Strategic Capability Response Analysis

David Walters • Deborah Helman

Strategic Capability Response Analysis

The Convergence of Industrié 4.0, Value Chain Network Management 2.0 and Stakeholder Value-Led Management

 Springer

David Walters
University of Technology Sydney
Sydney, NSW, Australia

Deborah Helman
DeVry University
North Brunswick, NJ, USA

ISBN 978-3-030-22946-7   ISBN 978-3-030-22944-3   (eBook)
https://doi.org/10.1007/978-3-030-22944-3

This Springer imprint is published by the registered company Springer Nature Switzerland AG
The registered company address is: Gewerbestrasse 11, 6330 Cham, Switzerland

In memory of Helen Amelia Walters,
1937–2016 and everything in between

Acknowledgments

We would like to acknowledge Dan Gowler for his inspirational and insightful work on revolutions and revisions in postindustrial societies – insights that surfaced at Templeton College Oxford in 1984 – that foretold the impacts of changes in technology that would impact organizations and consumers.

We would like to thank Guy Helman for his contribution to this book – generously sharing his considerable publishing experience in general and his talents with Illustrator in particular.

Introduction

What Is Different?

Three developments impacting the business environment in recent years have been digitization and connectivity, the acceptance of the wider application of the value chain, and the importance of stakeholder value management as a more equitable corporate response than shareholder value management. Digitization, in the shape of Industrié 4.0, has resulted in an extensive application of "connectivity" across the manufacturing and servitization delivery of product hardware and serviceability software. The value chain has matured and has enhanced collaboration, coordination, and communication among value chain partners. Stakeholder value-led management is becoming important to enlightened CEOs who are publicly mindful that they are answerable to more than just shareholders; their role is one of being answerable to a much broader group as pressure from customers, suppliers, employees, government, and other regulators, environmental groups, as well as shareholders make them answerable to "everyone" – the outcome is they must deliver a balanced performance. Capex (high capital intensity) business models are becoming Opex (low capital intensity-intangible fixed assets) based models focusing on organizational intangible assets (brands, developing supplier, and customer relationships) that contribute to Network Value Advantage.

Industrié 4.0

The business environment is changing rapidly. Specifically, the changing and challenging business environment of Industrié 4.0, the focus on stakeholder value considerations, customer-centricity, shared value, and value-creating systems, are changing the perspectives of value. The required contribution from the value chain network as a business model is to provide a strategic response role, one which identifies capabilities required for success, their sources, and their coordination.

The "connectivity characteristics" of Industrié 4.0 offer opportunities to increase the strength of relationship management in the value chain network and the effectiveness and efficiency with which value-based strategies are designed and value is delivered. Other characteristics offer equally interesting opportunities. The confluence of technological capabilities has expanded the production and productivity options available from flow line mass production and batch manufacturing (mass customization) to a batch size of one. Data management/analytics (collection, analysis, sharing, and collaborative decision-making) is increasing "organizational effectiveness and efficiencies"; component suppliers locate closer to their assembler customers, abandoning the one-time advantage of low costs and increasing quality reliability and time4response. Connectivity has brought M2M (machine2machine) facilities, M2P (machine2people – employees and customers), and M2NC (machine2network collaboration) enabling information transparency.

Value Chain Network 2.0

The conceptual notion of the value chain network has moved on into the age of strategic digital capabilities. Changes to the original Porter (1985) concept have been traced and explored in Fig. 1. Porter's value chain was very much a cost model, one that offered individual organizations the opportunity to manage the flow of inputs/materials through the production conversion process and into distribution channels. There was a suggestion of "leverage," whereby suppliers could extend their processes towards those of their customers and perform some of the tasks of their customers. For example, in industrial markets, suppliers to wholesalers would deliver very large orders to wholesalers' customers and share the cost savings.

Normann and Ramirez (1994), McHugh and Wheeler lll (1995), Parolini (1999), and others subsequently developed a more meaningful approach to the value chain based upon partnership networks. Parolini went so far as to suggest the "… basic difference between Porter's original value chain and the approach taken by Normann and Ramirez (and indeed subsequent approaches) is that the former (Porter) takes the company value chain as his starting point, whereas the latter (Normann and Ramirez 1994) underline the greater importance of the value-creating system." (Parolini 1999)

It became:

> A set of companies that acts integratedly and organically; it is constantly re-configured to manage each business opportunity a customer presents. Each company in the network provides a different process capability…. (McHugh and Wheeler lll 1995)

Figure 1 illustrates this argument by focusing upon the customer as the first consideration in the value chain network process. It also suggests organizational flexibility enabled the network to recruit (and reposition) organizations within the network to improve the customer value proposition. As the value chain network developed, it became more closely connected and collaborative; information was

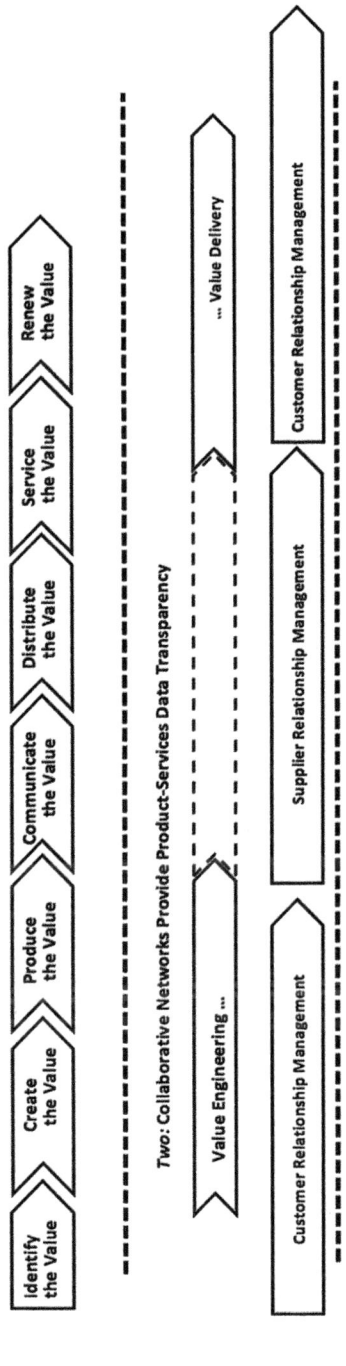

Fig. 1 The evolving interactive customer-centric value chain

made available throughout the network, and the networks became increasingly strategically *effective* and operationally *efficient*. As a result, product-service development became a shared activity, and the notion of customer-led stakeholder value management optimization took effect; network activities and processes are needed to be managed optimally rather than totally biased towards a specific member or group of participants.

Collaborative Networks Provide Product-Service Data Transparency

Serviceability (the intra- and inter-organizational activities that create service as value throughout the value chain network) has become an essential feature of the value proposition. Capital-intensive organizations have adopted an *increased quality and reliability* approach towards customer satisfaction, often focusing on intangible product-service characteristics in their value proposition. This required a reappraisal of investment policies resulting in a negotiated restructuring of the traditional *product*-service into a product-*service*; capital-intensive organizations opted to purchase suppliers' product performance output rather than the capital hardware:

> Producibility ("strategic and operational infrastructure"); The fusion of; research design and development, with manufacturing and distribution operations, "serviceability", and of product-service renewal activities, creates a value producing infrastructure that is a seamless and continuous process.

As a result, customer and supplier relationships became integrated, and customer performance management is becoming a value-adding product-service "enabled" by connectivity based upon sensors and monitoring equipment.

The integration and collaboration have increased in depth as well as application:

> Industrié 4.0 is deepening the relationships between manufacturing, customers, and suppliers. Industrié 4.0 shifts manufacturing from isolated, optimized cells of business processes, systems, and resources to fully integrated data and product flows across corporate borders.

Stakeholder Value-Led Management

The value-based management "fad" of the 1980s/1990s largely focused on shareholder value management (strategies to enhance the wealth of shareholders) – which largely ignored the interests of the other groups. The promoters of shareholder value management argued that the responsibility of senior management was (and to an extent remains) to maximize the value of the organization, but this failed for two primary reasons: first due to their requirements for very detailed information and the cost in time and managerial effort to produce the data, and second because most of the models ignored the need for customer interests to be considered, let alone to consider customer satisfaction a focal issue. Interestingly, the emphasis is now on the argument that customer value drives shareholder value. Payne and Holt (2001) cite Cleland and Bruno (1996) who suggest that an organization should ensure that its customer value strategies are successful in generating revenues that deliver levels of profitability (margins) that exceed its cost of capital sufficiently such that it (the organization) consistently builds wealth for the shareholders. Doyle (2008) was more succinct in emphasizing that shareholder value maximization requires a focus on delivered customer value and proposes this as a primary objective for marketing. However, an increasing number of organizations now see the role of the CEO as defining the business in terms of "a broader purpose," that is, one that benefits a larger set of *stakeholders* (customers, employees, suppliers, and future generations) than would otherwise gain from simply increased profits and shareholder value and indeed shared value. It follows that the *broader purpose* approach requires strong data support; Industrié 4.0 makes this possible. Long-time observers are questioning the validity of what appears to have been a focus on short-term value for the shareholder as opposed to the long-term prospects of the business – in other words – stakeholder value management.

A shareholder-only approach usually has several characteristics such as the following:

- A narrow focus.
- Objectives are driven by quantifiable metrics.
- Executive management may react to poor evaluations with extreme responses (cost reductions, restructuring, etc. that are likely to make situations worse).
- Tangible.
- Financially focused performance evaluation with little emphasis on intangible performance drivers.
- Concept of "value" (and who it is generated for and how) is not clearly understood or articulated.
- New ideas and methods are not easily understood and accepted.
- Management tends to embrace quick fix solutions, sometimes adopting irrelevant solutions.
- People who create relevant "value" often are viewed as "too radical".
- Lack of understanding of the organizations' value proposition.

The profit focus is on earnings-based metrics, and the performance reflects traditional approaches to growth that allocate resources to marketing, acquire other companies, and control costs, rather than considering partner stakeholder expectations. In other words, the business success is measured by what we create for our *shareholders*.

By contrast, stakeholder thinking tends to be deeper and broader and may include the following characteristics: a sustainable, visionary, and competitive-thinking approach, involving a multidimensional view of the organization. Performance evaluations are used to measure strategic achievements, as well as the operations delivery systems that extend across the entire value chain. Stakeholders can be classified as primary and secondary (external) stakeholders. Primary stakeholders are customers, employees, management, shareholders and investors, and partners, suppliers, and distributors. Secondary stakeholders comprise of competitors, the community, and government. Primary stakeholders welcome open, close, and ongoing relationships involving the exchange of creative ideas, innovation, management approaches to change, and performance improvements. The interests of secondary stakeholders have specific focus: competitors monitor progress and have interests in opportunities for collaboration (typically co-opetition, licensing patents, and using any excess capacity they have to manufacture and distribute competitors product-services); the community is interested in employment opportunities and expansion of revenues (and, therefore, in local taxes); and the government has interests in employment, innovation, exporting activities, and a contribution to GDP. An extension of this shift towards a wider perspective of "responsibility" is the B Corporation, a label that reflects an organization's ethical, social, and environmental practices. B Corporations are certified by an independent movement called "a Lab," founded by Jay Coen Gilbert who argues "We need to correct an error on source code of capitalism: shareholder primacy." B Corporations promote better governance and better serve the interests of workers, suppliers, and wider society, in addition to investors (Economist 2018). In other words, business success is measured by the value we create for all *stakeholders*.

Sayers (2019) identifies a corporate change. He cites research suggesting trust in business is increasing. We are still not shooting the lights out, but at 52%, we are more trusted than either governments or the media, and people trust their own employer more than any other institution. There is still a significant discontent in society, and there is an appetite for change. But importantly, the barometer shows people want business to lead that change. An astonishing 79% of Australian respondents want CEOs to lead on social issues, and 76% of global respondents think CEOs should take the lead on change rather than waiting for governments to impose it.

Industrié 4.0 + Value Chain Network 2.0

Industrié 4.0 is a unique opportunity. One specific benefit is the connectivity facility that creates a data-based model capable of evaluating the implications of actively pursuing market opportunities across a structured mix of interrelated organizational response capability centers (i.e., performance, profitability, productivity, partnerships, producibility, and preservation (resources and sustainability)) that are basic in their contribution to an analytical approach planning and decision-making and encouraging management to ask "what-if" questions concerning the interactions across the capability centers. Accepting the premise that the overriding objective of the business is being able to meet the value expectations of *all* value chain network stakeholders and is typically spelt out by its value proposition, it is essential that the network organization identifies the optimal use of its capabilities. Given a market opportunity, an organization must first evaluate the end-user customers' expectations. Often, this will require an exploration of the potential customers' "essentials and the desirables," their basic value expectations, i.e., product-service performance, asset management (cost-service-price optimization delivery), time uses, and locational availability (product-service and service support, e.g., online purchasing facilities). Delivering the value proposition where the product-service is required successfully requires it to achieve "total stakeholder satisfaction." Being a networked-based model, it has the potential for considering both internal and external trade-off characteristics. Should there be roles and/or tasks or other hurdles to be overcome, requiring specialist expertise in the overall process of delivering the promised value to the satisfaction of each of the stakeholder partners, the network model encourages collaborative activities. Artificial Intelligence (AI) is being applied to Value Chain 2.0 increasingly to manage routine operations (inventory levels and flows, network transportation), but it is building algorithms for strategic and operational futures. Blockchain is being incorporated into Value Chain 2.0, increasing transaction visibility and traceability and reducing risk (Fig. 2).

Industrié 4.0 + Value Chain 2.0 + Stakeholder Value-Led Management: A Convergence

Collectively, Industrié 4.0, Value Chain 2.0, and stakeholder value-led management are making a significant impact on strategy, structure, and competing business models. Connectivity, data availability, analysis, and application made available by the digital capabilities now deliver increased response agility, thereby enabling more accurate and timely responses to customer demand and separating organizations that are rapidly adopting the benefits of connectivity, collaboration, and transparency from those attempting to compete with "traditional thinking" and methods. It could be argued this is Industrié 5.0 – a connected and collaborative development that has the capability of increasing the effective and efficient activities of industrial activity with ongoing incremental capability changes!

General Electric, under the leadership of Jeff Immelt, proposed that the added value realized from a synergistic approach of combining activities is focused on long-term stakeholder value management and therefore invested in acquisitions and developed strategic alliances with specialist organizations in emerging industries such as energy, healthcare, renewable energy, transportation, and aviation. By contrast, Siemens are suggesting traditional conglomerates do not have a future. The Siemens CEO is focusing on laying the foundations of the next generation by developing digital approaches to solve current and future value creation activities. General Electric has organized their approach in a subsidiary activity with *Predix* and Siemens with *TeamCenter*. Bosch is going to great lengths to become more of an intangible focused organization, reflecting the current shift towards a world that sees value coming increasingly from software, services, and data – not tangible, hardware products. Bosch is aware of the problems when software and hardware meet in industries such as automotive manufacturing, in the world of "connectivity," they are at risk of being viewed as commodity suppliers. Bosch is positioning itself in the value chain as a "trusted custodian of data."

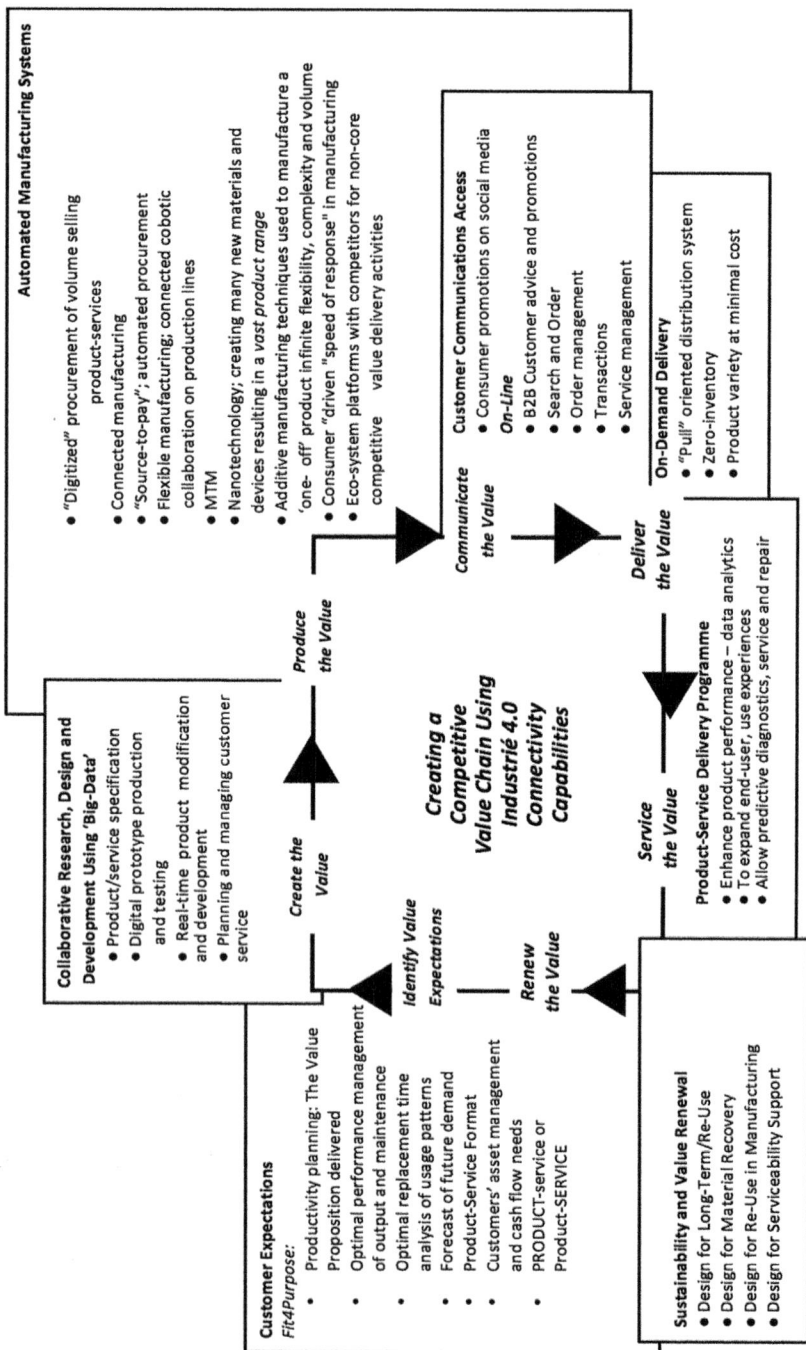

Fig. 2 The "connected value chain approach" to value delivery: Value chain network 2.0

This has not been without problems; it is interesting to note that General Electric, an organization responsible for many of the recent advances in digital manufacturing, ended 2017 experiencing a cash flow problem and saw the loss of more than $100 billion in the market value since the previous year and recently cut its dividend for only the second time since the financial difficulties of 2008/2009. Mr. Immelt left the company in 2017 and was replaced by another long-term GE executive, John Flannery. It was suggested that Mr. Immelt's acquisitions of industrial businesses, worth some US$126 billion (plus Wall Street fees of some US$6 billion), were excessive and appeared not to be demonstrating a cash return on the investment. Some 14 months later, Mr. Flannery was replaced by an outsider, Larry Culp; it is too soon to comment on the success of the changes. Culp has made structural changes and has divested a number of activities (assets) that are beginning to reduce the debt (Culp's Ability 2018; Blame Game 2018).

What Is the Rationale and Logic?

Business strategy and structure decisions are becoming increasingly holistic; the growth of collaborative networked organizations presents strong evidence of this trend. The principles and processes addressed are now being influenced by the digital direction of Industrié 4.0 and the convergence of collaboration and integration in Value Chain 2.0. The impact and implications of the convergence of Industrié 4.0, Value Chain 2.0, and stakeholder value-led management are being identified and discussed in the professional press, among the prominent major consulting organizations and business dailies such as the London *Financial Times* and *The Economist*. To date, there does not appear to be a published contribution that provides material discussing the impact of the changes, challenges, and opportunities on the strategic and operational aspects of management decision-making.

It is arguable that this convergence offers organizations an opportunity to be more analytical when making strategic decisions by increasing the amount of relevant data available concerning strategic options *and* their implications for, and their impact on, operational management activities. Our research embraces the "value expectations" of *all* stakeholders in a business enterprise. The notion of a business enterprise is currently better expressed as a "value-based network" since almost all organizations are linked to demand-supply-response relationships with other organizations. Such networks deliver a focused "agile-rolling-value proposition" that optimizes the expectations and the resources of its stakeholder constituents based on real-time data. This permits an evaluative, rather than prescriptive, approach allowing each of the pathway options to achieve a total stakeholder satisfaction to be considered across the range of value-based network responses and resources available for their implementation.

Therefore, the purpose of this ongoing research was (and remains) to consider the changing environment within which value chain management now exists. Examples from the large corporations suggest that the traditional structural model based upon functions (AKA silos) is rapidly losing support in favor of holistic structures reflecting *response capabilities*; these are characteristics that reflect an understanding of the market place and of the opportunities it offers and the characteristics of capabilities essential for successful engagement.

Industrié 4.0 has presented numerous opportunities across all industries to improve both the effectiveness of strategic decisions and the efficiency of their implementation for the network stakeholders. To this end, a new approach to business model design is suggested. The "big data" feature of Industrié 4.0 changes the characteristics of decision-making; the availability of real-time operating data facilitates the identification of solutions to problems and the decision alternatives. This in turn suggests the need for a model that considers the impact of alternative solutions on the core business model capability components of *performance* (the value engineering and value delivery); *profitability, productivity, partnerships*, and *producibility* (the strategic and operational infrastructures – the processes and activities of value engineering and value delivery); and *preservation* (sustainability issues: organizational, economic, social, and environmental) (see Fig. 3). This contrasts with the traditional view that there are key features comprising a business model; while they are interdependent, they do not have the facility of exploring cross impacts of decision alternatives to arrive at an optimal solution for the stakeholders. We use the conceptual notion of the value chain network but move this on into an age of strategic digital capabilities and introduce the broader perspective of stakeholder value-led management.

Our approach is to identify consistency and convergence in the practices of leading organizations and to build the material into what is developing into best practice methodology. We have identified a value-based network business model approach that incorporates the core connected business model resource capability components (of performance, profitability, productivity, partnerships, producibility, and preservation) that has evolved in response to the changing business environment of Industrié 4.0., Value Chain 2.0, and stakeholder value-led management. The reasoning behind the model is simply that the characteristics and facilities of the convergence are such that data outputs can now incorporate the interests of the entire spectrum of stakeholders and can be structured to consider topics such as the impact of value engineering and delivery decisions on investment decisions.

Process

Capability Responses: the approach views the firm as a portfolio of capabilities that evolve in response to the (perceived) demands of the business environment. An "enterprise" comprises one organization or collaborative network of firms. Response capabilities are characteristics that reflect an understanding of the market place, the expectations of its stakeholders and of the opportunities it offers, and, therefore, the characteristics of the capabilities essential for successful engagement. The core detailed business response capability model components are the

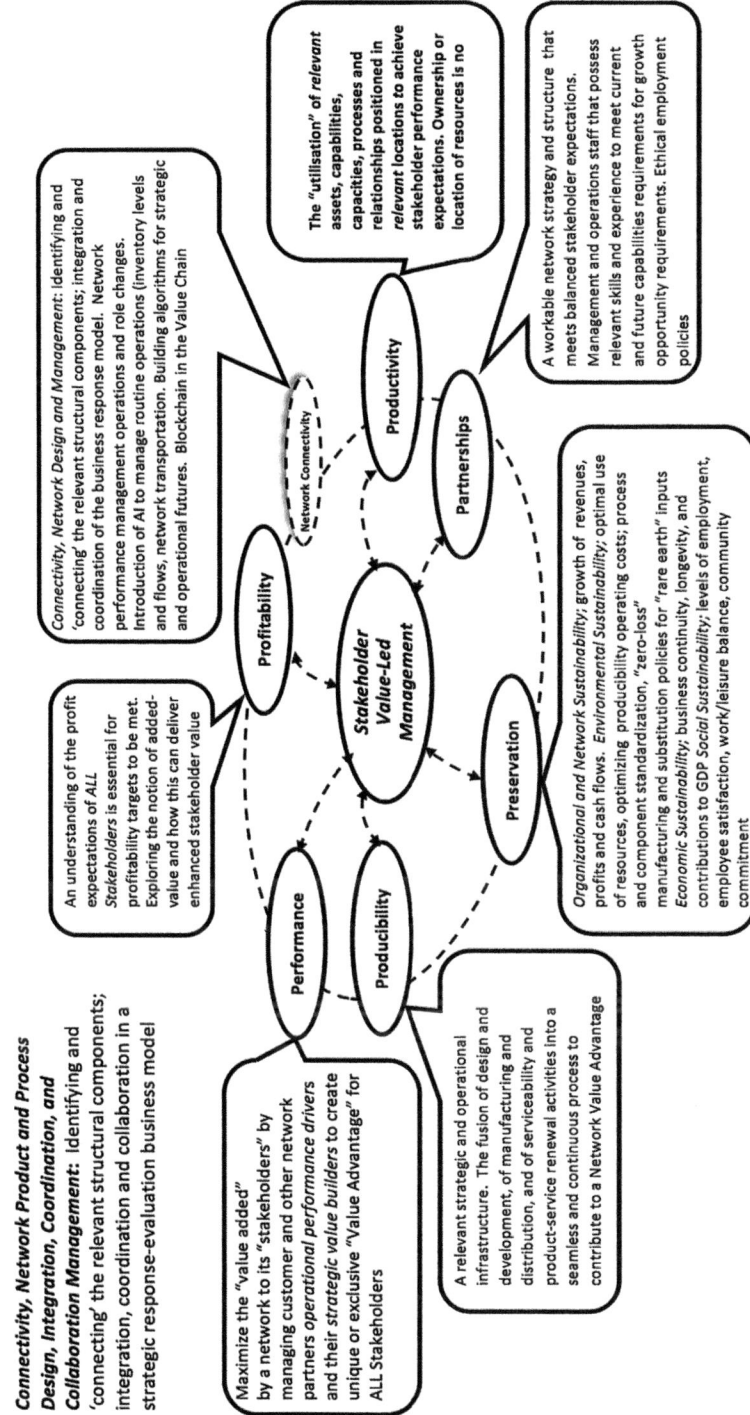

Fig. 3 The components of a connected strategic capability response model

Connectivity, Network Product and Process Design, Integration, Coordination, and Collaboration Management: Identifying and 'connecting' the relevant structural components; integration, coordination and collaboration in a strategic response-evaluation business model

Connectivity, Network Design and Management: Identifying and 'connecting' the relevant structural components; integration and coordination of the business response model. Network performance management operations and role changes. Introduction of AI to manage routine operations (inventory levels and flows, network transportation. Building algorithms for strategic and operational futures. Blockchain in the Value Chain

The *"utilisation" of relevant* assets, capabilities, capacities, processes and relationships positioned in *relevant* locations to achieve stakeholder performance expectations. Ownership or location of resources is no

A workable network strategy and structure that meets balanced stakeholder expectations. Management and operations staff that possess relevant skills and experience to meet current and future capabilities requirements for growth opportunity requirements. Ethical employment policies

An understanding of the profit expectations of ALL *Stakeholders* is essential for profitability targets to be met. Exploring the notion of added-value and how this can deliver enhanced stakeholder value

Maximize the *"value added"* by a network to its "stakeholders" by managing customer and other network partners *operational performance drivers* and their *strategic value builders* to create unique or exclusive *"Value Advantage"* for ALL Stakeholders

A relevant strategic and operational infrastructure. The fusion of design and development, of manufacturing and distribution, and of serviceability and product-service renewal activities into a seamless and continuous process to contribute to a Network Value Advantage

Organizational and Network Sustainability; growth of revenues, profits and cash flows. *Environmental Sustainability;* optimal use of resources, optimizing producibility operating costs; process and component standardization, "zero-loss" manufacturing and substitution policies for "rare earth" inputs *Economic Sustainability;* business continuity, longevity, and contributions to GDP *Social Sustainability;* levels of employment, employee satisfaction, work/leisure balance, community commitment

Productivity

Partnerships

Profitability

Network Connectivity

Performance

Producibility

Stakeholder Value-Led Management

Preservation

following: *performance* (value engineering, value delivery, and value proposition, stakeholder-performance expectations, fit4purpose, etc.), *profitability* (financial viability, sustainable economic profit, positive cash flow), *productivity* (total factor productivity; optimal utilization of capital, labor, materials, and service inputs; and EVA (economic value added)), *partnerships* (expertise and skills of value-generating employees, workplace cultures, and management styles in all network partners), *producibility* (the seamless intra- and inter-organizational infrastructure of sequential processes and activities of value engineering and value delivery that creates, produces, delivers, and captures value), and *preservation* (the socioeconomic and responsible use of environmental resources in creating stakeholder value and the concerns of environmental and corporate sustainability and responsibility).

Capability Response Analysis: is a management process that identifies the relevant capabilities that will be required to undertake a successful response to a market opportunity. The response capability approach can reinforce a market (and value chain) positioning, as well opening up alternative market opportunities. The analysis is conducted and based on the expectations of potential customers in order that these are fully understood and can be used as input for planning a response. However, the vendor organization also has expectations based upon the same capability criteria (not the same characteristics), but they are determined by the expectations of the organizations' stakeholders.

Capability Response Management: is the ability to structure, combine, and leverage internal and external resources for the purpose of creating a capability-led response to a market opportunity to create value for stakeholders by maximizing competitive value advantage. Organizational policy, not strategy, can establish prerequisites or ground rules concerning response capability levels, such as a margin of the amount by which an EVA return should exceed the annualized weighted cost of capital *(profitability),* or perhaps the amount of incremental increase in capital intensity the organization is prepared to undertake as further investment in pursuit of an opportunity *(producibility and productivity).*

Example One: Bosch – Aligning Capabilities for Repositioning in the Value Chain Network

Bosch is a 130-year-old hardware giant tackling the transformation of *product*-service hardware products into a software-driven product-*service* company.

Bosch is a very large organization. It has 440 subsidiaries and employs 400,00 people. Market analysts identify it as a car parts manufacturer, making everything from fuel injection pumps to windscreen wipers. It is also well known for its white goods and power tools. It is very proud of its "Made in Germany" labels. The organization identifies itself as a "supplier of technology and services" or an IOT (Internet-of-Things) company. The lawns of its Stuttgart headquarters location are mowed by robotic mowers indicating its emerging direction.

Bosch is going to great lengths to become more of an "intangible assets" (product-*service*) based organization, reflecting the current shift towards a world that sees value coming increasingly from software, services, and data – not tangible, hardware products. Bosch is aware of the problems when software and hardware meet in industries such as automotive manufacturing, in the world of "connectivity," they are at risk of being viewed as commodity suppliers. Bosch is positioning itself in the value chain as a "trusted custodian of data."

Bosch is 92% owned by a foundation, freeing it to invest heavily in RD&D €73 billion in 2016. It recently opened a research campus and has a funding of €420 million. The CEO continues to see Bosch as a product company but is becoming a product-*service* company with investment in software and what Volkmar Denner considers to be "middleware," a portfolio of supporting services. Software investment has resulted in a platform which runs IOT services and applications, allowing other companies access. Bosch has its own cloud and data center unlike other global competitors, like General Electric and Siemens, who use Amazon's cloud to run their platforms. Bosch argues that this offers greater speed and flexibility and data security. More are planned.

Bosch plans to increase the value developed by hardware with a strategy that is based upon sensors, software, and services. Over 50% of its electrical product-services groups have digital capability, and the company has invested €1 billion in a semiconductor plant to manufacture chips and sensors to act as the "eyes and ears" of the IOT. Bosch's long-term objective is to extend the connectivity into teaching things to think, e.g., training equipment to respond to unexpected events. An AI (artificial intelligence) activity has been established and is functioning; machines are limited by the data they operate with. Bosch currently manages data from British Gas customers to anticipate energy maintenance needs. Data from manufacturing and agriculture "sensors" is managed in a similar manner for the same purpose. Bosch head of AI foresees product updates in the future which will prompt pricing changes focused on the volume of data managed.

The Company has an electric scooter-sharing scheme in Berlin and Paris. Bosch provides the platform and purchases the scooters from Taiwan. It is their first direct-to-consumer business in their "electro-mobility" business in which Bosch invests €400 million per year. Bosch considers this a growth area. It has some 3000 people working on driver assistance systems and has some 1000 patents. By 2019, it anticipates earnings of €2 billion – twice the amount earned in 2016.

Bosch has had a reputation for being conservative and secretive. However, it is making connections with a range of partner organizations. These include a map-building partnership (Chinese), working with Tesla on autonomous vehicles, and Amazon's voice-controlled computer for input into Bosch smart-home systems. The company considers innovation partnerships make sense in the precompetitive stages of product-service development. A question concerning network activity in the competitive stage requires to be addressed. McKinsey's Eric Lamarre (2017) suggests that as platforms develop (horizontal platforms such as Amazon and vertical manufacturing platforms such as Bosch and General Electric), the questions concerning the "value" of data will become critical. Bosch considers that customers will soon value data more than hardware products. That said, they have a strong reputation for quality and reliability; it could be argued that with their increasing expertise, data analytics will make them very competitive (see Fig. 4).

Example Two: The Application of "On-Demand Manufacturing"

An alternative example demonstrates the response capability model being applied to a manufacturing process in the automotive industry. The application of additive manufacturing (3D printing) continues to expand as design capabilities and materials are developed. Figure 4 illustrates the work ongoing at General Electric in developing a new component system, the redesign fuel injection system, to reduce number of components using additive manufacturing as an OEM and service parts which results in (Fig. 5):

Reduction of components
Decrease in overall weight
Increased fuel efficiency
Increased engine power output
Increases maintenance interval times
Components that are stronger and more durable

These examples are representative of what is occurring on a wide international front. Our purpose is to consider the impact of a rapidly changing environment within which value chain network management now formally exists and its long-term emphasis on stakeholder value. Industrié 4.0 has presented numerous opportunities across all industries to improve both the effectiveness of strategic decisions and the efficiency of their implementation to the network stakeholders. It has brought about significant changes to the value chain network; customer demand and supply response time have been condensed, and in some, product-market digitization is making the response simultaneous. In capital goods markets, it is increasingly the case that it is the service output that is purchased, not the hardware; fixed costs become variable costs as organizational investment focuses on intangible assets such as brand reinforcement and customer services such as bespoke data analytics that identify performance management improvements and "flag-up" serviceability requirements.

Convergence

Industrié 4.0, Value Chain 2.0, and stakeholder value-led management are having significant (and far reaching) impact on strategic opportunity. It is interesting to note that relevant output from the major consulting organizations are identifying the need to consider Industrié 4.0, the reimagined Value Chain 2.0, and stakeholder value-led management as a collaborative coalition rather than a collection of

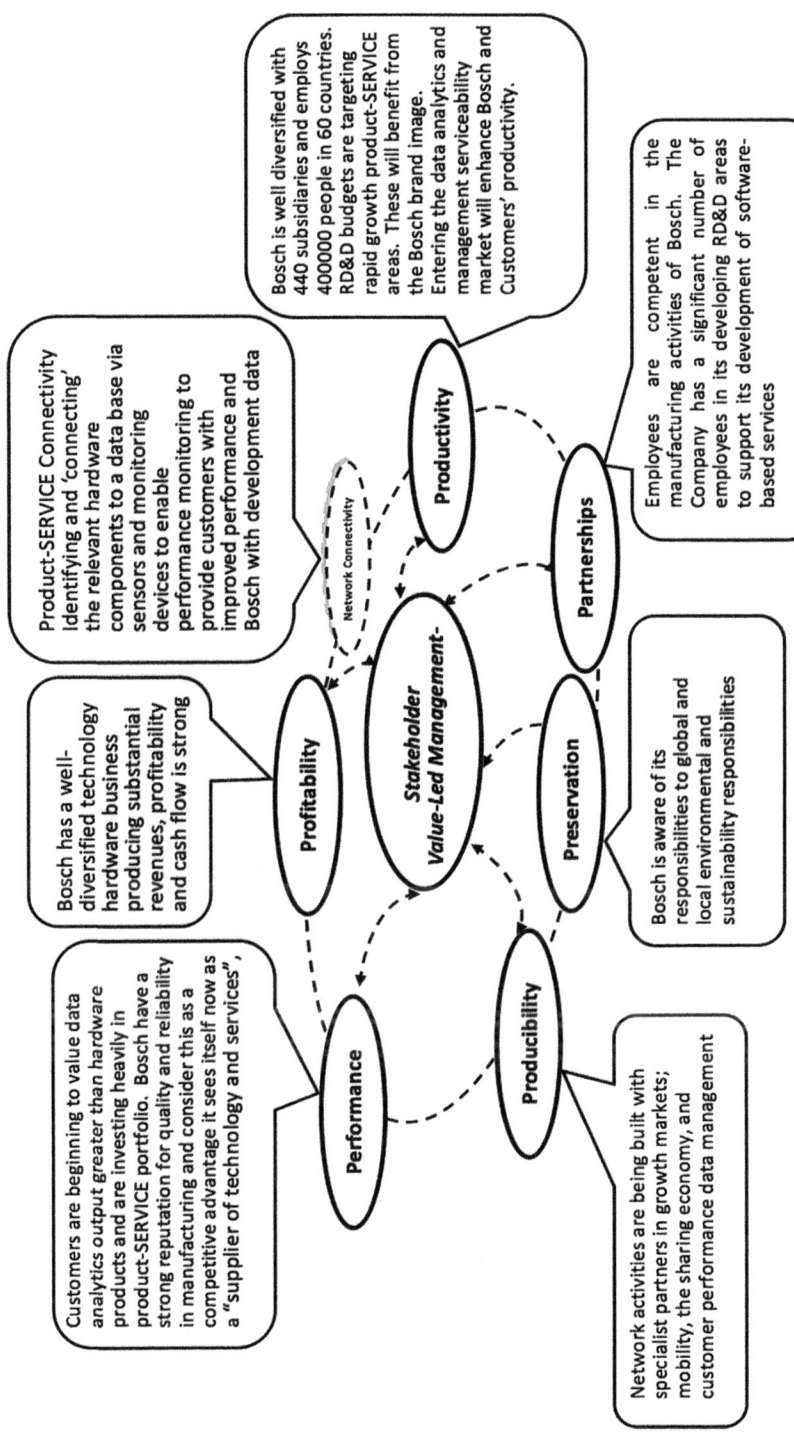

Product-SERVICE Connectivity Identifying and 'connecting' the relevant hardware components to a data base via sensors and monitoring devices to enable performance monitoring to provide customers with improved performance and Bosch with development data

Bosch is well diversified with 440 subsidiaries and employs 400000 people in 60 countries. RD&D budgets are targeting rapid growth product-SERVICE areas. These will benefit from the Bosch brand image. Entering the data analytics and management serviceability market will enhance Bosch and Customers' productivity.

Employees are competent in the manufacturing activities of Bosch. The Company has a significant number of employees in its developing RD&D areas to support its development of software-based services

Bosch has a well-diversified technology hardware business producing substantial revenues, profitability and cash flow is strong

Bosch is aware of its responsibilities to global and local environmental and sustainability responsibilities

Customers are beginning to value data analytics output greater than hardware products and are investing heavily in product-SERVICE portfolio. Bosch have a strong reputation for quality and reliability in manufacturing and consider this as a competitive advantage it sees itself now as a "supplier of technology and services",

Network activities are being built with specialist partners in growth markets; mobility, the sharing economy, and customer performance data management

Network Connectivity

Productivity

Partnerships

Profitability

Stakeholder Value-Led Management-

Preservation

Performance

Producibility

Fig. 4 Bosch: A supplier of technology and services – strategic response capabilities

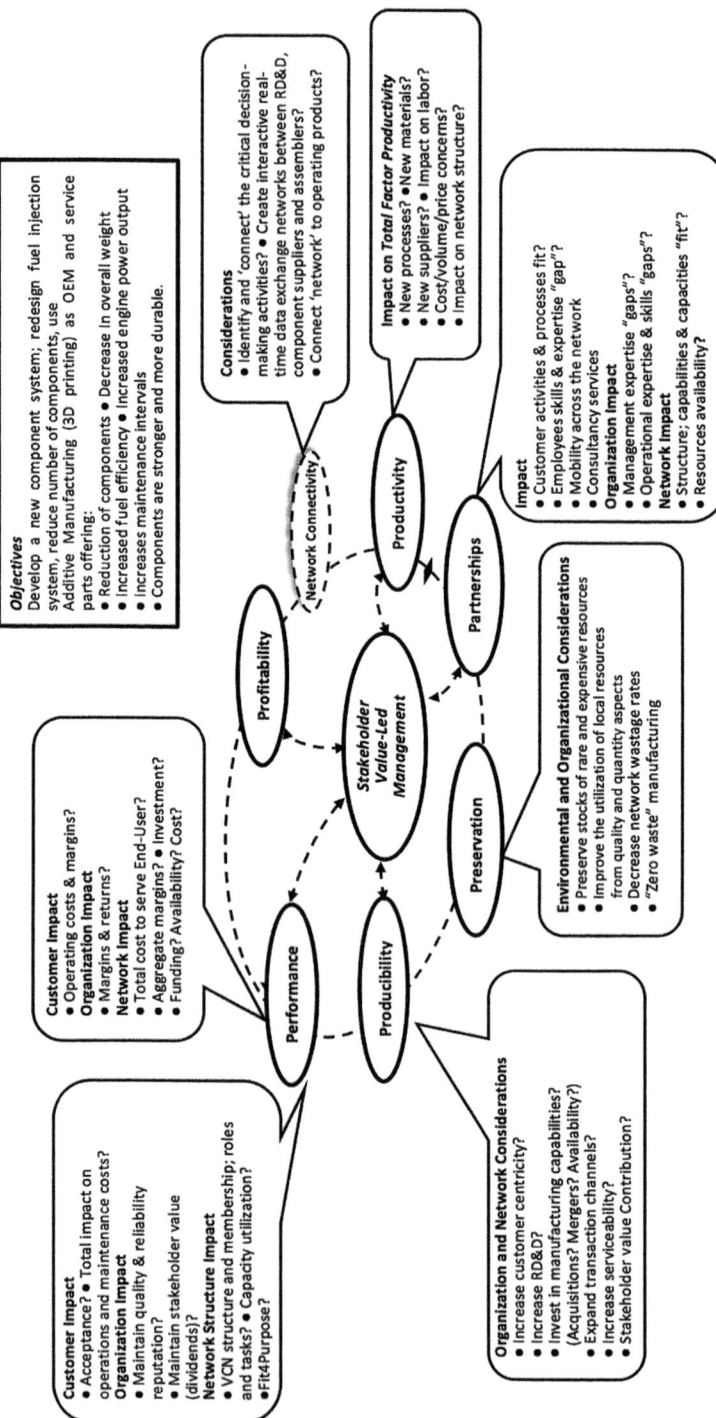

Objectives
Develop a new component system; redesign fuel injection system, reduce number of components, use Additive Manufacturing (3D printing) as OEM and service parts offering:
- Reduction of components ● Decrease in overall weight
- Increased fuel efficiency ● Increased engine power output
- Increases maintenance intervals
- Components are stronger and more durable.

Considerations
- Identify and 'connect' the critical decision-making activities? ● Create interactive real-time data exchange networks between RD&D, component suppliers and assemblers?
- Connect 'network' to operating products?

Impact on Total Factor Productivity
- New processes? ● New materials?
- New suppliers? ● Impact on labor?
- Cost/volume/price concerns?
- Impact on network structure?

Impact
- Customer activities & processes fit?
- Employees skills & expertise "gap"?
- Mobility across the network
- Consultancy services

Organization Impact
- Management expertise "gaps"?
- Operational expertise & skills "gaps"?

Network Impact
- Structure; capabilities & capacities "fit"?
- Resources availability?

Customer Impact
- Operating costs & margins?

Organization Impact
- Margins & returns?

Network Impact
- Total cost to serve End-User?
- Aggregate margins? ● Investment?
- Funding? Availability? Cost?

Environmental and Organizational Considerations
- Preserve stocks of rare and expensive resources
- Improve the utilization of local resources from quality and quantity aspects
- Decrease network wastage rates
- "Zero waste" manufacturing

Customer Impact
- Acceptance? ● Total impact on operations and maintenance costs?

Organization Impact
- Maintain quality & reliability reputation?
- Maintain stakeholder value (dividends)?

Network Structure Impact
- VCN structure and membership; roles and tasks? ● Capacity utilization?
- Fit4Purpose?

Organization and Network Considerations
- Increase customer centricity?
- Increase RD&D?
- Invest in manufacturing capabilities? (Acquisitions? Mergers? Availability?)
- Expand transaction channels?
- Increase serviceability?
- Stakeholder value Contribution?

Network Connectivity

Productivity

Profitability

Partnerships

Stakeholder Value-Led Management

Preservation

Performance

Producibility

Fig. 5 Exploring the introduction of additive manufacturing into component manufacturing: The impact on the VCN (Adapted from: General Electric Report (2017))

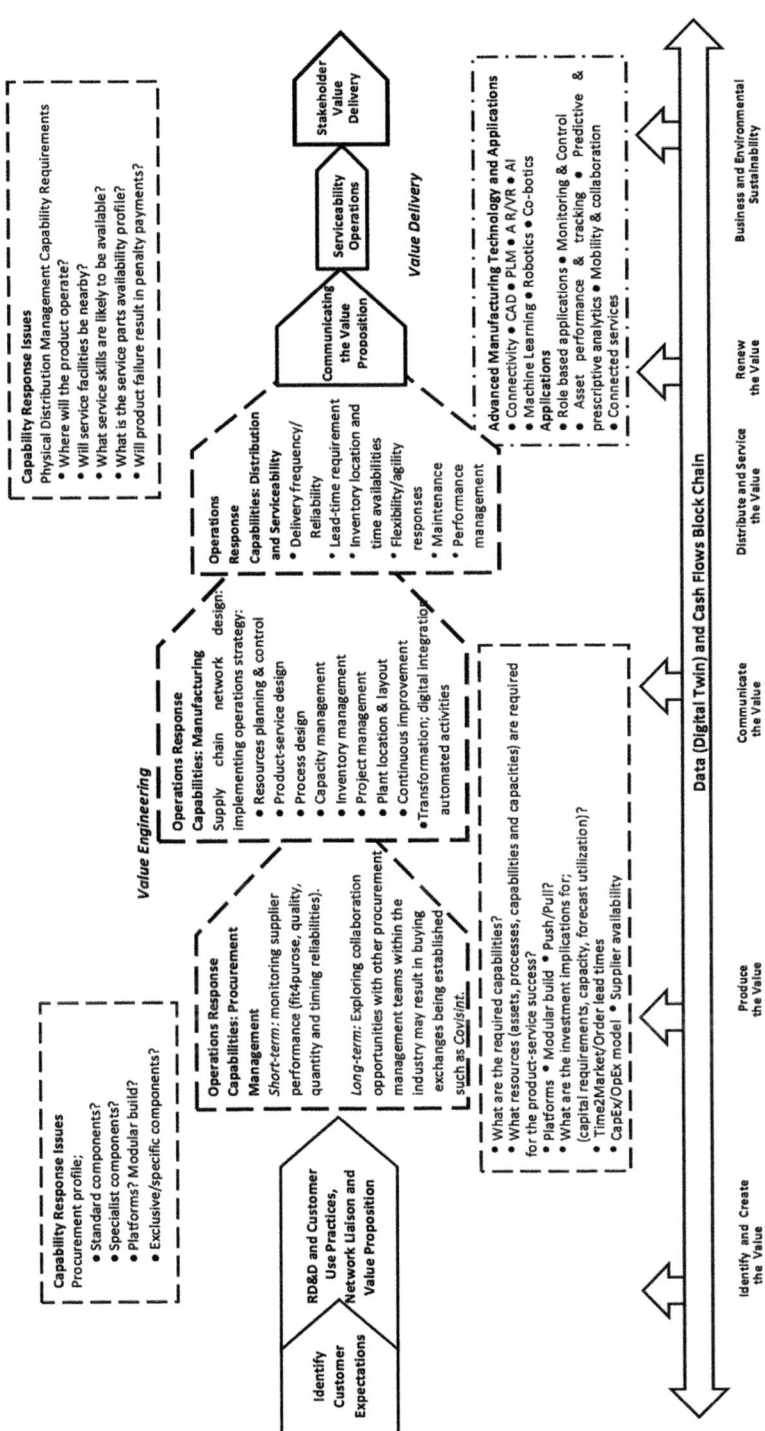

Fig. 6 Producibility: Producing the value

discreet models. The purpose of this research has been to consider the changing environment within which value chain management now exists and its long-term emphasis on stakeholder value. Industrié 4.0 has presented numerous opportunities across all industries to improve both the effectiveness of strategic decisions and the efficiency of their implementation to the network stakeholders. To this end, a new approach to business model design is suggested, a "response capability approach." The "convergence" of Industrié 4.0, Value Chain Network 2.0, and the expanded view of corporate responsibility, stakeholder value management, is leading towards the creation of a seamless data-led "producibility infrastructure" illustrated as Fig. 6.

References

Business, Blame Game, Economist (2018) 4 October.
Business, Culp-ability; General Electric powers downwards , Economist (2018) 3 November 2018.
Cleland, A., & Bruno, A. (1996). *The value based market process: Bridging customer and shareholder value*. Wiley/Jossey-Bass.
Danone rethinks the idea of the firm – Choosing plan B (2018, August, 9). The Economist.
Doyle, P. (2008). Value based marketing strategies for corporate growth and shareholder value. 2nd Edition. New York: Wiley.
General Electric Company Annual Report, 2017.
Lamarre, E. (2017). Making sense of internet of things platforms. McKinsey Quarterly, *McKinsey & Co*. May.
McHugh, P., & Wheeler lll, G. (1995). *Beyond business process reengineering*. Chichester: Wiley.
Normann, R., & Ramirez, R. (1994). *Designing interactive strategy: From value chain to value constellation*. New York: Wiley.
Parolini, C. (1999). *The value net*. Chichester: Wiley.
Payne, A., & Holt, S. (2001). Diagnosing customer value: Integrating the value process and relationship marketing. *British Journal of Management, 12*(2), 159–182.
Porter, M. (1985). *Competitive strategy*. New York: Free Press.

Contents

Part I
A Changing and Challenging Future: How the Future Is Developing

Chapter 1 explores perceptions of "value." Value has been suggested to be the residual of benefits received less the cost of acquisition, which were considered to extend across the life cycle of the purchase (its value-in-use) and which could focus the procurement decision on operating and service and maintenance costs as being critical in the procurement decision. More recent views of value have added the interests of the "owners" of value-producing organizations, the shareholders, and during the late 1980s/early 1990s, shareholder value management was favored and for many organizations was the principle concern of the organization. This has extended into stakeholders' concerns with the argument that value has a broader context, one that links customers to supplier organizations, well beyond that of the individual investor to those involved in the "value" creation and delivery activities. Stakeholders can be classified as primary and secondary (external) stakeholders. Primary stakeholders are customers, employees, management, shareholders and investors, and partner suppliers and distributors. Secondary stakeholders comprise competitors, the community, and government.

The discussion should include value in networks, as the holistic network structure is commonplace; value in networks is generated by collaboration on shared objectives and coordinated operations. By leveraging assets and allocating roles and tasks, a network approach optimizes its resource capabilities in terms of investment in PRODUCT-service/product-SERVICE offers resulting in a value proposition that meets customer expectations, optimizes capabilities, and is economically viable. Most successful networks are structured around a principle of interlocking financial engineering (resurrecting and updating existing assets).

"Shared value" and the sharing economy have emerged as significant and involve creating economic value in a way that also creates value for society by addressing its needs and challenges, creating economic value for the corporation through innovations that address society's needs and challenges. The "sharing economy" is an economic model often defined as a peer-to-peer (P2P) economy-based activity of acquiring, providing, or sharing access to goods and services that are facilitated by a community based online platform; Uber, Lyft, and Airbnb are examples. The sharing economy is one in which online platforms coordinate hundreds of thousands of freelancers to drive cabs, rent rooms (Airbnb), and clean laundry (Washio). While it has grown rapidly, and continues to do so, the sharing economy is not without risk. Gapper (2014) identified issues that are (or will require) some attention: "As companies recognize the threat, governments and regulators are struggling to adjust and consumers are unsure whether to trust the new type of business". However, the greatest uncertainty faces workers.

The *customer journey* is the complete sum of experiences that *customers* go through when interacting with a company and its brand. Instead of looking at just a part of a transaction or experience, the customer journey documents the full experience of being a customer. A McKinsey Executive Briefing in 2016 suggests it is necessary to understand the *total transaction as the customer* sees it, using their research and consulting experience to make some significant points. The data now being generated "enables" far more than customer transaction details, a progressive series of "touchpoints" that add up to the *experience* customers are given when interacting with staff and online facilities causing some rethinking and restructuring of the organization (and the value proposition), network processes, and activities. Related groups, or cohorts, usually share common characteristics or expectations within a defined time-span or an experience. Identifying cohorts has advantages. It increases customer loyalty and the communications with a specific customer interest group. Traditional segmentation focuses on demographics, socioeconomics, brand, and store loyalty, identifying cohort targeting is focused on an interest.

Value migration occurs when new business models enter an industry, and this may lead to industry disruption; thinking ahead is important, and while it is possible to design agile products and processes that can compete with changes as they occur typically, it offers short-term relief; ongoing monitoring industry competitor's activities is preferable. Value migration occurs when customer needs change or a new business design begins to encroach on an existing configuration of providers. Businesses which previously enjoyed a leadership position and overlooked potential competition are now caught in an outflow of value and declining revenue and customer bases.

Satisfying customer expectations requires an understanding of their product-service purchasing and their applications activities: a "value proposition" spells out a response to these. Webster (1994) contended that positioning and the development of the value proposition must be based on an assessment of the product offering and of the firm's distinctive competencies *relative to competitors*. Hence the value proposition should make clear its *value competitive positioning*, how it delivers a value contribution that increases revenues and profitability by enhancing the organization's own value proposition. Connectivity creates opportunity for greater involvement by stakeholders as the data generated by activities becomes wider and deeper. Increasingly we see the importance of data management as the relationship between stakeholder networks and end-user organizations.

The changes and challenges of the current business environment and capabilities presented by the convergence of Industrié 4.0, Value Chain Network 2.0, and the focus on stakeholder satisfaction favor the notion of *sustainable future competitiveness*; it is also proposed that a broader, macro framework is becoming necessary to structure the identification of *capability response factors* rather than simply a list of the latest industry or marketplace drivers; the future (given the current developments in some industries) suggests some industries may change the way customers' needs are addressed (e.g., healthcare using connectivity to monitor patients, reducing hospital and specialists visits). It is also possible that "industry purposes" may change (e.g., recent developments in the automotive industry are adding non-ownership-based offers to their mobility value propositions).

Chapter 2 identifies and suggests that monitoring and exploring "industry dynamics" for opportunities is essential if *sustainable future competitiveness* is to be *satisfied*. Such activity revolves around a framework comprising *six* generic *influencers*: knowledge management, technology management, process management, relationship management, regulatory compliance, and managing geopolitical events. But much has changed. There are numerous examples of investments in assets, processes, and indeed capabilities that have proven to be short lived (often because the company has been slow to react to process innovation or, perhaps, even worse there has been a lack of awareness of ongoing developments that have undermined existing processes); e.g., the rate at which consumer acceptance of online transactions would develop, and as a result these organizations are left with resources that are not obsolescent but totally obsolete. This chapter develops an argument around the notion that organizations should respond to changing industry dynamics and seek to develop related capabilities. The successful organizations are those that identify the characteristics that are essential for creating value advantage ahead of their competitors.

Response capabilities: the topics of Chap. 3 are specific characteristics derived from an industry dynamic (or dynamics) that are exclusive, possibly unique, to an organization. They are focused on to a specific asset, process, or activity within an organization to facilitate the delivery of the organization's value proposition *and* to create value advantage. Typically capabilities are created from combined industry dynamics; the value proposition is *enabled* by the convergence of a direct selling approach to customers (*relationship management*), digital order and transactions management systems (*technology management*), JIT inventory management/production assembly methods (*relationship management and process management*), current computing capabilities through collaboration with specialist component suppliers (*knowledge management and relationship management*), and the enhancement or "halo" effect of using components manufactured by leading industry brand owners (*relationship management*) and by collaborating with other members of the value chain network.

Capability management seeks to integrate the network organization's capabilities to ensure it achieves a value competitive position in an industry/market value chain. Capability management has become a method of creating *enterprise architecture* in recent years. It seeks to build a model of an enterprise that identifies its component parts and their relationships for planning the evolution (and continuity) of the enterprise. A capability management perspective (such as Leonard's model or Teece's Dynamic Capabilities Theory) suggests the firm is viewed as a collection of capabilities (rather than functions or silos) comprising the means by which an organization responds to market opportunities. Typically, response capability is based upon tangible and intangible assets of the firm comprising the traditional business functions that have become "silos."

A response capabilities approach views the firm as a portfolio of capabilities that evolve in response to the (perceived) demands of the business environment. An "enterprise" comprises one organization or collaborative networks of firms. Examples from the large corporations suggest that the traditional structural model based upon functions (silos) is rapidly losing support in favor of holistic structures reflecting *response capabilities*; these are characteristics that reflect an understanding of the market place and of the opportunities it offers and the characteristics of capabilities essential for successful engagement. The evident changes in management approaches of narrowly defined customer centricity; a changing view of the enterprise, stakeholder value management rather than just shareholder value management; a positive approach to network collaboration; network "connectivity"; and the availability of real-time data analysis and management suggest the demise of the silos and replacement by a holistic, interactive approach that enables an organization to identify the capabilities that market opportunities require and the ability to structure a relevant response that meets the market demand precisely rather than attempting to mold the market need into an opportunity that fits the silo's requirements.

Three developments impacting the business environment in recent years have been digitization, the acceptance of the wider application of the value chain, and the importance of stakeholder value management as a more equitable corporate response than shareholder value management. Digitization, in the shape of *Industrié 4.0*, has resulted in an extensive application of "connectivity" across the manufacturing and servitization delivery of product hardware and serviceability software. The *value chain* has matured, and *Value Chain Network 2.0* has enhanced collaboration, coordination, and communication among value chain partners. Stakeholder value-led management is becoming important to enlightened CEOs who are publicly mindful that they are answerable to more than just shareholders; their role is one of being answerable to a much broader group as pressure from customers, suppliers, employees, government, and other regulators, environmental groups, as well as shareholders is making them answerable to "everyone" – the outcome is they must deliver a balanced performance balanced performance.

The "big data" feature of Industrié 4.0 changes the characteristics of decision-making; the availability of operating data facilitates the identification of solutions to problems and the decision alternatives; in many situations the customer offer (the value proposition) has become an agile response, a "rolling value proposition," one that can be matched to the specific, current needs of customers.

The dynamics of these changes suggest the need for a model that considers the impact of alternative solutions on the core business response capability model components of *performance management* (the value engineering and value delivery proposition; stakeholder-wide, performance, fit4purpose, etc.), *profitability* (financial viability, economic profit), *productivity* (total factor productivity, optimal utilization of capital, labor, materials, and service inputs), *partnerships* (building and managing network partnerships possessing relevant skills, experience to meet current and future capabilities requirements for future growth opportunity requirements, and ethical business policies throughout the network), *producibility* (the seamless intra- and inter-organizational infrastructures of processes and activities of value engineering and value delivery that creates, produces, delivers, and captures value), and *preservation* (a socially responsible use of environmental resources used in creating stakeholder value and the concerns of environmental and corporate sustainability). The chapters comprising part two identify the capability areas and explore each of them in depth.

References

Webster, F. E. (1994). *Market driven management*. New York: Wiley.
Gapper, J. (2014). Pharma needs an injection of financial engineering. *The Financial Times* May 4.

Chapter 1
Changing Perspectives of "Value"

Abstract Value is now being addressed in a much broader context as the longer-term implications of a customer focus is explored. While the view of value and the entire process of value creation, production, and delivery downstream are important, there are now emerging views concerning the overall activities and relationships of the value creation and delivery process. Many of the perspectives are a resurrection of earlier work: the stakeholder concept can be dated back to work undertaken and published by the Stanford Research Institute (SRI) in the 1960s and that had been influenced by earlier work in Lockheed's planning department. SRI argued that management needed to take cognizance of the concerns and interests of shareholders, employees, customers, suppliers of materials, suppliers of capital, and society to plan an acceptable business strategy. This has extended into stakeholders' concerns with the argument that value has a broader context, one that links customers to supplier organizations and the investors involved in "value" creation and delivery. It is important to consider what we mean by "value" and to review recent contributions to its expanding use. "Value" in the context of consumer and organizational transactions implies value to be a positive net result of benefits received by adding value to both customer and supplier, less the costs involved.

Keywords Value · Value proposition · Value migration · Shared value · The sharing economy · Customer journey · Customer experience · Customer cohorts

Introduction

From a simple but clear view of "value", value can be viewed as the residual of benefits received, less the cost of acquisition, but other dimensions of value have recently been added to the vocabulary. Concurrently the use and development of the notion of value has expanded. From a somewhat vague inference of a "quality/quantity-monetary" transactional expectation, it has taken on a more meaningful context.

© Springer Nature Switzerland AG 2020
D. Walters, D. Helman, *Strategic Capability Response Analysis*,
https://doi.org/10.1007/978-3-030-22944-3_1

Value is now being addressed in a much broader context as the longer-term implications of a customer focus is explored. While the context of the Slywotzky and Morrison (1997) view of value and the entire process of value creation, production, and delivery downstream is important, there are now emerging views concerning the overall activities and relationships of the value creation and delivery process. Many of the perspectives are a resurrection of earlier work: the stakeholder concept can be dated back to work undertaken and published by the Stanford Research Institute in the 1960s and that had been influenced by earlier work in Lockheed's planning department. SRI argued that management needed to be cognizant of the concerns and interests of shareholders, employees, customers, suppliers of materials, suppliers of capital, and society to plan an acceptable business strategy. This has extended into stakeholders' concerns with the argument that value has a broader context, one that links customers to supplier organizations and the investors involved in "value" creation and delivery. Pine and Gilmore (1998) suggested the economic context of value can be considered as a progression commencing with low price-low differentiation extracted commodities (raw materials), manufacturing, service delivery, and staged experiences, high price-high definition.

It is important to consider what we mean by "value" and to review recent contributions to its expanding use. While "value" in the context of consumer and organizational transactions implies value to be a positive net result of benefits received by adding value to both customer and supplier, less the costs involved in acquiring the value, additional perspectives should be considered and discussed.

Corporate Perspectives of "Value"

Shareholder Value-Led Management

The *value-based management* "fad" of the 1980s/1990s largely ignored the SRI thinking focusing instead on an emphasis on shareholder value management (strategies to enhance the wealth of shareholders) – which largely ignored the interests of the other groups. They argued that the responsibility of senior management was to maximize the value of the organization, but this failed for two primary reasons: the first was due to their requirements for very detailed information and the cost in time and managerial effort to produce the data and the second because most of the models ignored the need for customer interests to be considered let alone to consider customer satisfaction a focal issue. Interestingly the emphasis is now on the argument that customer value drives shareholder value. Payne and Holt (2001) cite Cleland and Bruno (1996) who suggested an organization should ensure that its customer value strategies are successful in generating revenues that deliver levels of profitability (margins) that exceed its cost of capital sufficiently such that it (the organization) consistently builds wealth for the shareholders. Doyle (2000) was more succinct in emphasizing that shareholder value maximization requires a focus on delivered customer value and proposes this as a primary objective for marketing.

Freeman (1984) proposed the notion of identifying stakeholder management as a means by which rigor could be put into managing the relevant groups and their interests that impact and influence an organization's pathway towards reaching its objectives. This text develops a cross-activity evaluation approach to strategic management; it is based upon observation of the behavior of a range of organizations by size and response to market and industry dynamics, such that both market and value chain positioning are variables. Possibly one of the largest international industrial activities, the automotive industry is demonstrating changes that only a few years ago would not have been considered, let alone seen as a competitive necessity.

A shareholder-only approach usually has several characteristics, such as: narrow focus, driven by quantifiable metrics, executive management may react to poor evaluations with extreme responses (cost reductions, restructuring, etc.), tangible, financially focused – performance evaluation with little emphasis on intangible performance drivers, the concept of "value" (and who it is generated for and how) is not clearly understood or articulated, new ideas and methods are not easily understood and accepted, management tends to embrace quick fix solutions too quickly, sometimes adopting irrelevant solutions, people who create relevant "value" often are viewed as "too radical" and lack understanding of the organizations value proposition, the profit focus is on earnings based metrics, on traditional approaches to growth that allocate resources to marketing, acquiring other companies, and controlling costs, rather than considering partner stakeholder expectations. In other words, business success is measured by what we create for our *shareholders.*

However, an increasing number of organizations now see the role of the CEO as defining the business in terms of "a broader purpose," that is, one that benefits a larger set of *stakeholders* (customers, employees, suppliers, and future generations) than would otherwise gain from simply increased profits and shareholder value and indeed shared value. It follows that the *broader-purpose* approach requires strong data support; Industrié4.0 makes this possible.

Stakeholder Value-Led Management

By contrast, stakeholder thinking tends to be deeper and broader and may include the following characteristics: a sustainable, visionary, and competitive-thinking approach, involving a multi-dimensional view of the organization. Performance evaluations are used to measure strategic achievements, as well as the operations delivery systems that extend across the entire value chain. Stakeholders can be classified as primary and secondary (external) stakeholders. Primary stakeholders are customers, employees, management, shareholders and investors, and partners suppliers and distributors. Secondary stakeholders comprise competitors, the community, and government. Primary stakeholders welcome open, close, and ongoing relationships with organizations, involving the exchange of creative ideas and innovation, management approaches to change, and performance improvements. The interests of secondary stakeholders have specific focus: competitors monitor

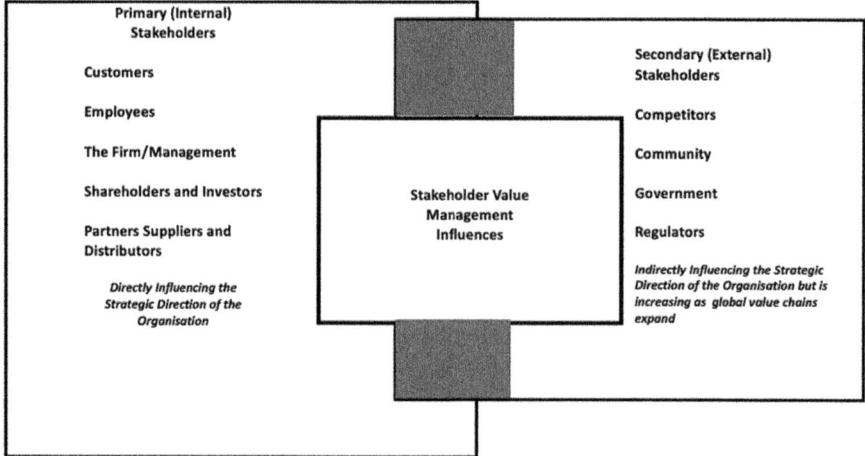

Fig. 1.1 Stakeholder hierarchy

progress and have interests in opportunities for collaboration (typically co-opetition, licensing patents, and using any excess capacity they have to manufacture and distribute competitors product-services); the community is interested in employment opportunities and expansion of revenues (and therefore in local taxes), and government has interests that include employment, innovation, exporting activities, and a contribution to GDP. In other words, business success is measured by the value we create for all *stakeholders*. See Fig. 1.1.

Stakeholder theory is, essentially, a theory of organizational management and addresses morals and values in managing the "total organization." It was originally detailed by Freeman's (1984) article on stakeholder theory in the *California Management Review* in late 1983 and is largely attributed for the development of the concept based upon internal discussions in the Stanford Research Institute. This was followed by the publication of *Strategic Management: A Stakeholder Approach* by Freeman in 1984. His book identified and models the groups which are stakeholders of an organization and recommended methods by which an organization's management can (should) consider the interests of those groups. The traditional view of the firm is that shareholders are the owners of the company and the firm has a legally binding duty to put their needs first, to increase the value of the firm for them. The stakeholder view argues that there are other parties involved, including customers, employees, shareholders/investors, partners (suppliers and distributors), competitors, the community, and government which are counted as stakeholder, their status being derived from their capacity to affect the firm and its other morally legitimate stakeholders. Drawing on crowd-sourced knowledge – a Wikipedia entry suggests two categories of influential stakeholders; *primary stakeholders* can be defined as those having formal, official, or transactional relationships and have a direct and necessary impact on the organization. *Secondary stakeholders* can be defined as those who in the past, present, or future influence or might be influenced by the

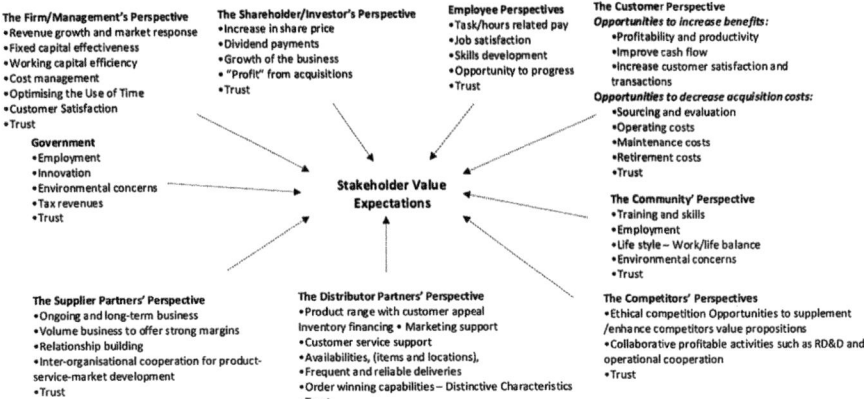

The Firm/Management's Perspective
• Revenue growth and market response
• Fixed capital effectiveness
• Working capital efficiency
• Cost management
• Optimising the Use of Time
• Customer Satisfaction
• Trust

Government
• Employment
• Innovation
• Environmental concerns
• Tax revenues
• Trust

The Shareholder/Investor's Perspective
• Increase in share price
• Dividend payments
• Growth of the business
• "Profit" from acquisitions
• Trust

Employee Perspectives
• Task/hours related pay
• Job satisfaction
• Skills development
• Opportunity to progress
• Trust

The Customer Perspective
Opportunities to increase benefits:
• Profitability and productivity
• Improve cash flow
• Increase customer satisfaction and transactions
Opportunities to decrease acquisition costs:
• Sourcing and evaluation
• Operating costs
• Maintenance costs
• Retirement costs
• Trust

Stakeholder Value Expectations

The Community' Perspective
• Training and skills
• Employment
• Life style – Work/life balance
• Environmental concerns
• Trust

The Supplier Partners' Perspective
• Ongoing and long-term business
• Volume business to offer strong margins
• Relationship building
• Inter-organisational cooperation for product-service-market development
• Trust

The Distributor Partners' Perspective
• Product range with customer appeal Inventory financing • Marketing support
• Customer service support
• Availabilities, (items and locations),
• Frequent and reliable deliveries
• Order winning capabilities – Distinctive Characteristics
• Trust

The Competitors' Perspectives
• Ethical competition Opportunities to supplement /enhance competitors value propositions
• Collaborative profitable activities such as RD&D and operational cooperation
• Trust

Fig. 1.2 Identifying network stakeholder expectations

firm's strategic and operational decisions without being directly engaged in transactions with the firm in question and thus are not essential for its survival.

Figure 1.2 identifies the varied expectations of stakeholders. Clearly these differ based upon the perspectives of each stakeholder group. The relative importance of the shareholder group's influence is likely to differ depending upon its purpose and structure, for example, consider the differences between for-profit and not-for-profit organizations. Furthermore, their impact on decision-making will also differ. We suggest *primary stakeholders* will have an important influence on the strategic direction of the organization. By contrast *secondary stakeholders* are more likely to influence the strategic direction of the organization indirectly. A rider to this would be the legislative role of government concerning health and safety issues; these we suggest are taken as given aspects of the organization's business environment and accepted as constraints to all organizations. Figure 1.2 suggests a positioning of primary and secondary stakeholders, accepting and indicating there to be an overlap of influences.

Jensen (2001) identified the potential conflicts presented by stakeholder value management: attempting to maximize multiple objectives. Because stakeholder theory does not specify how to make the necessary trade-offs among competing interests, it leaves managers with a theory that makes it impossible for them to make purposeful decisions. Jensen described *enlightened value maximization*, and it is identical to what he identifies as *enlightened stakeholder theory*. Enlightened value maximization utilizes much of the structure of stakeholder theory but accepts maximization of the long run value of the firm as the criterion for making the requisite trade-offs among its stakeholders. Enlightened stakeholder theory specifies long-term value maximization (or value seeking) as the firm's objective and, as the author suggests, solves the problems that arise from the multiple objectives that accompany traditional stakeholder theory. This also suggests a problem concerning the differing time horizons of each of the stakeholder groups. Supplier and distributor stakeholders are likely to have shorter time spans that are based upon period-based operating (and cash2cash) cycles. Customer expectation time spans will vary depending upon product-service characteristics such as frequency of purchasing, the item's share of total spend, and commitment to the brand/supplier of the item.

Stakeholder recognition is becoming increasingly important for several reasons. The significant increase in partnerships, particularly those occurring between large multinational organizations and much smaller specialist companies, emphasizes the importance of producibility and people in the business model. The application of the benefits of connectivity, transparency, reach, richness, and time offered by Industrié 4.0 is becoming widely accepted and implemented.

The business media is consistently bringing focus onto stakeholder management by its reporting of militant shareholders and large individual investors contesting executive pay scales and bonuses. Large organizations that once might have ignored "messages" are now responding. For example, GE's CEO announced reductions and said it would cut management bonuses if it failed to meet financial targets adopted after talks with activist investor Nelson Peltz. CEO Jeffrey Immelt adopted the goals after pressure from Peltz's Trian Fund Management, which took a stake in GE in 2015 after the company slimmed down lending operations to refocus on making jet engines, gas turbines, and oil field equipment. The shares dropped 5.5% during a 12-month period, while the S&P 500 Index advanced 14%. Immelt retired from GE mid-2017. This is but one example among many and emphasizes the growing importance of identifying and understanding this diverse group of "influencers." This is a particularly important development for network structured activities. BP, which had pledged to change the way it pays executives, asked investors to approve a new remuneration policy. A Bloomberg (2017) report commented that many BP shareholders voted against the CEO's 20 percent pay increase after the company reported a record net loss in 2015 and announced thousands of job cuts following the slump in oil prices. The revolt sparked investor discontent about compensation at other European companies.

Amazon is an interesting business in the context of stakeholder value management. Investors anticipate both an extraordinary rise in revenue, from sales of $136bn in 2016 to half a $trillion over the next decade, and a jump in profits. The hopes invested in it imply that it will probably become more profitable than any other firm in America ("Amazon, the World's Most Remarkable Firm" 2017). Amazon's focus is on the distant horizon. *Amazon emphasizes continual investment* to propel its two principal businesses, e-commerce and Amazon Web Services (AWS), its cloud-computing arm.

In e-commerce, the more shoppers Amazon attracts, the more retailers and manufacturers want to sell their goods on Amazon. That gives Amazon more cash for new services—such as two-hour shipping and streaming video and music—which entice more shoppers. Similarly, the more customers use AWS, the more Amazon can invest in new services, which attract more customers. Amazon's extensive platform structure is creating a commercial infrastructure that many of its competitors use. A third platform cycle is starting, Alexa, the firm's voice-activated assistant: as developers build services for Alexa, it becomes more useful to consumers, giving developers reason to create yet more services, and history repeats itself.

If shareholders maintain their current expectations, Amazon's model resembles a self-fulfilling prophecy. The company will be able to keep spending, and its spending will keep making it more powerful.

But, it is suggested there is a real problem with the investor stakeholder expectations surrounding Amazon ("Amazon, the World's Most Remarkable Firm" 2017) – If it gets anywhere close to fulfilling them, it will attract the attention of regulators. As it grows, so will concerns about its power. Even on standard antitrust grounds, that may pose a problem: if it makes as much money as investors hope, *The Economist* suggests its earnings could be worth the equivalent of 25% of the combined profits of listed, public Western retail and media firms. Amazon's business model will also encourage regulators to think differently. Investors value Amazon's growth over profits; that makes predatory pricing more tempting. In future, firms could increasingly depend on tools provided by their biggest rival. If Amazon does become a utility for commerce, the calls will grow for it to be regulated as one. Shareholders are right to believe in Amazon's potential. But success will bring it into conflict with an even stronger beast: "government." The value proposition now has the added function of demonstrating its ability to generate value for a stakeholder group, not just a segment of potential customers.

Network Organizations

Value in networks is generated by collaboration on shared objectives and coordinated operations. By leveraging assets and allocating roles and tasks, a network approach optimizes its resource capabilities in terms of investment in PRODUCT-service/product-SERVICE offers resulting in a value proposition that meets customer expectations, optimizes capabilities, and is economically viable. Most successful networks are structured around a principle of interlocking financial engineering (resurrecting and updating existing assets, General Motors used an existing assembly plant and outsourced (currently) expensive components (batteries) in the knowledge that their cost will decrease and enlisting debt investors who would be happy with lower returns, if they are steady, and to put investment into research projects rather than into companies and operational systems) (Gapper 2012; Stock 2016).

Value Chain Network 2.0 is described by Fig. 1.3 and includes the impact of the connectivity features of Industrié 4.0 and of the expanded notion of stakeholder-led value management. The significant developments of VCN 2.0 are those brought about by digitizing product-services and their production processes. Notable among these are the notions of *digital thread* data generation (and its use in connecting the physical world to the digital world) and the *digital twin* (a cyber (software) copy of a hardware product). Figure 1.3 suggests that digitization has been applied to most value chain network operations processes; production planning, procurement, production process management, and managing demand/supply response lead times. Digitization has impacted vendor/customer relationships by

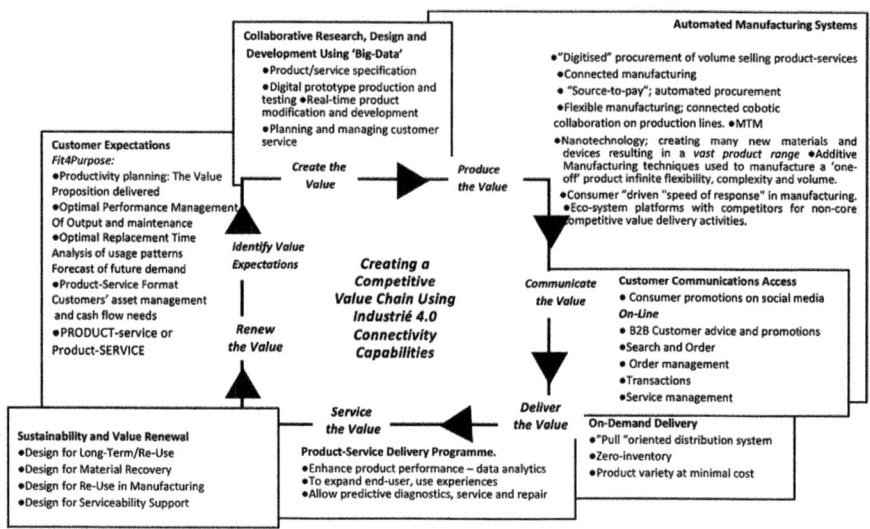

Fig. 1.3 The "Connected Value Chain Approach" to value delivery: Value chain network 2.0

using digital thread/digital twin combinations (Predix, General Electric, and TeamCenter, Siemens) to manage ongoing product-service modification and maintenance management programs.

Collaboration can be seen working in several ways in the global automotive industry. To maintain range credibility, major manufacturers need to offer specialist versions of their core brands; these typically are low volume vehicles; 20/30,000 vehicle per year, a small volume compared to the mainstream production vehicles. Usual practice is to outsource the assembly of the specialist vehicles to network partners who have the capability to manufacture the specialist parts required and to assemble this relatively low volume. Figure 1.4 illustrates the principle involved; Magna International (and other similar organizations), a global brand in manufacturing vehicle component parts for the high-volume manufacturers, undertakes this role within the industry.

Shared Value

Shared value is creating economic value in a way that also creates value for society by addressing its needs and challenges and involves understanding and building stakeholder expectations into the network organizations' value proposition. Shared value involves creating economic value in a way that also creates value for society by addressing its needs and challenges. *Shared Value* creates economic value for the

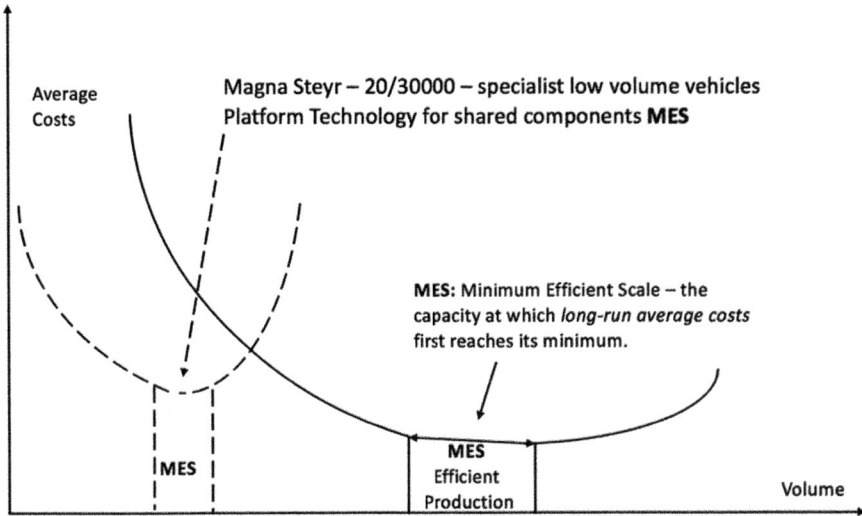

Fig. 1.4 Strategic costs: Economies of scale, MES and market-volume-profit

corporation through innovations that address society's needs and challenges. Porter and Kramer (2011) suggested companies create shared value in three ways:

- Reimagining products and markets (GE, Panasonic)
- Redefining productivity in the value chain (relevant positioning and inputs)
- Developing local business clusters (USA: restaurants using and promoting locally sourced produce)

If shared value involves creating economic value in a way that also creates value for society by addressing its needs and challenges, then businesses must reconnect company success with social progress. Shared value is not social responsibility, philanthropy, or even sustainability, but a new way to achieve economic success. It is not on the margin of what companies do but at the center. We believe that it can give rise to the next major transformation of business thinking.

Nestle has recently established a "Creating Shared Value Board" which acknowledges the company's remit includes shareholders, people, and the supply chain. It also considers the environmental footprint of factories and occupational health and safety (Thirkin 2017). Walmart and Johnson & Johnson are developing shared value initiatives that achieve new levels of social impact while improving corporate profitability. These involve product design that incorporates remanufacturing of selected components, packaging alternatives, and inputs into FMCG products. Dow AgroSciences developed a line of omega-9-rich canola and sunflower oils, with zero trans fats and the lowest levels of saturated fats. Since 2005 omega-9 oils have eliminated nearly a billion pounds of trans fat and 250 million pounds of saturated fat from North American foods. Companies can also improve the competitive context in which they operate by investing in their communities. Nestlé, for example, worked closely with the farmers of the Moga Milk District in India, investing in local infrastructure and transferring world-class technology to build a competitive milk supply chain that simultaneously generated social benefits through improved healthcare, better education, and economic development.

Consumer Perspectives of Value

The Sharing Economy

The sharing economy is an economic model often defined as a peer-to-peer economy (P2P)-based activity of acquiring, providing, or sharing access to goods and services that are facilitated by a community-based online platform. A peer-to-peer economy is viewed as an alternative to traditional capitalism, whereby business owners own the means of production and the finished product, hiring labor as necessary to carry out the production process the sharing economy, is clearly understood by the major automotive manufacturers as indicated by the partnerships entered by major automobile manufacturers (Toyota/Uber, VW/Gett, and GM/Lyft) and hotel and short-term accommodation, as demonstrated by the Qantas addition of Airbnb as an option on its website.

The sharing economy is one in which online platforms coordinate hundreds of thousands of freelancers to drive cabs, rent rooms (Airbnb), and clean laundry (Washio) has arrived. Successful clothes sharing activities are becoming popular. For a monthly fee, members can "visit" the store and hire clothes for specific occasions or simply because they "want a change." Availability is published and "wait lists" exist for "in demand" items.

While it has grown rapidly and continues to do so, the sharing economy is not without risk. Gapper (2014) identified issues that (or will) require some attention: "As companies recognize the threat, governments and regulators are struggling to adjust and consumers are unsure whether to trust the new type of business. However, the greatest uncertainty faces workers. As self-employment, start-ups and one-person 'micro-businesses' comprise a larger share of the workforce, workers are becoming freer and more at risk."

However, Gapper (2014) suggests the growth of freelance working has been accompanied by some concerns. The twenty-first-century freelance worker has no guaranteed work benefits and insurance, a hallmark of employment conditions of the twentieth century. There is a need to find new methods that can meet these changes in the work place that do not simply consider these perks belong to direct employment only, leaving people who work differently without protection. Gapper (2014) cites Sundararajan, at New York University's Stern School, who is concerned that such support could "just slip away." And: "Steady incomes and a social safety net are characteristics of a healthy economy which has moved past simply getting people to work for a living to creating a higher quality of existence."

The freelance economy brings two challenges. First, some freelance jobs are low-paid forms of direct employment. Companies label workers as "independent contractor", to avoid paying employment taxes and indirect benefits while treating them as employees: they must wear uniforms, obey rules, and so on; many are low-paid workers, such as delivery drivers or warehouse stackers. And second, even if workers are self-employed, the company or platform that "employs" them could choose to offer more than the minimum benefits. Employers traditionally provide health and pension plans, as well as training, to create and maintain a productive, reliable workforce. It is more expensive, but if it pays off in the standard of service they offer, then it will help them to beat lower-quality competitors. In the event of companies abdicating the role, then society needs to devise other ways to offer long-term support and security to the self-employed, as the Freelancers Union (USA) and others have been attempting to do (Gapper 2014).

Slee (2017) offers another perspective, suggesting what started as an appeal to community, person-to-person connections, sustainability, and sharing, has become the playground of billionaires, Wall Street and venture capitalists. He suggests: "The promise of a more personal alternative to a corporate world is instead driving a harsher form of capitalism: deregulation, new forms of entitled consumerism and a new world of precarious work. There is a lot of talk of democratization and networks, but what's happened instead is a separation of risk (spread among the service providers and customers) from reward, which accrues to the platform owners". And; "Despite the claims of ecological sustainability embodied in ideas like 'access over ownership' and the reuse of excess capacity, the on-demand sector is instead encouraging a new form of privileged consumption: 'lifestyle as a service'."

Clearly there are issues that may undermine the growth of the sharing economy. Carlos Ghosn, then Head of the Alliance (Renault, Nissan, and Mitsubishi), when asked questions related to the sharing economy during a BBC Business interview (Leggett 2018), suggested the automotive industry was aware that the growth in shared-use vehicles was gathering momentum. The Alliance has launched a car ownership product-service between several people based upon their social media profiles (Campbell 2018). Ride hailing apps such as Uber and Lyft have already proven very popular, as have short-term car rental schemes, and some experts believe that the development of automated taxis will accelerate this process and that the days of personal car ownership are numbered, at least in cities. However, car ownership in China, India, and Indonesia is very low in comparison with the rest of the world and this offered opportunity for continued growth for automotive manufacturers; adding, the industry is building systems designed for the sharing economy that will provide new avenues for growth: "Ownership of cars will continue. Maybe it'll be limited in mature markets to the benefit of shared mobility services. But we're going to be involved in both sides" (Leggett 2018).

Customer Journeys: A Sum of Experiences

The *customer journey* is the complete sum of experiences that *customers* go through when interacting with your company and brand. Instead of looking at just a part of a transaction or experience, the customer journey documents the full experience of being a customer. A McKinsey Executive Briefing: 2016 suggests it necessary to understand the *total transaction as the customer* sees it, using their research and consulting experience to make some significant points:

- A "journey" is a progressive series of "touchpoints" that add up to the *experience* customers are given when interacting with staff (and online facilities). Seeing and understanding their perspective helps structure the experience and helps structure the value proposition.
- Shaping the customer experience requires the organization to shape interactions into a coherent set of sequences and processes and the implications these may have for relevant activities that may involve other members of the value chain network as well as identifying an opportunity to apply a relevant contribution from the *Industry Dynamics Portfolio* (knowledge, technology, processes, relationships, regulatory compliance, and geopolitics; the topic of the next chapter).
- Rethinking and restructuring the organization (and the value proposition), network processes, and activities. This may involve culture change and changes in management style throughout the value chain network.

The advent of the data management and analysis facility has enabled both customer use behavior and product-service response data to be monitored, collected, analyzed, and "modelled." Artificial Intelligence has been developing algorithms that predict future customer journeys.

Creating Unique Customer Experiences as Expectations

"An experience occurs when a company intentionally uses services as the stage, and goods as props, to engage individual customers in a way that creates a memorable event. Commodities are fungible, goods tangible, services intangible, and experiences *memorable*". Pine and Gilmore (1998)

To build internal momentum for initiatives to develop a unique customer experience, a company must understand how that helps it perform distinctively in the market. The conviction and shared aspiration that stem from understanding the customer experience an organization wants to deliver can not only inspire, align, and guide it but also bring innovation, energy, and a human face to what would otherwise just be strategy. Lewis and Jacobs (2018) provide a number of examples from Japan that illustrate that consumers rate the experience of obtaining the product (the service) greater than the product. Their article provides a number of examples; one is in a pancake restaurant where the clientele is attracted by the experience of waiting (queuing) and the ambiance as their reason for patronizing the restaurant.

The customer experiences an organization wants to provide can vary widely. For some companies, this transformed experience represents a step change. For others, the aspiration may, at least in the short term, require only more modest changes. Either way, the aspiration will translate into an overall mission and, ultimately, into guiding principles for frontline behavior. Past performance and whatever helped satisfy customers in the past can often make small changes valid in the short term. But an understanding of the *fundamental* wants and needs of customers is an essential step in determining what a great experience for them should look like is necessary, if the organization is to create a *Value Advantage* (Boyarsky et al. 2016).

McKinsey (Executive Briefing: 2016) found that several key questions commonly underpin successful stories and strategies:

- How a company's approaches change in the short term. Does it have a goal to change the customer experience fundamentally or simply to improve it at the margins?
- What is the gap between the needs and wants of customers and what they do experience?
- How can the company gain a customer-experience advantage against competitors?
- At which point in the experience should the company concentrate to have a real impact?
- How do the *overall capabilities* of the staff (a people issue) support the customer experience the company wants to provide?

Customer aspirations are important inputs (their performance expectations) – as is that of the vendors, for whom the economic response affects the performance of the supplier business. To understand both interests, considerable research must be undertaken.

The McKinsey study suggests gathering and segmenting data are classic starting points in understanding customers. But data are not enough. Successful customer-experience efforts apply a human filter to the collected data to ask overarching questions. Exactly who are my customers as individuals? What motivates them? What do they want to achieve? What are the fundamental causes of satisfaction? Tackling these questions requires a concerted analytical effort, which helps an organization design and implement a more sophisticated program and, critically, persuade employees to embrace its goals. Four fundamental questions need to be answered if these questions are to be successfully addressed:

- What does the customer desire from the experience? Some needs are stated, but it is important to understand that many are not.
- What is the customer's underlying (possibly unstated) objective or purpose?
- What preconceived notions, positive or negative, does the customers have about the product-service experience?

- What emotions do customers have that are likely to experience, for example, anxieties concerning the product-service delivery (e.g., the time taken to "acquire delivery" reaching transport terminals)?

How employees deliver the experience is an important input: "A customer experience begins with employees who know about it, care about it, and are well positioned to deliver it" (Boyarsky et al. 2016). Organizations should ensure "front-line workers" experience the customers' experiences. Delivering the customer experience requires the understanding of an experienced employee, one capable of an empathetic response to customer questions.

A conclusion from the study suggests customer-experience programs require shared aspirations that serve as a direction for strategic decisions and execution. Otherwise, a corporate strategy to improve the customer experience will go only so far. Successful value delivery customer-experience transformation and the way stakeholders would achieve it require some basic inputs for success: *safety* (a priority for product-SERVICEs), *comfort* (*physical,* the quality of the experience in seating, entertainment, restrooms, interactions with employees, etc. and *non-physical* assurances of safety, reliability, etc.), *convenience* (providing time and effort saving service options), and *speed* (compressing time by introducing serviceability augmentation, reducing bottle necks in customer reception processes).

Customer Cohorts Rather than Categories

Related groups, or cohorts, usually share common characteristics or expectations within a defined time span or an experience. Its members share a significant experience at a point in time; a group of people who share a defining characteristic, typically who experienced a common event in a selected period, such as birth or graduation, forming a cross section at intervals through time.

Product-service providers can use a cohort approach to develop successful related businesses. Examples include airline frequent flyer groups, who target product-related interests, typically travel, but can be extended to include wine selections and restaurants.

Craft-based businesses that encourage continuity and increased involvement in product-service use by creating differentiated activities, such as achieved by TOFT, a UK alpaca wool producer who increased the sale of the basic product by offering applications classes (knitting and crocheting) and designing a range of animals that have captured the imagination of customers who purchase "kits" and build collections. Internet gaming has a range of theme-related interests in which cohort groups can be identified.

Identifying cohorts has advantages. It increases customer loyalty and the potential for communications with a specific customer interest group. Traditional segmentation focuses on demographics, socioeconomics, brand, and store loyalty; identifying cohorts targeting is focused on an interest.

Value Migration

New business models can lead to industry disruption and make it possible to design and be aware of these changes as they occur. Value migration occurs when customer needs change or a new business design begins to encroach on an existing configuration of providers. Businesses which previously enjoyed a leadership position are now caught in an outflow of value and declining revenue and customer bases. Slywotsky (1996) focused on how business designs (business models) can adapt to changing customer needs and create new demand. He suggested that creation of new technologies is no longer required for growth, nor is it necessarily the best path:

> In the age of business (model) design, success depends on the speed and skill with which competitors understand these designs and improve and adapt them to their particular customers.

Slywotsky (1996) argued that new models offer the same benefits to customers but at lower cost by changing the model structure. This change often results in a restructuring of profit sharing throughout the business model by restructuring activities in the value chain network. Customer-led value migration occurs with changes in their applications needs; for example, extracting fossil-based forms of energy (the introduction of fracking techniques) or a shift in consumer lifestyles involving time budgets and the preference for prepared, microwaveable meals.

Slywotsky (1996) suggested, in a value migration world, our vision must include two, three, or even four customers along the value chain. So, for example, a component supplier must understand the economic motivations of the manufacturer who buys the components, the distributor who takes the manufacturers products to sell, and the end-user consumer. Slywotsky (1996) provides some questions to aid awareness of value migration occurring in an industry. The automotive industry is undergoing significant changes, borrowing versus ownership (the trend in mature markets towards shared mobility services), and in aerospace the trend towards purchasing product-service outputs rather than the product. Customer and market value migration is increasingly being monitored on a regular basis to identify:

- Who is the customer? Are decision-makers and influencers changing?
- How many distinct new business designs have been introduced in the industry in the past 5 years?
- Identify the value migration event(s) that enabled the company to establish the present position. Who are the defeated incumbents?
- Hypothesize a date at which the current business design will be obsolete and why.
- Compare the current [business model] to that of competitors: traditional, newcomers, and non-traditional operators. Whose economic logic is more compelling?

Value migration is likely to accelerate as the benefits of the "connectivity" features of Industrié 4.0 become routinized and monitoring and comparison of competitive value offers of the organization with its competitors. The shift in emphasis away from the Capex business model (high fixed capital intensity) towards an Opex model (low fixed capital intensity) with stronger relationships with network partners is in evidence across a range of industries that include automotive manufacturing, healthcare (imaging interpretation), and education (the Emeritus Institute of Management programs now being offered by Columbia Business School, MIT Sloan, and the Tuck School of Business offering a collaborative network focusing on practicing managers).

A Value-Led Stakeholder Approach to the Value Proposition Using Industrie 4.0 and Value Chain Network 2.0

The genesis of the value proposition concept can be traced to a project undertaken by McKinsey & Co. in the 1980s. Brief mention only was made of the concept at that time (Bower and Garda 1985). Later, Lanning and Michaels (1988) defined value proposition as a statement of benefits offered to a customer group and the price a customer will pay. Their approach was presented as a "value delivery system" involving three steps which highlighted the critical importance of communicating value. The framework; choose the value; provide the value; and communicate the value were the basic characteristics of the McKinsey model. These authors argued that business success is constrained when companies adopt a value orientation that focuses on making and selling a product. The term "value delivery system" collectively referred to the formulation and implementation of a value proposition. The authors suggested it appeared to do so in a supplier-led manner. We would argue that this probably was, at the time, realistic, as it reflected an understanding of the customers' expectations and was modified by suppliers' expectations; however, the Amazon business model suggests maybe the value proposition should be more comprehensive and identify the expectations of other stakeholders.

Typically, a value proposition describes the bundle of product-services that create value for a specific market, segment, or customer. It is the reason why customers select one company over another – and often the reason why they return! The value proposition is a response, a solution, to an identified customer need. As such it should be afforded rigorous evaluation; the performance, profitability, productivity, partnership, producibility, and preservation model offer an opportunity for a "due diligence" approach. A value proposition consists of a package of products and/or services – or possibly a reimagined PRODUCT-service that has become a product-SERVICE because of changes in customers' business models, product-service applications, or a reformulation of the value delivery; an example is the transformation of a PRODUCT-services into a product-SERVICE capital goods as purchasers are considering the benefits of purchasing performance outputs rather than the physical product that produces the output.

The acceptance of the notion of "customer-centricity" implies that the value proposition is a purpose-led bundle of benefits to targeted customers. Some value propositions may be innovative solutions to problems; others may be tangible market entrants offering alternative performance management, asset management, profitability, productivity, or time and/or location management benefits – in other words an alternative business model application. They all should include reference to acquisition and other life cycle costs.

Satisfying customer expectations requires an understanding of their product-service purchasing and their applications activities: a "value proposition" spells out a response to these. Webster (1994) contends that positioning and the development of the value proposition must be based on an assessment of the product offering and of the firm's distinctive competencies *relative to competitors*. Hence the value proposition should make clear its *value competitive positioning*: how it delivers a value contribution that increases revenues and profitability by enhancing the organization's own value proposition to its customers or reduces the customers' acquisition and life cycle costs. In doing so it should communicate to the target customer the distinctive competence portfolio of the value chain network participants, demonstrating that it extends its collective skills and resources beyond the current dimensions of competitive necessity into creating competitive advantage that, in turn, offers customers an opportunity to do likewise.

Industriė 4.0 provides an opportunity to sharpen the focus of the value proposition. The provision of detailed, timely, accurate data together with data analytics techniques has enhanced the usefulness of the value proposition in the context of application (a wider target stakeholder and application focus). The use of gathering detailed data from customers, its analysis, and interpretation is proving to create greater value than ever before as capital equipment providers use the digital thread and twin concepts to assist their client companies with improving their operational performance. Figure 1.5 illustrates the structure of this relationship, demonstrating the interactive relationships between network partners that collaborate and coordinate optimal, efficient materials and cash flow management.

The value proposition is an integral activity within a strategy process. Figure 1.6 illustrates this point. Following the evaluation of the value proposition and modifications made following the appraisal by the *Response Capability Resources*

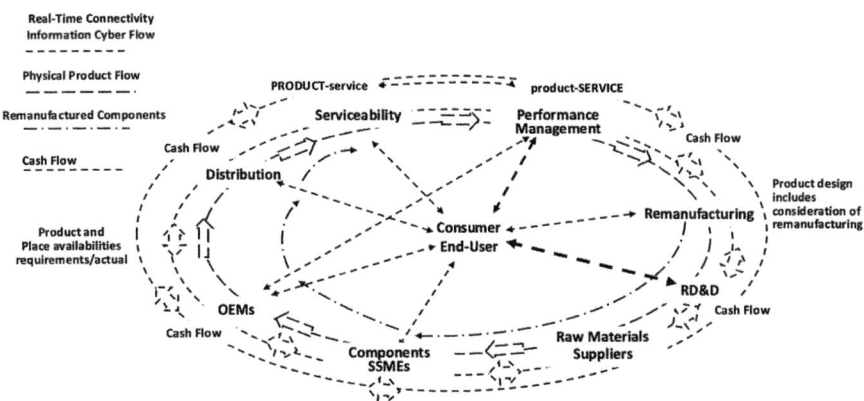

Fig. 1.5 The fusion of managing stocks and flows of information cash and product-services

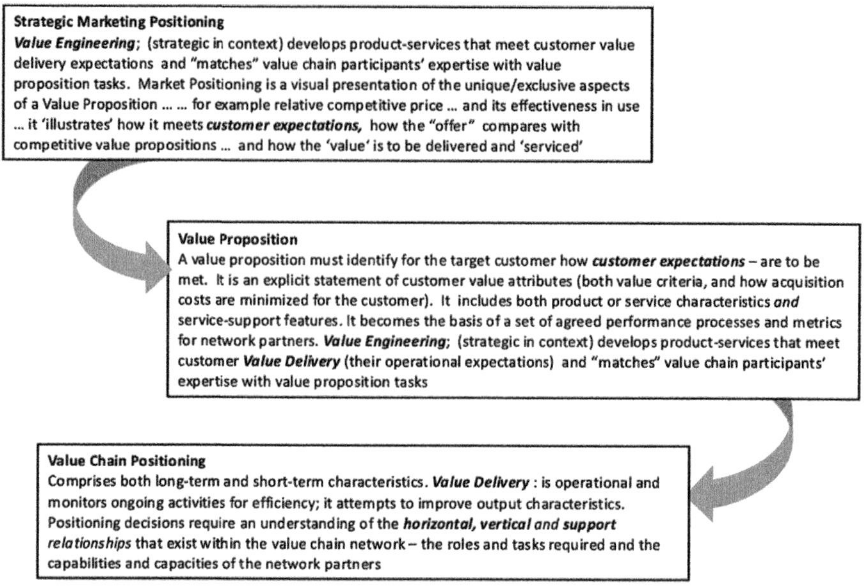

Fig. 1.6 The value proposition is an integral activity within a strategy process

Business Model the activity switches to a consideration of the network's *Strategic Marketing Positioning*. While this may be an ongoing feature of the network organization, this does not preclude regular reviews based upon market opportunities and capabilities and response capabilities. Given the outcome of this review, consideration can be given to the structure and the positioning of network partners, a review of overall value chain network positioning. The reluctance to move (or possibly retire) from the network is a real possibility. Some interesting examples from agribusiness activities are emerging as the application of connectivity, sensors, and machine-led decision-making is being applied to centuries of agricultural "nous," for example, the application of drones that survey farmland fertility and manage crop spraying accuracy and autonomous tractors using sensors and GPS to conduct specific tasks.

The process involves a collaborative involvement of all management activities. This is suggested by Fig. 1.7. The process identifies (and takes into consideration) the possible alternative responses given the changing approaches emerging concerning ownership; it also considers competitive concerns; these comprise a retaliation response or, alternatively, one of a collaborative response, which depends upon the competitors' capabilities to meet the target customers' requirements. Figure 1.7 creates a value proposition by first understanding the target customers' product-service expectations; these are matched with the organizations' resource capabilities (usually the network partnership) and often those of competitors who may have relevant unutilized resources. The offer is then reviewed with the target customer(s) and may need modifying to accommodate specific customer requirements or perhaps for changes in the vendor network. From this a "realistic" value proposition

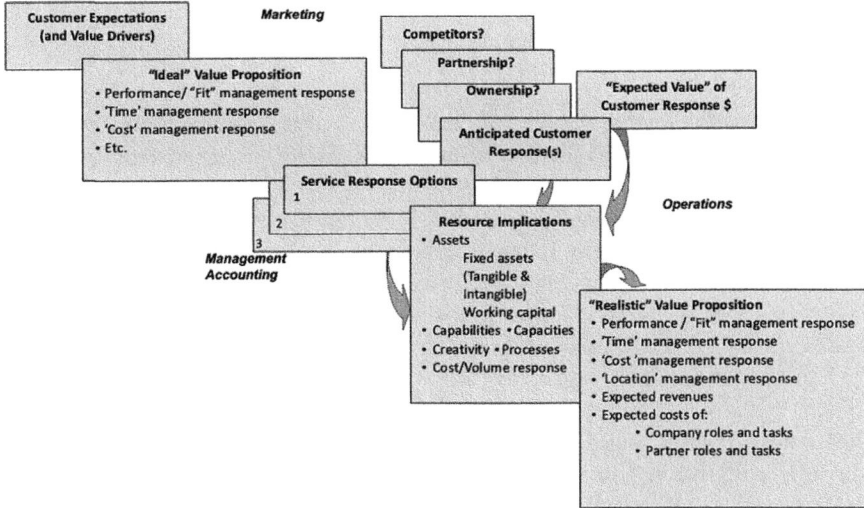

Fig. 1.7 Deriving a workable customer value proposition

will emerge, one that meets both customer and vendor stakeholders' expectations. Clearly the process can be a complex activity, and several alternatives may emerge before agreement is reached. Figure 1.7 suggests a procedure typically followed.

Value propositions have been focused on the primary stakeholder, predominantly the end-user customer. Given the capabilities offered by Industrié 4.0, the data linkages of digital thread connectivity and the digital twin (the software "copy" of the product-service offer) is beginning to create opportunities for "rolling value propositions," offering performance management and digitally managed modifications to the product and the surrounding service package. Barnes et al. (2009) approach developing the value proposition task by challenging the ability of the traditional business model approach that was in many ways a projection of the current strategic status and proposed a *value focused model* with more enlightened components:

Value-Focused Strategy
- Where will the organization be in 3/5 years?
- Who are the primary stakeholders? What are their expectations?
- Who are the secondary stakeholders? What are their expectations?

The Value Proposition
- What is/will be the VALUE PROPOSITION (market, customer benefits and acquisition costs, delivery model, distinctive differentiation, PRODUCT-service or product-SERVICE led)?
- What is the strategic "fit" (for the customer and for the vendor organization)?

A Valued-Focused Operating Model
- How will the value proposition be delivered?
- What structure will be required (internal or external resources CapEx or OpEx)?

Value-Creation-Production-Based Management and Execution
- How will this be managed and controlled to ensure optimal value delivery for *ALL stakeholders?*

The authors argue that a value-focused business model approach to strategy and operations requires a re-evaluation of the organization, its partnerships, and foremost the business environment confronting the organization. It should result in recognizing that profitability and productivity will depend upon building relationships with existing and potentially future stakeholders. While customers may be identified as primary stakeholders, they are linked to other primary and secondary stakeholders via a business environment that is increasingly becoming more transparent, more expansive, time responsive, and accurate.

Lawrence et al. (2002) argued that value is both generated and perceived by customers in their own internal processes. For B2C customers these processes are determined by their lifestyle profiles such as preparing meals, personal care, and activities within their homes. In the B2B markets, customers create value for their downstream customers and end-users. They argue both are using internal value generating to interact with suppliers and that these are inputs to customer value-generating processes rather than results of the sellers' processes. This transforms the relationship from a transactional process into one in which the vendor is supporting the buyer's value generating processes, i.e., increasing revenues (because of a superior product) or decreasing their prices (because of process improvements, reduced time, and/or improved cost management). The authors suggest that "all firms are, in effect, service firms" and that value occurs when the purchaser *uses* the goods and services provided by the vendor. Shared value extends the role of the value proposition to include the group of secondary stakeholders. It could be argued that "private-public-partnership" proposals are currently pursuing a shared value approach.

Value Proposition Responses to Primary Stakeholder Value Expectations

An example of the complexity of the process is provided by Robins, a Sydney-based, shoe manufacturer (Roberts 2005). J. Robins and Sons (Robins) operate a multi-skilled workforce of 250 employees who "move between work stations as part of a small-batch production process that gives Robins the flexibility to design, make and deliver a pair of shoes in only two weeks." The Robins response came about because of the inflexibility of China's production systems, requiring long production runs which excluded Robins from the fashion segments of the shoe market. Robins' managing director commented; "If you were looking at getting a new design from China you would be lucky to get the finished product back within 3 1/2 to four months. We test the market with small production runs and, if they are successful, we can come back and repeat the order." The Robins customers' value delivery expectations are shown as Fig. 1.8.

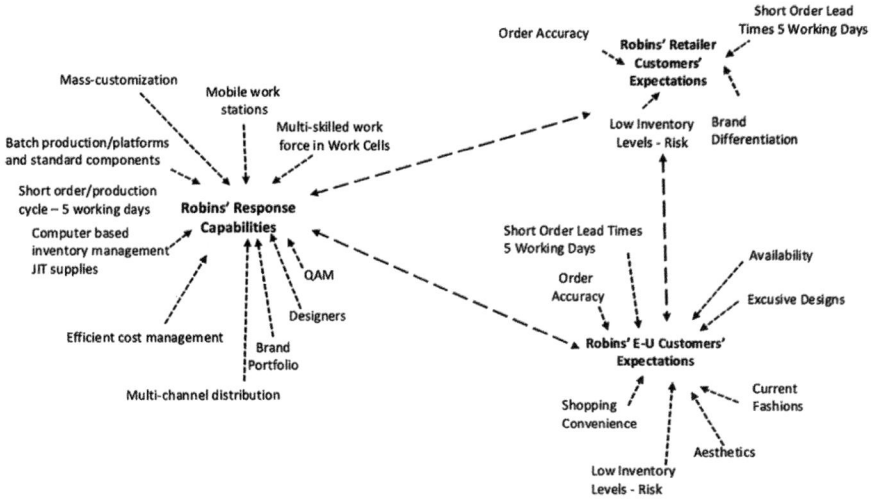

Fig. 1.8 J Robins an omni-channel shoe manufacturer has differing expectation

Newman (2005) found other successful organizations. Melba Industries (suppliers of protective fabrics to defense and firefighting activities) managed to expand by innovating in niche markets such as automotive seating and specialty fabrics for use in protective clothing for the police, military, and fire services. Melba has avoided the apparel sector because of the volume-price issues posed by China (Comment from Melba's CEO). Bruck Textiles remain competitive in mass markets by "being leaner, keener, more aggressive and more respectful of customers" (Comment from CEO, Alan Williamson).

While Robins, Melba, and Bruck Textiles were successful following a strategy of specialist products offering characteristics that attract higher margins or adopting lean principles, other companies are not so fortunate. Once global leaders, Actil (cotton sheeting) and Sheridan (a dye house) are examples of the not-so-fortunate companies. Clearly to be able to develop a meaningful capability response, network organization requires a thorough understanding of the customer organization's strategy, its objectives, operations, and its capabilities.

The Impact of the Industriè 4.0, Value Chain Network 2.0, and Stakeholder Value-Led Management

Connectivity creates opportunity for greater involvement by stakeholders as the data generated by activities becomes wider and deeper. Increasingly we see the importance of data management as the relationship between stakeholder networks and end-user organizations. Predix is General Electric's cloud platform offer that connects people and machines with "big data and analytics" on the cloud to help bring the Industrial Internet to customers' businesses and is changing the relationships between GE and its customers. The accessibility to more specific data, often in real time, now offers the possibility of a customized value proposition, and, further, this may have been made "task variable" by capturing operating data and making modifications online to improve productivity. The "cross activity" model can compare the important value performance features across the model to assess quantitative performance data and compare the actual performance impact on stakeholder expectations, resulting in suggested changes to the value proposition to exceed primary stakeholder expectations. Clearly this involves a thorough understanding of customers' expectations (their individual value performance drivers *and* their operating model) if this level of customer performance management is to be achieved. Figure 1.9 identifies the influence of Industrié 4.0, Value Chain 2.0, and Stakeholder Value-Led Management.

Until the full range of benefits offered by Industrié 4.0 became fully understood and applied, the value chain's role had been limited due to data limitations. Data was accurate and thought to be timely and rich (informative); however, it required its combination with information and control technology to offer far more benefits. Industrié 4.0 has introduced management decision-making to real-time data, more detailed data, and more focused data.

Combined, the value chain network and connectivity offer a formidable planning, production, and value delivery system. The value chain offers the means by which questions requiring answers can be asked of the right people and identify optimal operations, marketing (physical distribution and transactions channels), serviceability, and customer performance management. Connectivity collects and provides data for "analytics," at relevant times, with relevant coverage and distribution.

The value chain offers its members the opportunity to identify changes in an industry and its markets or in some instances attract new entrants in dynamic industries; it is necessary to monitor change and take full advantage of them to move into

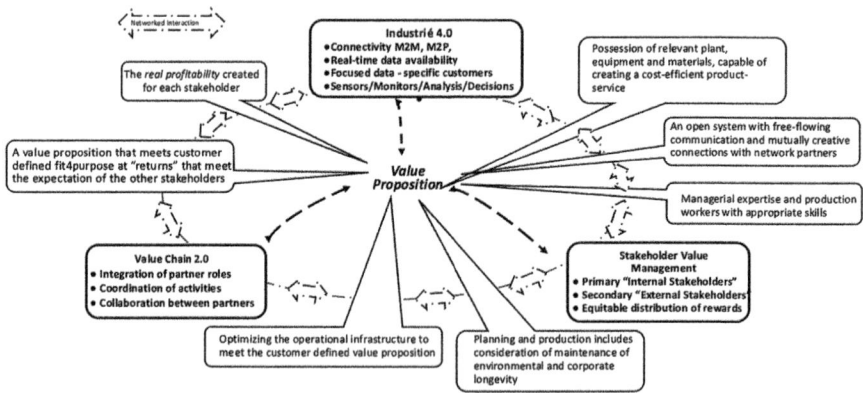

Fig. 1.9 Industrié 4.0 + Value Chain 2.0 + stakeholder value-led management and the "rolling value proposition"

a position within the value chain that offers an opportunity to increase stakeholder value. "Bodega" is an example of a positioning move within the value chain, non-perishable goods in an "unmanned pantry box." The company was founded by two former Google employees, Paul McDonald and Ashwath Rajan. Customers can unlock the box and purchase products via an app. It brings products "to where people already are so that they can access them immediately, when they need them. This beats out any two-hour delivery - or even half-hour delivery - alternative." McDonald suggests it could make corner stores obsolete. The aim of Bodega is to take the boxes nationwide soon, envisioning them on college campuses and as an amenity in the lobbies of hotels and big apartment buildings (Judkis 2017). Metcash, an Australian food and drinks wholesaler, is adding a connection point in the value chain. Metcash services an independent group of retailers trading under an identity known as IGA (Independent Grocers Association). Typically, it supplies about 70 percent of their requirements, the remaining 30 percent they purchase directly from suppliers. Metcash is putting the final touches on a new online portal known as "indieDirect" which will enable independent retailers to link up directly with suppliers and suppliers to contact a wider range of independent retailers. Metcash says the portal will enable independent retailers to source a wider range of products to differentiate themselves from Coles and Woolworths (the two dominant FMCG chains in Australia) and better compete with Amazon (active in Australia in 2018). It will also enable Metcash to play a role by negotiating better deals with suppliers on behalf of retailers and earn a small commission or margin on additional sales (estimated at AUD$ one billion business annually) without having to distribute or hold the stock. See Fig. 1.10.

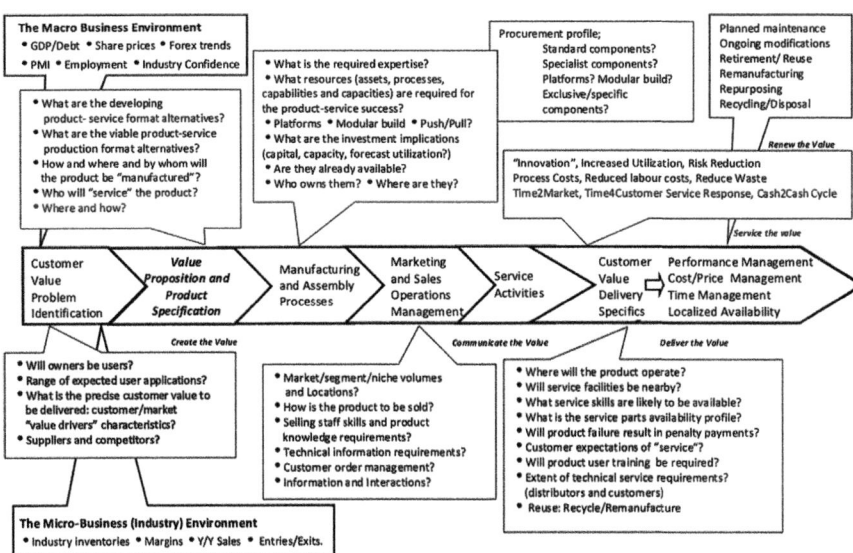

Fig. 1.10 Using connectivity and the value chain network for developing a value proposition

Aerospace is an example of a rapidly changing industry; not only has it been responsible for developing technology change; it has also encouraged technology transfer – specifically from the automotive industry. Technological developments offer existing members of the value chain opportunities to embrace the new technology (be it a process change or the application of new materials), or alternatively it attracts new entrants. Aerospace has adopted carbon fiber composite technology and with-it global sourcing, just-in-time scheduling, stringent quality, and cost controls. The Australian company, Quickstep, has specialized in carbon fiber composite pieces made in patented processes to have very low weight and metallike strength and durability (Abernethy 2017). Quickstep commenced business some 15 years ago, providing carbon fiber technology to the automotive industry; due to the growth of aerospace demand, it has a diversified portfolio of clients and is well established in the defense aerospace industry. The CEO suggests the entry requirements to this "burgeoning" market will be capabilities based upon developing cost-efficient processes that enable niche suppliers to produce at commercial levels of rate and flow rather than in batches.

UK-based Icon Aerospace Technology has been on a 10-year transition from low value output to products of a more complex nature, supported by a multi-million-pound investment in continuous improvement. Icon's product range has mirrored the physical business' evolution, steadily producing the more innovative and technologically advanced engineered products demanded by aerospace technology. Prior to the global financial recession of 2008–09, 70% of Icon Aerospace Technology's output was specific products for a customer application, while customized items were at the lower end of the value chain. Management realized that most aerospace and defense companies were looking for a supplier able to provide something of a far greater engineering complexity as a capability response. For specialist suppliers such as Icon, this represents an opportunity to shift towards high added value activities (Abernethy 2017; Williamson 2017).

Concluding Comments

The expansion of the understanding and application of value by organizations brings the notion of alternative product-service formats into the vendor-customer relationship and is changing marketplace attitudes towards product-service delivery. Furthermore, the expansion of value to include stakeholder value management, shared value, and sharing value has acknowledged the changing market opportunities towards product-service delivery. Optimizing the focus of interests to consider all stakeholders as a group entity, rather than as discrete, individual entities, widens the ability of a networked organization to achieve competitive advantage based upon a "value advantage," one which can be performance related (which we will discuss in later chapters).

The development of the concept of the value proposition has occurred alongside that of "value," and it is arguable that both have resulted in a more robust approach to understanding the broader aspects of customer satisfaction. *Strategic Marketing Positioning* reflects the value competitive differentiation of the network organization. *Market Positioning* is a visual presentation of the unique/exclusive aspects of an organization's value proposition, for example, its strategic implications, its relative competitive price, its effectiveness in use, and the support serviceability the network organization offers. It "illustrates" how it plans to meet *customer expectations*, how the "offer" compares with competitive value propositions and how the "value" is to be delivered and "serviced." It may be a product-led PRODUCT-service or a service-led product-SERVICE, thereby acknowledging the role of investment and asset management of the customer. It is an explicit statement of customer value attribute requirements (both value criteria, being offered and how acquisition costs are minimized for the customer). It includes both product or service characteristics *and* service-support features. It becomes the basis of a set of agreed performance processes and metrics for network partners and other, secondary, stakeholders. Value production, delivery, and serviceability decide upon the *Value Chain Positioning* of each of the network partners. These comprise both long-term and short-term characteristics. *Value Engineering* (strategic in context) develops product-services that meet customer value delivery expectations and "matches" value chain participants' expertise with value proposition tasks. *Value Delivery* is operational and monitors ongoing activities for efficiency; it attempts to improve output characteristics. Positioning decisions require an understanding of the *horizontal and vertical relationships* that exist within the value chain network – the roles and tasks required and the capabilities and capacities of the network partners.

Figure 1.11 serves to combine the interests of customers-investors-suppliers-employees-distributors-the firm-community-competitors-government and the notion that progressive organizations see stakeholder value-led management is relevant to

Fig. 1.11 The integrating role of data management and analysis in the value proposition

successful growth. It suggests the value proposition has a key role in developing the growth of individual and networked organizations' operating profit and cash flow management. The diagram "connects" the expectations of stakeholder expectations and their capability contributions through the data connectivity facility of Industrié 4.0 that provides transparency and accuracy in real-time enabling management to match the expectations of all stakeholders.

References

Abernethy, M. (2017, August 31). "Quickstep" carves a niche in aerospace, The Australian Financial Review.

Amazon, the world's most remarkable firm, is just getting started. (2017, March 25). The economist.

Barnes, C., Blake, H., & Pinder, D. (2009). Creating & delivering your value proposition: Managing customer experience for profit. Kogan Page.

Bower M and R Garda (1985) The role of marketing management, McKinsey Quarterly, Autumn.

Boyarsky, B., Enger, W., & Ritter, R. (2016). Developing a customer experience vision. McKinsey and Company. March.

Campbell, P. (2018, July 6). Architect of Renault's (Alliance) new transport push steps down. *Financial Times (London)*.

Cleland, A. & Bruno, A. (1996). The value based market process: Bridging customer and shareholder value. Wiley/Jossey-Bass.

Doyle, P. (2000). How shareholder value analysis redefines marketing. *Market Leader.* Spring.

Executive Briefing. (2016). The CEO guide to customer experience. *McKinsey Quarterly*, August.

Freeman, R. E. (1984). *Strategic management: A stakeholder approach*. Boston: Pitman.

Gapper, J. (2012, May 4). Pharma needs an injection of financial engineering. Financial Times *(London)*.

Gapper, J. (2014). Pharma needs an injection of financial engineering. *The Financial Times* May 4.

Jensen, M. (2001, October 29). Value maximization, stakeholder theory, and the corporate objective function. *The Monitor Group and Harvard Business School*.

Judkis, M. (2017, September 15). Bodega, a corner shop disruptor has already become America's most hated start-up. Washington Post/Australian Financial Review.

Lanning, M.J., & Michaels, E.G. (1988). A business is a value delivery system. McKinsey.

Lawrence, T. B., Hardy, C., & Philips, N. P. (2002). Institutional effects of Interorganizational collaboration: The emergence of proto-institutions. *The Academy of Management Journal, 45*(1), 281–290.

Leggett, T. (2018 July 1). Carlos Ghosn: Why carmakers must adapt - or die, BBC News.

Lewis, L., & Jacobs, E. (2018, July 13). How business is capitalising on the millennium Instagram obsession, *FT Services Millennium Moment: The business of a Generation, Financial Times*.

Newman, G. (2005, October 22). How textile survivors fashion a profit, The Australian.

Payne, A., & Holt, S. (2001). Diagnosing customer value: Integrating the value process and relationship marketing. *British Journal of Management, 12*(2), *159.*

Pine, B.J., & Gilmore, J. (1998). Welcome to the experience economy. *Harvard Business Review.* July-August.

Porter, M., & Kramer, M. (2011). Creating shared value. *Harvard Business Review*, Jan-Feb.

Roberts, P. (2005, December 6) How textile survivors fashion a profit. *Australian Financial Review.*

Slee, T. (2017). *What's your is mine: Against the sharing economy*. New York: OR Books.

Slywotsky, A. (1996). *Value migration*. New York: Free Press.

Slywotzky, A. J., & Morrison, D. J. (1997). *The profit zone*. New York: Wiley.

Stock, K. (2016). The chevy bolt is the ugly car of the (very near) future. Bloomberg, 19. December.

Thirkin, P. (2017). Shareholders versus Stakeholders. *BOSS*. March.

Webster, F. E. (1994). *Market driven management*. New York: Wiley.

Williamson, J. (2017, August 8). Aerospace supplier advises how to move up the value chain. Manufacturer.

www.bloomberg.com>opinion>articles 12 April 2017.

Chapter 2
Industry Dynamics

Abstract Beginning with the notion that is underlying competencies rather than their current manifestation in the latest product that is important, it is proposed that *survival* be replaced with *sustainable future competitiveness*; it is also proposed that a broader, macro, framework is becoming necessary to structure the identification of *capability factors* rather than simply a list of the latest industry or marketplace drivers. It is suggested that such a framework revolves around *six* generic focused characteristics of *industry dynamics*: knowledge management, technology management, process management, relationship management, regulatory compliance, and managing geopolitical events. The established notion that industry dynamics and value builders indicated "pathways" to success for organizations, and that vendor value propositions should reflect these expectations, is explored. This chapter develops an argument around the notion that industries should respond to industry dynamics and organizations seek to develop related capabilities and that the successful organizations are those that identify the characteristics that are essential for creating value advantage. The increasing presence of value chain networks emphasizes the need, and the available facility, to develop customer-centric solutions. We will provide examples of organizations that have identified industry dynamics, capabilities, and support capabilities in other industries and have imported them with considerable success.

Keywords Industry dynamics · Strategy dynamics

Introduction: Industry Dynamics, Capabilities, and Being Value Competitive

Grant's (1995) introduction of the notion of "survival" into critical success factors raised an interesting and important distinction between necessity and advantage. Arguably a lot of what are often identified as key success factors are essentially *competitive necessities*. Unless they are addressed an organization will be unable to compete successfully in a selected market or market segment. While they may appear to be sources of advantage to outsiders, in fact they are more akin to a simple

© Springer Nature Switzerland AG 2020
D. Walters, D. Helman, *Strategic Capability Response Analysis*,
https://doi.org/10.1007/978-3-030-22944-3_2

market entry status than true sources of significant advantage over otherwise evenly matched competitors. Arguably, all of the key success factors identified by Leidecker and Bruno (1984) and many identified by Grant above can be seen as competitive necessities. Much of the work in this area has focused on the divination of critical success factors in particular industries. It is suggested however that what is required is some broader view of what constitutes a key success factor in a digital network economy context. Just as Rumelt (1987) suggested, it is the underlying competencies, rather than their current manifestation in the latest product, that are important. It is proposed that *survival* be replaced with *sustainable future competitiveness*; it is also proposed that a broader, macro, framework is becoming necessary to structure the identification of *capability* factors rather than simply a list of the latest industry or marketplace drivers. It is suggested that such a framework revolves around *six* generic focused characteristics of *industry dynamics*: knowledge management, technology management, process management, relationship management, regulatory compliance, and managing geopolitical events. But much has changed ....

To create *competitive advantage*, or perhaps, a *value advantage*, something has always been required. An organization such as Walmart has clearly mastered the competitive necessities of retailing, but what has truly set it apart from its competitors has been location, with the company historically locating in regional areas away from traditional competitors in city or suburban centers. Initially developing this involved skill and thinking that were at the time more akin to a real estate developer, but which are now firmly entrenched in the retail environment. Product-service availability (localized availability as well as instore availability), an essential feature, was ensured by establishing a robust response distribution support system to deliver both. Similarly, though now widely copied, Walmart built a franchise around "everyday low prices" by enticing established brands to lower costs (and enhance price/brand-led "value") through *technology*-driven supply chain efficiencies.

It is also arguable that in a rapidly changing business environment, it is possible for a network-led organization to possess a *competitive disadvantage*. There are numerous examples of investments in assets, processes, and indeed capabilities that have proven to be short-lived, often because the company has been slow to react to process innovation, or, perhaps, even worse there has been a lack of awareness of ongoing developments that have undermined existing processes, for example, the rate at which consumer acceptance of online transactions would develop, and as a result these organizations are left with resources that are not obsolescent but totally obsolete. Often there is no alternative other than to write off the equipment. Sainsbury (the UK former market leader in food retailing) noted in the late 1990s in an annual report the positive impact on overall profitability of its increased logistics productivity and saw this as a key corporate strategy and therefore a value advantage. In November 2004 Sainsbury admitted to not meeting customer expectations. CEO Justin King commented at the time: "they (Sainsbury) had built an unsustainable business model, (it had) become too driven by profit margins and was too heavily focused on short-term profits. Furthermore, it had invested in infrastructure that had failed to deliver the intended benefits; the customer offering was neglected"

(Gluyas 2004). In short, its perceived value advantage had become a value-destroying disadvantage!

Perhaps the more contemporary suggestions of Slywotzky and Morrison (1997), and more recently those of Phelps (2004), are now more relevant for a customer-centric business environment. In the previous chapters, the work of both sets of authors argued that customers have expectations that can be met by exploring developments in industry dynamics and value builders (see following chapter) and utilize these to form the basis of their value propositions. Slywotzky and Morrison (1997) identified value migration as a reason for change. They argued that consumers' preferences change and that competitors challenge the existing "way of doing business" with more efficient business models. Phelps (2004) suggested that industry dynamics and value builders indicated "pathways" to success for organizations and that vendor value propositions should reflect these expectations. This chapter develops an argument around the notion that industries should respond to industry dynamics and organizations seek to develop related capabilities and that the successful organizations are those that identify the characteristics that are essential for creating value advantage. The increasing presence of value chain networks emphasizes the need, and the available facility, to develop customer-centric solutions. We will provide examples of organizations that have identified industry dynamics, capabilities, and support capabilities in other industries and have imported them with considerable success.

An Alternative Approach: Industry Dynamics, Capabilities, and Support Capabilities as "Future Critical Success Factors" and Industrié 4.0

The introduction suggested the notion that *Industry Dynamics* are "*Macro*" *characteristics*, resulting often from external research and development activities that can be applied to improving value engineering and value delivery characteristics that will better meet customers' value expectations and provide a pathway to creating competitive value advantage. *Industry Dynamics* are not unique; but individual organizations (and networks) may create unique opportunities by identifying value creation and production and capturing opportunities their competitors miss.

It is suggested that such a framework revolve around the six generic, but focused, industry dynamic frameworks. These are knowledge management, technology management, relationship management, process management, regulatory compliance, and geopolitics. Many of the contributions predate Industrié 4.0 but nevertheless remain relevant.

Blumentritt and Johnston (1999) identified the importance of *knowledge management* for value chain network (virtual organization) managers, suggesting that "….the ability to identify, locate and deliver information and knowledge to a point of valuable application is transforming existing industries and facilitating the emergence of entirely new industries." Furthermore, they suggested that while knowledge is recognized as a key to *value advantage* and is indeed an asset, there are challenges awaiting managers when attempting to "manage" this intangible asset. The linkages that are created between customer behavior information and demand management through customer loyalty programs is an example of the importance of knowledge management as an aspect of critical industry success factors in retailing. Currently we are witnessing the corporate view that knowledge management (intellectual property) is being considered to be an intangible asset and extends beyond RD&D into insights into supplier and customer performance management, as data analysis and management increases from product-service applications from customers via a digital thread link to a software representation of the product-service, the digital twin.

Technology management is the process of planning and strategic technology management and particularly the *integration of technology strategy with business strategy* in the context of engineering and technological organizations. The operational aspects of managing technology, including the relationship with production, marketing RD&D, and human resources activities, are a new framework for understanding current technology management issues. Irani and Love (2001) considered *"technology management"* within an organization as a "business process that facilitates the development of a comprehensive and robust techno-centric infrastructure, consequently enhancing the delivery of accurate, timely and appropriate services within an organization, which in turn increases the economic viability of the business." This is not necessarily pure RD&D; rather it is applied RD&D, and it includes the innovative thinking involved in combining new and recent technologies into new "value-engineered," deliverable agile (rolling) value propositions, or perhaps the "technology transfer" of existing processes and equipment to take advantage of other inputs (e.g., low-cost labor and/or raw materials but increasingly data outputs from customer product-service application). They referred to a "technology management gap" that may result in a value advantage being jeopardized. The linking of digitized manufacturing (e.g., additive manufacturing) with automated distribution and manufacturing processes to reduce order response times in the aerospace and defense industry is an example of a technology management-based critical industry success factor – the closing of the technology gap.

Relationship management is essentially the effective (and efficient) coordination of the different internal and external parts of a business. Some considerable time ago, Jarrillo (1993) suggested the plethora of articles and books on topics such as networking, value-adding network relationships, modularization, etc. all point in one direction – the need for companies to nurture long-term relationships in all spheres; "companies must look at their boundaries with new eyes - things that have traditionally been 'inside' should perhaps be outside, and outsiders might perhaps deserve the treatment of 'insiders'." The current business environment is arguably changing such that an organization can only grow effectively by developing strategic networks with its suppliers, distributors, and customers and collaborating with joint venture developments in product-service and process developments; often these activities are inter-industry and international.

The fourth "dynamic" is *process management*. Processes have defined business outcomes for which there are recipients that may be either internal or external to the organization. They cross organizational boundaries enabling virtuality to become reality. Virtuality defines the ability to create partnerships across companies using value chain network structures with complementary companies that work together to maximize the value delivered to all stakeholders, not only customers. The scope of process management is such that it may be applied to strategic and operational tasks and structures alike. The process management skills of the new process model are the coordination of customer-"designed" products with the "just-in-time" (JIT) delivery of the components required to meet the assembly of the computer to meet the order. Digitization of the model minimizes inventory holding but meets three very important, customer expectations – product specification, availability, and improved performance management. An important point to be made concerns the rapidly expanding availabilities of offshore facilities. Emerging supply markets are appearing in the developing economies of Asia and Eastern Europe. An interesting aspect of this development concerns the specialist nature of the developing countries' expertise; for example, India has an efficient labor force in software engineering, skilled linguistics (as deployed into call center operations), and expanding healthcare facilities. Korea offers process (production capacity) management expertise, while the PRC (China) that once dominated the supply of low-cost labor is now becoming significant in digital software. Industrié 4.0 extends the proposition with development of digital-led manufacturing that is broadening the product range and compressing the time offers to customers, thus reducing the supply response to demand expectations dramatically. Processes can overlap organizations, for example, plastic bottle packaging. A packaging supplier to soft drinks companies developed a process by which the size and final shape of the container were completed when the bottle was filled; this reduced costs but also eliminated a problem of misleading content levels that was a feature of the replaced filling process and was often a source of customer queries and complaints.

The conduct of business, work, and leisure is becoming increasingly complex. Legislation concerning health, workplace safety, and food standards ingredients and storage (in commercial and institutional locations) and in a wide range of vendor/customer relationships are making *regulatory compliance* an industry dynamic. Recent actions by governments and industrial organizations suggest *regulatory compliance* offers opportunities as well threats. They suggest that foresight will be a necessary component in strategy process. For example, the recent decision by *Transport for London* to withdraw/review Uber's operating license is an indication of difficulties confronting businesses when developing product-service offers within a changing landscape. This response suggests that organizations agree an industry-wide set of obligatory standards they enforce rather than make it the responsibility of government. The "sharing economy" offers suppliers and consumers economic and convenience benefits, but both Uber and Airbnb have found there are problems that, if they had been identified ahead of the product-service launch, would prevent downstream problems.

Lack of regulatory compliance has high-risk applications implications. Food retailing is one important example. Product recalls, if missed or ignored, can have far-reaching implications. GS1 Australia has a number of response supply chain data management services, one of which coordinates (on an industry basis) data management changes throughout an industry. The GS1 Global Data Synchronization Network® (GDSN®) is a network of interoperable data pools enabling collaborating users to securely synchronize master data based on GS1 standards. GDSN supports accurate, real-time data sharing and trade item updates among subscribed trading partners. This means organizations can have confidence that when one of their suppliers or retailers updates their database, their own database is similarly updated as a result. Everyone has access to the same continuously refreshed data. For this to happen, each organization needs to join a data pool certified and tested by GS1, who connect to the GS1 Global Registry®, a central directory which keeps track of connections, guarantees the uniqueness of data, and ensures compliance with shared GS1 standards. There are many data pools spread across the world (see www.gs1au.org).

At an international level, regulatory compliance is often difficult to predict in the value chain when third-party activities compromise multinational organizations; there have been a number of problems with contaminated infant milk products and health supplements resulting in the collapse of the offshore entrants' markets. Social media and "search engine" product organizations have been confronted with unexpected operating restrictions. The connectivity characteristics of Industrié 4.0 can be used to monitor product-service and market activities. Sensors are available to track "sensitive" products such as infant formula and similar products. For longer-term protection, scenarios that ask "what if?" questions may be built into the capability audit.

Geopolitics has a range of definitions. Initially (early in the twentieth century) it was used to consider the impact of power relationships in international relations. However, socioeconomic considerations, and the implications of politics, monopoly, geographical position, climate, and topography have been issues since the days of the Greek and Roman empires. Access to seaports, water, arable land, energy sources, minerals, etc. are topics that have at one time or another been the subject of disputes and, as such, are geopolitical in their nature. Current issues in commercial geopolitics concern the activities of multinational organizations and the global value chain networks that have strong influence on the economic well-being of developing nations. It is also an important issue within developed countries as the growth of *populism* expressed in the UK referendum to leave the UK and the Trump presidential election demonstrates. At the time of writing, the UK/Brexit issue is unclear; however, reaction to the US situation is becoming apparent by the response of large automotive assemblers and component suppliers with German and Japanese suppliers announcing intentions to build manufacturing facilities in the USA.

The foregoing paragraphs can be described as "macro" considerations. "Micro" aspects are demonstrated by the decisions of large multinationals to enter into joint venture investment activities with "local" organizations in Asia. Other characteristics include the development of specific activities that reflect the specialist skills and resources of offshore locations. General Electric has located R&D hubs in India and China; Dyson (consumer durables) relocated its manufacturing and distribution facilities in Malaysia. Clearly the other incentives, tax regimes, the ability to move profits to home locations, and labor and materials costs are also important. This suggests that geopolitical risk be identified and measured. See Investopedia website for full description and sources of services available.

Geopolitical risk is the risk investment returns could suffer as a result of political changes or instability in a country; returns from offshore investment activities could stem from a change in government, legislative bodies, other foreign policy makers, or military control and become more of a factor as the time horizon of an investment gets longer. Geopolitical risks are difficult to quantify because there are limited sample sizes or case studies when discussing an individual nation. Some political risks can be insured against through international agencies or other government bodies. The outcome of a political risk could significantly reduce investment returns or even go so far as to remove the ability to withdraw capital from an investment.

There are a range of decisions governments make that can affect individual businesses, industries, and the overall economy. These include taxes, spending, regulation, currency valuation, trade tariffs, labor laws such as the imposition of minimum wage, and environmental regulations. Regulations can be set at all levels of government, including federal, state, and local, as well as in other countries. Companies that operate internationally, known as multinational businesses, can purchase political risk insurance to remove or mitigate certain political risks. This allows management and investors to concentrate on the business fundamentals while knowing losses from political risks are avoided or limited. Typical actions covered include war and terrorism.

Walmart Stores Inc. outlined certain political risks it faced in its fiscal 2015-year filing with the SEC. Under its operating risk section, risks associated with suppliers, Walmart identified potential political and economic instability in the countries that foreign suppliers operated as likely to be labor problems and foreign trade policies and tariffs that could be imposed. In its *regulatory*, *compliance*, reputational and other risks section, the company outlined risk associated with legislative, judicial, regulatory, and political/economic risks. Risk factors mentioned included political instability, legal and regulatory constraints, local product safety and environmental laws, tax regulations, local labor laws, trade policies, and currency regulations. Walmart mentioned Brazil specifically and the complexity of its federal, state, and local laws (www.Investopedia.com).

These "macro" frameworks should be used to help obtain a profile of industry characteristics against which it is then possible to identify "organizational" characteristics that influence *industry success* – capability requirements. Having identified them the next task is to identify where they are located, who owns them, and how they can be accessed. A major skill in the management of a value chain network is to establish the most effective position for itself as a partner within the value chain network. For this it is necessary to take a more detailed view of the organization and its environment, to look for *capabilities*.

Partnership Leverage: Capabilities, Support Capabilities, and Collaboration

Capabilities

Capabilities then are specific characteristics derived from an industry dynamic that are exclusive, possibly unique, to an organization. They are focused onto a specific asset, process, or activity within an organization to facilitate the delivery of the organization's value proposition *and* to create value advantage. Typically capabilities are created from combined industry dynamics; the value proposition is *enabled* by the convergence of a direct selling approach to customers (*relationship management*), digital order and transactions management systems (*technology management*), JIT inventory management/production assembly methods (*relationship management and process management*), current computing capabilities through collaboration with specialist component suppliers (*knowledge management and relationship management*), and the enhancement or "halo" effect of using components manufactured by leading industry brand owners (*relationship management*). By collaborating with other members of the value chain network, it has become possible to develop a value proposition that offers its customers equipment that can be specified to a customers' requirements (within the limits of a predetermined "menu"), at a competitive price and within a controlled delivery time frame. Digitization enables a value proposition that manages performance as well as maintenance. Caterpillar is an interesting organization in this respect, one that has extended its use of both technology and relationship management by *skillful application of capabilities*. Caterpillar and its distributor network is an example of creating a total relationship approach. Using product technology as an enabler (a remote serviceability diagnostic), ICT networks and a committed dealer network, the "total" Caterpillar network, offer a guarantee of reliable and rapid global serviceability to end-users. Innovative (entrepreneurial) management identifies capabilities that are successful in one industry application and *exports* them into an unrelated industry, often achieving remarkable success.

The foregoing identifies a link between industry dynamics and capabilities for *sustainable stakeholder value* delivery characteristics. The argument here is based upon the changes in direction that have been observed in very large organizations. IBM led the change in its "raison d'être" by exiting the personal computer sector (selling this to Lenovo) and identifying itself as a "solutions provider." The SaaS Solution Provider initiative from IBM enables business partners to sell full-service offerings supported by IBM SaaS offerings to their customers. The "Business Partners" bring business acumen and related added value services together with IBM SaaS to create and build long-term value with customers. General Electric moved from being a financial services provider organization to an industrial capital equipment organization based around platforms and then into a digital-systems-data-led business in which customer performance systems management is becoming its primary product-SERVICE. Predix is a digital data platform. Digital industrial transformation offers immediate cost benefits and long-term strategic advantages that facilitate rapid innovation, optimal asset performance, cost-effective investment (by applying machine learning analytics and controls at the edge, in the cloud, or anywhere in between), future planning flexibility, scalability, and security as the foundation of an entire digital industrial transformation.

Existing industries may well change; they may well be absorbed, or they may absorb other, parallel, industries. New industries are expected to develop out of current, successful parent industries; the interest of the large automotive manufacturers in "connectivity" companies and in vehicle sharing is an example of this beginning to happen. Healthcare is developing into preventative medicine, into self-monitoring and individual well-being programs. Precision medicine, automated regulation, digital currency, food provenance, and real-time freight tracking were suggested by Eyers (2004). Clearly the aware organization will be seeking the linkages between relevant industry dynamics and the capabilities required for successful entry through the value advantage they can create by successfully coupling industry dynamic(s) and the relevant capabilities. They will also be aware of the opportunities offered by partnership leverage.

Support Capabilities

Support capabilities are organizations operating successfully in other industries and markets but are parallel with an organization's core activities (see Normann and Ramirez 1993; Christensen 1997; Stabell and Fjeldstad 1998; Allee 2003). Often, they have characteristics that can be combined and provide an opportunity to create value advantage and/or expand a market. For example, the introduction of credit by General Motors in the 1920s (a concept limited in its use to B2B relationships at that time) had a significant impact on the growth of sales in the automotive industry. During 2010 GM announced the acquisition of AmeriCredit for about $3.5 billion dollars, suggesting it "supports our efforts to design, build and sell the world's best vehicles by expanding the financing options we can offer to consumers who want to buy GM vehicles" ("GM to Acquire Auto Finance Firm in $3.5 Billion Deal" 2010). AmeriCredit already had links with about 4000 GM dealers; the acquisition improved sales penetration rates through coordinated GM branding and targeted customer marketing initiatives. Similarly, the combination of insurance and healthcare resulted in a rapid expansion of private healthcare services in countries where healthcare was administered by government-managed services.

Retailing organizations are actively seeking to provide support capabilities services to their product-service ranges. IKEA has acquired US start-up firm TaskRabbit, which allows users to hire people to help them assemble furniture as well as a host of other chores like house cleaning or lawn mowing ("Ikea acquires TaskRabbit" 2017). IKEA President and Chief Executive Jesper Brodin justified the acquisition as being a move to "make our customers' lives a little bit easier." The

John Lewis Partnership (a successful UK department store and food retailing group – Waitrose) has followed a similar strategy. The company partnered with Ocado, a technology-based organization that has built a successful online food retailing structure that has recently expanded overseas into 23 international markets. Subsequently Waitrose developed its own online activity.

Collaboration

Collaboration, well established in the automotive industry, is becoming a feature of defense industry manufacturing (Ferguson 2017). Modern defense equipment is complex; it usually integrates technologies from several different domains, it being impossible for any one organization to be an expert in all of them. As a result, collaboration is increasing, and a value chain approach is emerging. Dr. Jens Goennemann, managing director of the Advanced Manufacturing Growth Centre (AMGC), suggests the term "manufacturing" is being redefined (Ferguson 2017). Goennemann believes it must embrace the whole spectrum, from R&D and design to logistics, production, distribution, sales, and after-sales service. This more holistic view on manufacturing encourages firms to own more of the value chain in the manufacturing process – tangibles such as production itself and the less tangibles that precede and follow it. These are the areas where collaboration and research partnerships can make a massive difference to Australian businesses and help them to compete on value rather than being forced to compete on price.

Industrié 4.0 provides opportunities to create "additional added value" by facilitating collaborative integration and coordination to create shared capabilities within value chains, between partners and across processes. It offers solutions to problems that not long ago were difficult to resolve. For example, recent (mid-2017) geopolitical stances being pursued by the Trump government attempting to protect the US automotive investment and employment have resulted in Toyota and Mazda collaborating to build a manufacturing unit within the USA. However, this example extends collaboration well beyond the current agreements, typical within the automotive industry; Mazda CEO Masamichi Kogai suggested it will "transcend the boundaries of previous cooperation [...] to evaluate how best to utilize each company's respective strengths" (Greimel 2017). Lippert and Trudell (2017) add detail to the agreement that Toyota Motor Corp. and Mazda Motor Corp. have agreed to buy stakes in each other and jointly build a $1.6 billion US-located factory, the auto industry's first new assembly plant to be announced under President Donald Trump. The deepening ties between Toyota and Mazda also are driven in part by the enormous costs automakers are facing to develop connected, self-driving, and battery-powered vehicles. The two companies will work together to develop electric vehicle technology.

Industry Dynamic "Collaborative Overlap"

Clearly "overlap" is to be expected across the range of *Industry Dynamics*. For example, if a manufacturer of apparel items chooses to outsource the entire production of the product range to a supplier offering acceptable finish quality at a much-reduced unit cost (*process management*), it follows that without an expression of confidence and trust (*relationship management*) the partnership is unlikely to be successful. The Dell business model could not have survived without strong *relationship management* skills being exercised between the company and its suppliers. It is these intersections that are the *enabler factors*; and it is the innovative application of capabilities that creates value advantage – the unique (perhaps exclusive) way in which the industry dynamics are positioned. Partnerships are reached on a mutual basis; organizations seek to restructure (or transform) themselves in order to meet the challenges of the marketplace, and typically this is a two-way process. These examples tend to be permanent and as such result in organizational structures that develop inter-organizational response capabilities that achieve long-term strategic value advantage and growth. Organizations manufacturing healthcare equipment and pharmaceutical products (*knowledge management* (*intellectual property*), *technology management*, and *process management* (*manufacturing and applications*)) are required to meet the healthcare protocol and practice standards required by governments (advised by local healthcare organizations).

Figure 2.1 is a typology of these component concepts

Successful network partnerships are those in which leadership, in the form of an "entrepreneur," identifies how end-user satisfaction (value) can be delivered and structures, integrates, and coordinates the network capabilities and activities to achieve a predetermined objective.

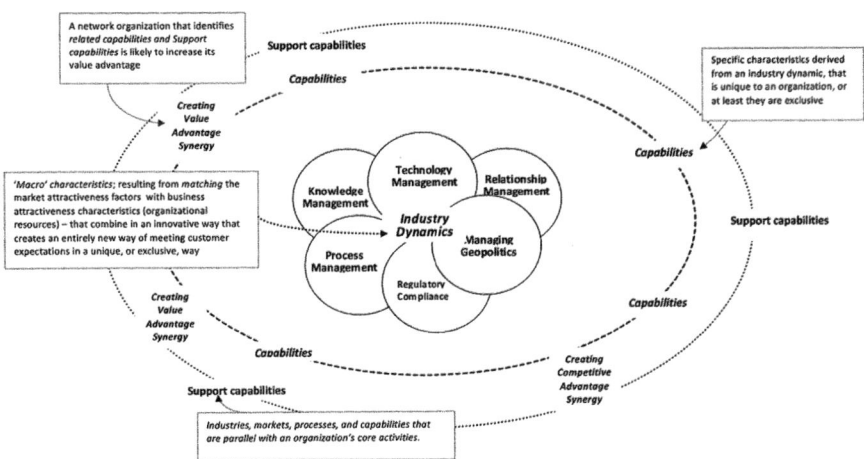

Fig. 2.1 Industry dynamics, capabilities, and support capabilities

Recent Examples of Industry Dynamics, Capabilities, Support Capabilities, and Collaboration

The Automotive Industry

In Fig. 2.2 the industry dynamics and capabilities that have been developed in recent years of the automobile industry are illustrated. The automotive industry has responded to the opportunities presented by the primary drivers: knowledge management, technology management, process management, and relationship management.

Examples of the application of *knowledge management* are seen in the reading of supply and consumer markets (e.g., the development of offshore production by VW and Peugeot in China where initially the low-cost labor gave an advantage, but so too did the increasing disposable income and vehicle ownership moved from aspiration to become reality). The application of electrical power and battery technology is gathering momentum with the major manufacturers announcing success in their efforts to move away from traditional fuel sources. Notable *technology management* applications can be observed in the ability to integrate output from global manufacturing sources. The development of shared product platforms has facilitated the offer to the customer of differentiation and mass customization at attractive end-user prices (this has also been facilitated by the use of "standard" industry products and procurement systems). The application of "external technology" developments, e.g., ABS braking systems and GPS indicators from their

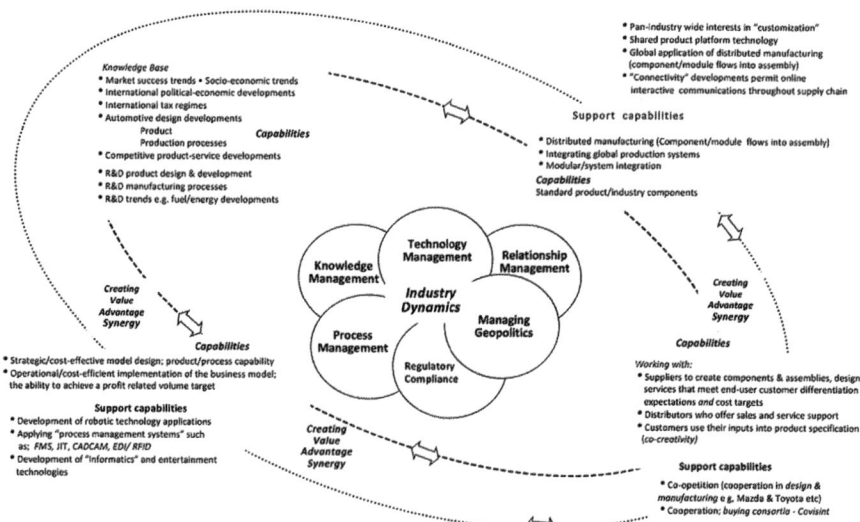

Fig. 2.2 Identifying the industry dynamics, capabilities, and support capabilities: An automobile industry perspective

military application, has enhanced the industry's value proposition. Within the area of *relationship management*, equally significant developments have occurred; the development of shared platforms between competitors and the cooperation between manufacturers and suppliers (Toyota) have resulted in cooperative efforts to improve quality *and* the cost/price profile of components. The recent preferences of end-users for smaller vehicles have resulted in cooperative moves between manufacturers in which the marketing channels of one producer will be used to sell the vehicles of another (offshore)-based company. *Process management* also qualifies as an important driver. Vehicle design now incorporates consideration of where and how the vehicle is produced – an interesting combination of effective strategic design for operational implementation. The "Save Your Factory" campaign has provided input into the automotive (and other labor repetitive industries) in an attempt at reducing variable costs by exploring the increased application of robotics to assembly production processes.

In 2009/2010 the "Toyota Model" of collaboration (specifically relationship management) failed. A view was offered by Takahiro Fujimoto (an economics professor from the University of Tokyo and a leading authority on the Toyota production system and automotive product development) in an interview in 2016 with John Paul MacDuffie from Wharton's Management Department (Jacobides et al. 2016). Fujimoto said: "Toyota was growing very rapidly. But other companies, like Hyundai, were growing even faster than Toyota. Growth multiplied by other factors was probably the real cause of the problems – other factors like the number of production lines, production facilities, number of models sold in the global market, and the growing complexity of each individual vehicle due to social pressure and market demand. All these things multiplied together created explosive expansion of the workload for handling quality problems. Although Toyota usually had the capability to handle these kinds of things, even they couldn't really handle this tremendous expansion of complexity." And, significantly, "I saw some of my friends complaining about how Toyota treated people when there was a problem. You know, they go to the dealers and then the dealer reports to Toyota about these problems. But [the customers] tended to get a reaction from the company that said, 'The product must be good. We are confident of the quality of the products. So logically it must be your driving problems.' That made many people very angry." This point was also made by Tabuchi (2010) in an article published in *The Australian Financial Review* in which the author cited Toyota among many other who badly handled customer complaints. Tabuchi quoted a lawyer who has handled a number of complaints related to Toyota vehicles: "In Japan there is a saying: 'if something smells put a lid on it'." A quote from the head of the Japan Automobile Consumers Union suggested that the "State" sided with the auto makers and not the consumers, in an action against the automobile manufacturers. This suggests that relationship management relates to the supply chain and not the demand chain!

Champy (2008) recalls and reflects on some 30 years' experience. In his recent book, Champy offers descriptions of very real and practical strategies that are working for successful businesses, offering strategies that will sustain their success into the future. Champy reviews the strategy decisions of a number of interesting companies, most of them little known outside the USA; however, their strategy decisions share a logic that can be explored using this model. The following example illustrates how by identifying the industry dynamics of an organization's business environment, the capabilities required for success, and support capabilities/ activities MinuteClinic has been able to create a response to a marketing opportunity that other organizations had overlooked or ignored.

MinuteClinic (Now a Component of CVS Caremark)

MinuteClinic is an example of an alternative healthcare delivery strategy. Champy (2008) identifies this:

> a classic example of a company whose founders recognized a significant unmet consumer need and seized the opportunity by borrowing the idea from a seemingly unrelated industry.

He describes the application of quick service auto-maintenance to healthcare. MinuteClinic has *added* a value-adding component in the healthcare value chain network by introducing a complementary component – not a service that is directly competitive and threatening to others. MinuteClinic offers a relatively low-cost, conveniently accessible method of identifying and treating a range of common ailments. Champy (2008) argues: "you don't need a fully trained mechanic to change the oil in your car." MinuteClinic operates a kiosk-based service (usually based in shopping malls) that is open 7 days for a total of 72 hours. The clinics clearly post the ailments that can be treated together with fees applicable. The kiosks are staffed by qualified nurses who are licensed to prescribe drugs. Each kiosk has an on-call physician to whom the nurse can refer in cases of doubt. Treatment typically takes 10–15 minutes. If the waiting time is likely to exceed 5 minutes, patients are given an electronic pager that informs them of their appointment.

The customer base includes individuals and organizations; the customer base rapidly expanded by associating with trade unions and employers who saw the MinuteClinic operation as a means of keeping members healthy and working! Health insurance organizations also benefit from the MinuteClinic model in the context of lower cost claims to be met. A new CEO in 2005 brought marketing expertise into the organization resulting in an approach that focused on changing how healthcare was delivered rather than the traditional focus on solving the problems of diagnosis. He undertook an expansion strategy and the group went from 19 clinics to 83 between 2005 and 2007.

Acquisition of MinuteClinic in 2007 by CVS Caremark, the largest retail pharmacy in the USA, resulted in the opportunity of further expansion; CVS operate 6000 retail outlets and specialty stores in 43 states. The chairman, a surgeon, built bridges with the medical community such that a few hospitals have actively embraced the concept by inviting the company to set up kiosks within the hospitals.

As Champy (2008) comments: "it's a boon to have nurse practitioners take over the simple chores that otherwise would tie up doctors and adds to their long lines of 'emergency' patients."

Champy (2008) summarizes the success of MinuteClinic as being based upon an understanding of customer/patient expectations and the application of philosophies and practices from outside of the industry. Applications of customer service techniques from retailing and the adoption of operating processes from quick service organizations have given MinuteClinic a strong position in the healthcare value chain network. Figure 2.3 explores the success of the organization's approach using industry dynamics and capabilities. MinuteClinic has identified a niche area in healthcare delivery and applying a "borrowed" *process management enabler* – a quick response technique to a relatively simple, but frequently occurring problem – together with a relationship management enabler, working with the endorsement and cooperation of medical practitioners.

Convergence …..

Observation of the behavior of organizations suggests a structured value chain network response to interpretation of business environment changes. For example, a number of national governments have established target dates for 100 percent of electrical vehicles. This can be seen as either a geopolitical or regulatory compliance dynamic. It also suggests significant changes for each of the other industry

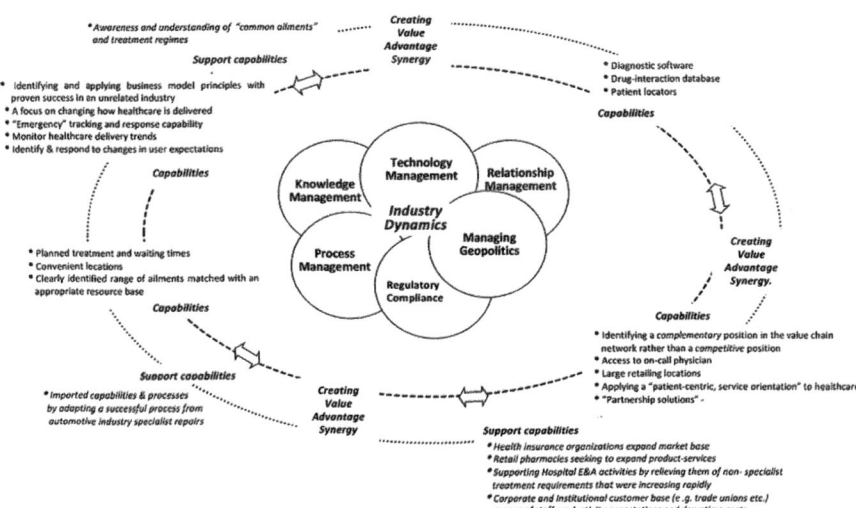

Fig. 2.3 Identifying the industry dynamics, capabilities, and support capabilities: MinuteClinic

dynamics. By contrast the vehicle exhaust emissions problems are requiring technology management, process management, and network relationship management changes which are required in the immediate future. A structured approach is apparent. The longer-term development of industry dynamics (not only the switch to an EV automotive industry but one that considers vehicle use and ownership and regulated access) is apparent in the strategic alliances and merger and acquisition activities currently taking place in the industry and that reflect a consensus of future strategy and structure decisions. A pattern is discernible, one that we suggest should become standard practice for monitoring opportunity impact/urgency matrix analysis. It also adds emphasis to the ability of the value chain network model to provide the basis for a stakeholder-value-led response. The observed, suggested, approach is based upon answers to a number of questions:

Value Chain Network Performance

How are individual organizational and network capability centers responding? Are value driver objectives being met?

If not, where are problems occurring?

Are there any changes requiring a review of specific capabilities? Are they sufficiently important to warrant evaluation? Will changes improve individual organizational and/or network performance? What are the cost implications?

If no, are there problems with a lack of understanding of the industry dynamics?

Impacts from Industry Dynamics for the Network

What is/are the critical Industry Dynamics for the network?

Are there identifiable changes taking place in the relevant Industry dynamics? Are these significant for the short-term (incremental adjustments) and/or the long-term (major shifts in one or more)?

Impacts from Industry Dynamics on the Network's Customers

If significant change is identifiable, how have these impacted/influenced the network capability centers, i.e., performance, producibility, productivity, profitability, partnerships, and preservation, that contribute to business-intelligence capabilities and key planning decisions?

Do these changes require changes in supplier organization responses? Will these changes impact on suppliers' relative value advantage positions? How? Will it require organizational change(s)?

Impacts from Industry Dynamics on the Network's Customers' Customers

Identify the Impact of Industry Dynamics changes on customers and customers' customers.

What are the network organizations' existing customers' capability center responses to current industry dynamics?

References

Allee, V. (2003). *The future of knowledge: Increasing prosperity through value network*. Oxford: Butterworth-Heinemann.

Blumentritt, R., & Johnston, R. (1999). Towards a strategy for knowledge management. *Technology Analysis & Strategic Management, 11*(3), 287–300.

Champy, J. (2008). *Outsmart*. Upper Saddle River, NJ: Pearson Education/FT Press.

Christensen, C. (1997). *The innovator's dilemma: When new technologies cause great firms to fail*. Cambridge, MA: Harvard Business School Press.

Eyers, J. (2004, September 18). P/E tool not sharp enough say experts. *Australian Financial Review*.

Ferguson, G. (2017, September 27). Collaboration the key to advanced defence manufacturing. *The Australian Financial Review*.

Gluyas, R. (2004, November 15). Article in the Australian.

GM to acquire auto finance firm in $3.5 billion deal (2010, July 22). *Agence France-Presse*.

Grant, R. (1995). *Contemporary strategy analysis*. Oxford: Blackwell.

Greimel, H. (2017). Toyota, Mazda form partnership to share technologies, confront cost challenges. *Automotive News*.

Ikea acquires TaskRabbit (2017, September 29), *BBC News*.

Irani, Z. & Love, P. (2001). The propagation of technology management taxonomies for evaluating investments in information systems. *Journal of Management Information Systems*, Winter.

Jacobides, M., MacDuffie, J. P., & Tae, J. C. (2016). Agency, structure, and the dominance of OEMs: Change and stability in the automotive sector. *Strategic Management Journal, 37*(9), 1942–1967.

Jarrillo, J. C. (1993). *Strategic networks: Creating the borderless organization*. Oxford: Butterworth-Heinemann.

Lippert, J. & Trudell, T. (2017, August 4). Toyota Motor Corp. and Mazda Motor Corp. agreed to buy stakes in each other and jointly build a $1.6 billion U.S. factory. *Bloomberg*.

Leidecker, J.K. & Bruno, A.V. (1984). Identifying and using critical success factors. *Long Range Planning*. February.

Normann, R., & Ramirez, R. (1993). *Designing interactive strategy: From value chain to value constellation*. New York: Wiley.

Phelps, R. (2004). *Smart business metrics*. Harlow: Pearson Education.

Rumelt, R. P. (1987). Theory, strategy and entrepreneurship. In D. Teece (Ed.), *The competitive challenge: Strategies for industrial innovation and renewal*. Cambridge, MA: Ballinger.

Slywotzky, A. J., & Morrison, D. J. (1997). *The profit zone*. New York: Wiley.

Stabell, C., & Fjeldstad, O. (1998). Configuring value for value advantage on chains shops and networks. *Strategic Management Journal, 19*(5), 413–437.

Tabuchi, H. (2010, March 9). Consumers come second in Japan. *The Australian Financial*.

Chapter 3
Using Capabilities to Build a Response-Led Strategic Decision Model

Abstract The dynamics of the business environment are now changing rapidly, and we are witnessing a rate of change that 15/20 years ago could not have been imagined. The digital and connectivity features of Industrié 4.0, the increased integration, coordinated collaboration of Value Chain Network 2.0, and the broader accountability of stakeholder value-led management has brought a requirement for a new approach; individual organizations and networks can no longer rely on existing capabilities; they need to be both agile and future focused, when identifying strategic futures and the *capability responses* required for success. A response capabilities approach views the firm as a portfolio of capabilities that evolve in response to the (perceived) demands of the business environment. An "enterprise" comprises one organization or collaborative networks of firms. Examples from the large corporations suggest that the traditional structural model based upon functions (silos) is rapidly losing support in favor of holistic structures reflecting *response capabilities*; these are characteristics that reflect an understanding of the marketplace and of the opportunities it offers and the characteristics of capabilities essential for successful engagement. The evident changes in management approaches of narrowly defined customer centricity, a changing view of the enterprise-stakeholder value management rather than just shareholder value management, a positive approach to network collaboration, network "connectivity," and the availability of real-time data analysis and management suggest the demise of the silos and replacement by a holistic, interactive approach that enables an organization to identify the capabilities that market opportunities require and the ability to structure a relevant response that meets the market demand precisely rather attempting to mold the market need into an opportunity that fits the silo's requirements.

Keywords Response capabilities · Support response capabilities · Capability planning · Value contribution · Value advantage

© Springer Nature Switzerland AG 2020   53
D. Walters, D. Helman, *Strategic Capability Response Analysis*,
https://doi.org/10.1007/978-3-030-22944-3_3

Introduction

Capability Management seeks to integrate the organization's capabilities to ensure value competitive positioning in an industry/market value chain. Capability management has become a method of creating *Enterprise Architecture* in recent years. It seeks to build a model of an enterprise that identifies its component parts and their relationships for planning the evolution (and continuity) of the enterprise. A capability management perspective (such as Leonard's model or Teece's Dynamic Capabilities Theory) suggests the firm is viewed as a collection of capabilities (rather than functions, or silos) comprising the means by which an organization responds to market opportunities. Typically, response capability is based upon tangible and intangible assets of the firm comprising the traditional business functions that have become "silos."

A response capabilities approach views the firm as a portfolio of capabilities that evolve in response to the (perceived) demands of the business environment. An "enterprise" comprises one organization or collaborative networks of firms. Examples from the large corporations suggest that the traditional structural model based upon functions (silos) is rapidly losing support in favor of holistic structures reflecting *response capabilities*; these are characteristics that reflect an understanding of the marketplace and of the opportunities it offers and the characteristics of capabilities essential for successful engagement.

Leinwand et al. (2016) identify Haier's four differentiating capabilities. These are depicted in Fig. 3.1. Haier has focused upon *customer-responsive innovation*, initially seen as a *customizer*, focusing on local customer needs but moving into the role of a *solutions provider* (helping consumers manage domestic issues such as water quality and home design). *Operational excellence* is pursued by creating a workplace culture of internal competition through a zero-defect continuous improvement. *Management of local distribution networks* has resulted from overall market size that has created a decentralized value chain. *Demand production and delivery*

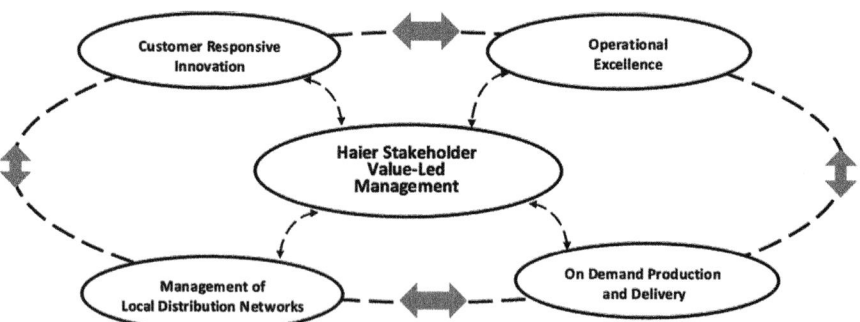

Fig. 3.1 Haier delivers value by excelling at four differentiating capabilities. (Adapted from: Leinwand et al. (2016). Source: *Strategy That Delivers*, Paul and Cesare Mainardi with Art Kleiner, Harvard Business Press, 2016)

is based upon a "pull-oriented" distribution system and a zero-based inventory management policy. This has been achieved by "being a consistently coherent and capable company: staying true to its core identity as a company dedicated to solving problems for consumers, while continually reinventing itself with imagination and verve" Leinwand et al. (2016). Haier demonstrates a requirement for success that is based upon a policy of identifying a framework of core capabilities that underly its strategic responses. These have clearly been successful by modifying them to meet changing customer needs and competitive challenges.

Three developments impacting the business environment in recent years have been digitization, the acceptance of the wider application of the value chain, and the importance of stakeholder value management as a more equitable corporate response than shareholder value management. Digitization, in the shape of *Industrié 4.0,* has resulted in an extensive application of "connectivity" across the manufacturing and servitization delivery of product hardware and serviceability software. The *value chain* has matured, and *value chain network 2.0* has enhanced collaboration, coordination, and communication among value chain partners. *Stakeholder value-led management* is becoming important to enlightened CEOs who are publicly mindful that they are answerable to more than just shareholders; their role is one of being answerable to a much broader group as pressure from customers, suppliers, employees, government and other regulators, environmental groups, as well as shareholders makes them answerable to "everyone" – the outcome is they must deliver a balanced performance.

It is arguable that an approach to response capability can be identified by considering response capabilities that are consistent with the changing and challenging business environment that while reflecting an underlying strategic opportunity, response capability reflects the awareness of value migration which identifies both changes in consumer demand *and* the means by which they can be successfully responded to. It encourages constant vigilance to identify opportunities and how they may be addressed effectively. An additional, new approach to business model design is suggested, a "response capability approach," matching strategic opportunities to existing or network available response capabilities. The "big data" feature of Industrié 4.0 changes the characteristics of decision-making; the availability of operating data facilitates the identification of solutions to problems and the decision alternatives; in many situations the customer offer (the value proposition) has become an agile response, a "rolling value proposition," one that can be matched to the specific, current needs of customers.

The extent and the significance of these changes suggest the need for a model that considers the impact of all possible, and viable, alternative solutions on the core business response capability model components of *performance management* (the value engineering, value delivery, and the value proposition; stakeholder-wide, performance, fit4purpose, etc.), *profitability* (financial viability, sustainable economic profit, positive cash flow), *productivity* (total factor productivity, optimal utilization of capital, labor, materials, and service inputs and EVA (economic value added)), *partnerships* (expertise and skills of value-generating employees, workplace cultures, and management styles), *producibility* (the seamless intra- and inter-organizational infrastructures of sequential processes and activities of value engineering and value delivery that creates, produces, delivers, and captures value), and *preservation* (a socially responsible use of environmental resources used in creating stakeholder value and the concerns of environmental and corporate sustainability and responsibility).

This contrasts with the view that there are key features comprising a business model canvas of interdependent decision-based strategic components; while they are interdependent, they do not have the facility of exploring cross impacts of decision alternatives to arrive at an optimal response solution for the stakeholders. Resource response capabilities are grouped in six centers, each responding to a specific, current capability that can add differentiation to the customers' value proposition, and at the same time, to the responding value chain network.

Figure 3.2 illustrates the connectivity components of the emerging digitized capability response model. The application of the model requires an understanding of the benefits of digital operations – the benefits of networked strategic and operational activities and a focus on stakeholder value management. Industrié 4.0 introduces (and facilitates) a need for faster, focused responses to market opportunities. Market responses require the orchestration of a combination of capabilities offering an optimal response solution to all stakeholders within the value chain by offering a value advantage over competing value proposition offers in terms of fit4purpose, relevance to the customers strategic direction, and time and location availability and at an acceptable customer value proposition comprising a cost/price/margin/serviceability combination.

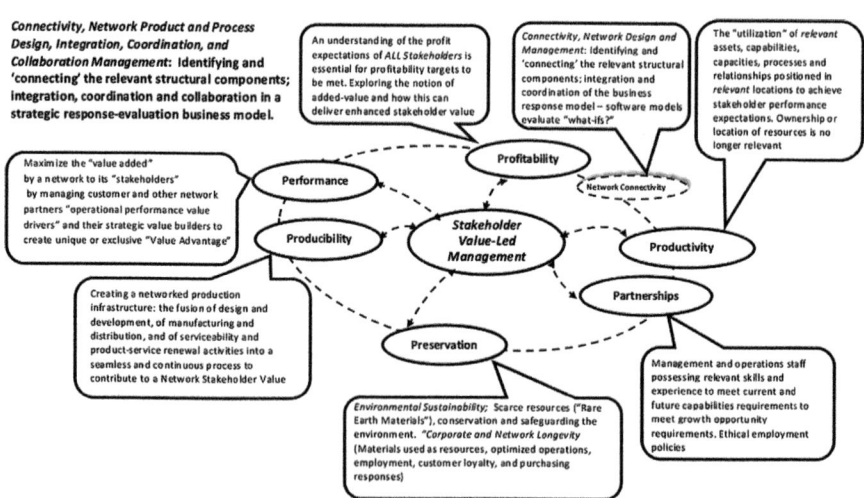

Fig. 3.2 The components of a connected strategic capability response model

Resources as Capabilities

The *resource-based view of the firm* was an early view of competitive advantage and was proposed by Penrose (1959) in *The Theory of the Growth of the Firm*. The fundamental thesis of this theory was that firms derive their profitability from their control of resources – and are in competition to secure control of resources. Resources were considered to be tangible and intangible. *Tangible resources* are physical resources (i.e., property, plant, equipment, and cash). *Intangible resources* are resources such as the knowledge and skills of employees, a firm's reputation (its brand), and a firm's culture. Intangible resources are more likely to meet the criteria for strategic resources (i.e., valuable, rare, difficult to imitate, and not substitutable) than are tangible resources. Industrié 4.0, Value Chain 2.0, and the broader based accountability of stakeholder management have shifted this emphasis. Organizations now seeking to achieve long-term competitive value advantage should place a premium on developing their firms' intangible *capability response resources*, i.e., performance, profitability, productivity, partnerships, producibility, and preservation relate to the tangible and intangible assets identified.

Prior to the emergence of capability management, i.e., the management of a portfolio of capabilities to ensure the organization is market competitive and as such will achieve profitable revenues, the dominant theory explaining the existence and competitive position of firms, was based on the resource-based view of the firm (RBVF); the fundamental thesis of this theory is that firms derive their profitability from their control of resources – and are in competition to secure control of resources. Other contributions to the resource-based view of the firm include Teece and Pisano (2003), Barney (2001), Ulrich and Lake (1990), Leonard-Barton (1992), Hamel and Prahalad (1994), and Kay et al. (2003).

Leonard-Barton (1992) analyzed the nature of the business concluding that "core capabilities" comprise at least four interdependent dimensions as follows:

- Physical technical systems; machinery, databases, software systems etc.
- Managerial systems; systems for the management of operations, including operation of technical systems
- Skills and knowledge (systems); systems for the maintenance of personnel and team skills and knowledge
- Values and Norms; systems for the regulation of behaviors and objectives in organizations

Leonard-Barton's (1992) "core capabilities" relate closely to the resource capabilities proposed above and relate to implications they have for the tangible and intangible assets identified in the proposed model.

Teece and Pisano (2003) developed a macro-level theory of *dynamic capabilities* and framework for their management. A *dynamic capability* is defined as "the firm's ability to integrate, build, and reconfigure internal and external competences to address rapidly changing environments. Ulrich and Lake (1990) make a distinction between capabilities and competencies: individuals have competencies while organizations have capabilities. Both competencies and capabilities have technical and social elements. Ulrich and Lake's (1990) distinctions are helpful as they identify a demarcation between individuals and organizational networks.

This would appear significant in the current business environment in which "intrapreneurship" (the management task of structuring and coordinating collaborative value adding networks) identifies the opportunity to create value and add vision and leadership by building organizational networks with relevant capabilities for value creation, production, and delivery.

However, the organization requires a "means to this end": Hamel and Prahalad (1994) succinctly described a company network's competitiveness as being derived from its core competencies and core products (the tangible results of core competencies). Core competence is the collective learning in the organization, especially the capacity to coordinate diverse production skills and integrate streams of technologies. It is also a commitment to working across organizational boundaries, the notion of structured networks involved in value production and delivery. To maintain consistency, we suggest that competitiveness is derived from the *network capability responses* we have identified earlier in this chapter. The inclusion of connectivity as a performance value driver adds emphasis to the requirements of ongoing competence among partners, the management, and the labor force.

Haier demonstrates a requirement for success that is based upon a policy of identifying a framework of core capabilities that underly its strategic responses. These have clearly been successful by modifying them to meet changing customer needs and competitive challenges. They have responded to opportunities as they saw and interpreted them, and their capability response can be mapped upon the proposed model.

An Industry in Transformation: Selling Mobility Capability Services Not Automotive Hardware Requires a Review of Organizational Capabilities

The automotive industry is undergoing rapid change, and it is suggested that this will continue for some significant time in the future. Two interrelated changes are occurring; these are ride hailing and autonomous driving technology. It is estimated that every day some 10 million partners use Uber. Uber operates in more than 600 cities in 82 countries. Autonomous vehicles are not quite ready for widespread introduction but are making rapid progress.

The automotive industry's reaction can be seen in the response of its major organizations (vehicle assemblers and component suppliers): they have been very active in establishing relationships with ride hailing and software and hardware automatic developers. The combination of developing user behavior (ride hailing) with autonomous, electrically powered vehicles would appear to have major significance for not only the automotive manufacturers and users but for public transport and urban planners. Car ownership would appear likely to be impacted by ownership cost per mile. Currently ride hailing costs per mile are high (at US$2.50

per mile) compared with private cost ownership (at US$1.20 per mile); the projected costs of automated robotaxis travel is around US$0.70 per mile. Estimates of the impact of autonomous vehicles suggest major changes in the "passenger economy." They suggest that by 2025 "robotaxis" will have 80 percent of partners using them in cities and by 2030 some 25 percent of passenger miles travelled in the USA will be in shared, self-driving electrical vehicles, reducing vehicle numbers on city streets by 60 percent, emissions by 80 percent, and road accidents by 90 percent (UBS and Strategy Analytics). General Motors announced (March 15, 2018) a $100 million capital investment program for two plants in Michigan as it plans to launch production of its Cruise AV self-driving vehicle. GM has also announced production of the fourth-generation Cruise AV is expected to begin in 2019. The electric-powered automated vehicle is projected for use in autonomous ride-sharing fleets. The Orion plant is currently producing the Chevrolet Sonic and Bolt EV vehicles, with a total output of 5.1 million per year. Clearly GM intends to play a major role in developing these applications.

A global estimate suggests the convergence of autonomous vehicles and ride hailing will be a market worth US$7 trillion, perhaps as much as US$ 10 trillion. There can be no surprise in the knowledge that carmakers, technology companies, and ride-hailing companies are deeply engaged in attempts to create dominant market positioning. The combination of ride hailing and autonomous vehicles is seen as an unprecedented challenge for the automotive manufactures, hence the moves being made to change their business models; their current *strategic response capability portfolios* are woefully inadequate to meet this challenge to asset management. A significant cost concern is that the automotive technology of autonomous vehicles is high, very much higher than the total cost of the vehicle, making individual ownership difficult, added to the fact that usage rates and return on investment of a privately owned automated vehicle would be difficult to achieve as private vehicles are stationary for 95 percent of the time. Some automotive manufacturers have created subsidiary companies in shared vehicle and ride hailing, others with alliances to supply vehicles for mobility organizations. Investment in ride-hailing companies may be attractive to large investment organizations; the forecast "24-7" utilization of the vehicles promises high returns.

There is also suggestion of problems for some national economies such as Germany with large numbers of employees within the assembly and components supply activities. The premium price sector may be threatened by both the electric vehicle (EV) and the autonomous vehicles. This may prove not to be as daunting as suggested; Volkswagen (Audi), BMW, Daimler Mercedes-Benz, and Jaguar-Land Rover have announced launch dates for EV versions of their vehicles.

The automotive industry is clearly expecting to undergo challenging changes. Two important issues will influence their decisions concerning a likely transition from vehicle manufacturing to mobility services providers. One is government influence; the autonomous vehicle is expensive to manufacture; private ownership will require subsidization, either of the vehicle cost or, possibly, its power supply in terms of convenience of accessibility and the cost of re-charging, particularly as electricity prices have experienced large and rapid increases internationally.

Government is also involved in broader contexts; it has been argued that urban planners have built cities around the automobile and EV and autonomous vehicles enable planners to take a more holistic view. The second issue concerns the likelihood, and timing, of consumer change and acceptance of ride hailing, autonomous vehicles, and electrically powered vehicles. This involves a significant level of trust. The diesel-powered vehicle was supported by government by an argument based on economy of use and cleanliness: in many countries the taxation incentive was reduced, and emissions problems have made the consumer question the cleanliness argument.

These examples are representative of what is occurring on a wide international front. The purpose of this proposed text is to consider the impact of a rapidly changing environment within which value chain management now exists and its long-term emphasis on stakeholder value. Industrié 4.0 has presented numerous opportunities across all industries to improve both the effectiveness of strategic decisions and the efficiency of their implementation to the network stakeholders. It has brought about significant changes to the value chain network; customer demand and supply response time has been condensed, and in some product-markets digitization is making the response simultaneous. In capital goods markets, it is the service-output that is purchased not the hardware; fixed costs become variable costs as organizational investment focuses on intangible assets such as brand reinforcement and customer services such as bespoke data analytics that identify performance management improvements and "flag-up" serviceability requirements. Capex (capital intensive) business models are becoming Opex (low capital intensity – intangible fixed assets)-based models focusing on organizational intangible assets (brands, developing supplier and customer relationships) that contribute to *network value advantage*, defined as a productivity-led efficiency advantage and/or a differentiation-led advantage derived from product-service characteristics that result in a superior value proposition (Christopher 1998).

Value Contribution and Value Advantage

Some capabilities directly contribute to the customer value proposition and have a high impact on network profitability and productivity. These "value advantage capabilities" are those assuring a positive *value contribution* that equals (preferably exceeds) the customer fit4purpose requirements, available from alternative competitive networks value propositions. Value advantage support capabilities make significant contribution in developing and delivering the value proposition, by assuring value contribution continuity. Kay et al. (2003) expressed a view on what is required to create value by suggesting "Business strategy is concerned with the match between the internal capabilities of the company and its external environment." Kay defines *Distinctive Capabilities* as "those capabilities a firm has which other firms cannot replicate even after they realize what the benefits are that owning the capability confers. These distinctive capabilities are the source or superior performance of successful firms. They provide potential access to a wide variety of markets, contribute to the customer benefits of the product, and are difficult for competitors to imitate." This is the network perspective (Value Chain 2.0). However, a planning and control model is required to consider operational and strategic time perspectives.

Capability resources management is based upon creative, collaborative, and coordinated combinations of specific responses. Typically, these evolve from network-based organizations utilizing the width, depth, accuracy, and immediacy of the data connectivity facility of Industrié 4.0, the collaborative connectivity of Value Chain 2.0, and the input of stakeholder expectations. Potential responses will require rapid assessment and evaluation by exploring alternative combinations of network capabilities. To manage the large range of potential options that will be available, each network partner will operate with a predetermined range of input capabilities. Examples of these facilitating the creation of *value advantage* characteristics are:

Performance
- Product-service fit4purpose specification
- Serviceability support
- Quality specifications
- Delivery frequency and reliability
- Availability (product-service and location)
- Customer-supplier-performance-management

Profitability
- Impact on "total-stakeholder-investment" effectiveness
- Economic profit and financial performance (EVA)
- Impact on optimal capital intensity and asset management
- Margin management and margin mix
- Cash flow management
- Impact on industry/market "profit pool" optimization (a shift in the value chain roles
- Network EVA (value contribution) distribution equity i.e., share of $ value of value advantage

Productivity
- Availability of relevant capabilities, capacities, and processes
- Network relationships (value chain network positioning)
- Network digitization and connectivity capability
- Availability ownership and location of resources

Partnerships
- Managerial capabilities requirements
- Labor force operations expertise and skills requirements
- Availability of relevant management if required
- Availability of relevant operations labor if required

Producibility
- The creation, production, delivery, and capture of value
- The impact of changes (and potential changes) to the value proposition
- Changes required to current and planned infrastructure model

- Implications for market and value chain network positioning
- Changes to product complexity and serviceability complexities
- Infrastructure connectivity and collaboration requirements
- Impact on the value contribution from an alternative infrastructure

Preservation
- Current implications on the market/industry of environmental sustainability requirements
- Impact of forecast changes on corporate and network processes and activities
- Implications for current and future costs

Support Capabilities: Characteristics

Support capabilities to facilitate the application of resource capabilities solutions are an essential component to the model. They will be structured around *leadership and professional capabilities*, identifying direction and applying managerial abilities that understand the nature of current market demand requirements and identifying the relevant response capabilities; *digital capabilities*, a responsive data management network and connectivity; *human capabilities*, ongoing recruitment and development and training of staff to match changes in market demand with appropriate responses; and *financial engineering capabilities*, the ability to repurpose existing assets for use in meeting changing market demand. They will vary between the strategic and operational activities of the network. For example, specific strategic support capabilities (building the business response) will identify opportunities and direct the planning and acquisition of relevant resource capabilities to address stakeholders' expectations such as:

- Product-service/product-SERVICE Fit?
- Economic profit targets (Customers/Network)?
- Productivity targets (Customers/Network)?
- Relationships fit (Stakeholders/Network)?
- Preservation (Stakeholders/Network)?
- Strategic operating infrastructure fit?

Operational support capabilities (managing the business response) will focus on implementation:

- Managing the operational operating infrastructure
- Digital SRM, CRM, and CPM (customer performance management)
- Increased application of digital capability
- Optimal productivity scheduling
- Improved data capture, analysis, and decision-making (accuracy and timeliness)
- Traceability data
- Enhanced security

Capabilities and Added Value: A Convergence of Qualitative Issues and Quantitative Performance

Capabilities and value advantage meet. Key to this is an appreciation that value itself is not necessarily a constant and that it needs to be measured regularly. The organization needs to identify and manage added value; this requires a quantitative means of doing so. Kay's (1993) approach to calculating added value is helpful in this regard:

> Added value is the difference between the (comprehensively accounted) value of a firm's output and the (comprehensively accounted) cost of the firm's inputs. In this specific sense, adding value is both proper motivation of corporate activity and the measure of its achievement.

Examples of added value in this context relate to supplier output that adds "features" to an assembler's product as it proceeds downstream towards an end-user customer. Features may be tangible (such as the use of low weight/stronger materials in construction that reduces fuel consumption and perhaps maintenance costs – the Boeing 787). They may be intangible features (such as changing the nature of the product; selling and servicing the output it provides – Rolls-Royce Engines – where product becomes *total serviceability*, the motive power and service support supported by digital maintenance). Clearly both forms of added value can be viewed in both qualitative and quantitative formats.

Kay (1993) calculates added value by subtracting from the market value of an organization's output the cost of its inputs:

> *Revenues*
>   *Less (wages and salaries, materials, and capital costs (i.e., the capital consumed in creating the value)*
>   *Equals*
>   *Added Value*

Kay (1993) suggested that added value is a measure of the loss that would result to national income and to the international economy if the organization ceased to exist:

> Adding value, in this sense, is the central purpose of business activity. A commercial organization which adds no value - whose output is worth no more than the value of its inputs in alternative uses - has no long-term rationale for its existence.

Or:

> *Revenues less (wages and salaries, materials, capital costs) equals Added Value*

An important aspect of Kay's work relates to a preference for measuring business profitability as economic profit rather than accounting profit. The commercial application of Kay's contribution was subsequently developed by Stern and Stewart (visit website). Stern Stewart Inc. offers an information service based upon calculated EVA (economic value added) details for organizations and industries.

Capability Planning: Management

Capability planning has been an active methodology in military strategy and operations for some time. Observation suggests that in business organizations, capability
planning and management is being approached by using a range of options.
Figure 3.3 identifies the components of a digital capability-led response model. The
model relies upon the connectivity capabilities delivered by Industrié 4.0, Value
Chain Network 2.0, and the broader responsibilities required by adopting stakeholder value management, rather than the narrower focus of shareholder value management. As the benefits of digital applications become more relevant, mergers and
acquisitions (mostly the latter) are increasingly considered in pursuit of competitive
capabilities. The automotive industry has seen numerous acquisitions of software
companies by the OEMs in the industry. In the aerospace industry, there has been
the recent acquisition by Boeing of KLX Inc. (a major supplier to Boeing) for $4.25
billion. KLX will become part of *Boeing Global Services* and will be fully integrated with Boeing's parts subsidiary, Aviall, with an anticipated annual cost savings of $70 million by 2021, according to the statement. Strategic alliances have
also been popular as methods of obtaining capabilities; again, the auto manufacturers have been notably active with GM and Ford. GM and Lyft agreed a long-term
strategic alliance to create an integrated network of on-demand autonomous vehicles in the USA, but the relationship between the two has grown increasingly com-
plicated as each company has pursued other partnerships with overlapping partners
as the scope of the industry expands.

It would appear they share basic criteria in deciding upon decisions on how to
obtain critical, value creating capabilities; the size of the investment is clearly

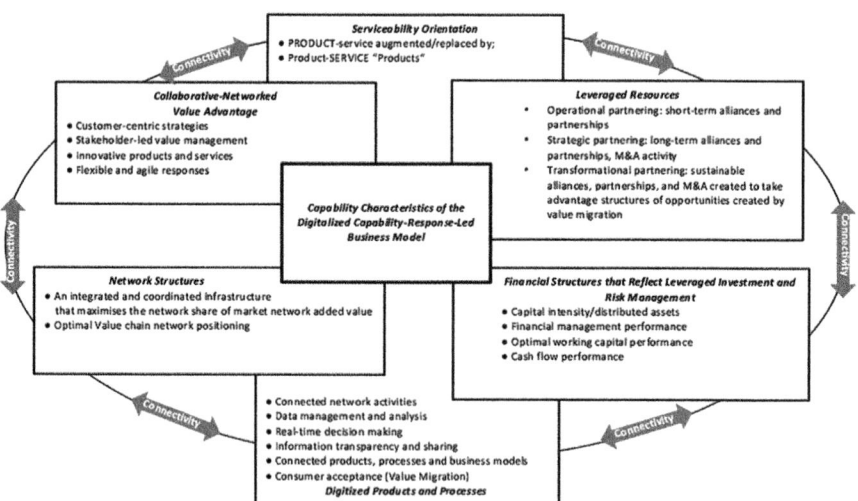

Fig. 3.3 The digitized business model: A "connected" network structure of relevant capability
resources

important as is the future return. Immediacy is also important; time2market is clearly an important issue. Talent and both management and technological competencies are significant in developing a product-service that can add value in the pursuit of superior fit4purpose. Operating efficiency (productivity) is being improved.

Value migration becomes important. Recent events in the media (news management – production and delivery) in Australia illustrate the impact of value migration. Two newspaper publishers dominate the newsprint industry, News Corporation and Fairfax; in July 2018 they agreed a consolidation move in which they shared each other's printing facilities on a regional basis. This was followed by a merger whereby a subsidiary of Channel Nine will own Fairfax and are expected to influence Fairfax editorial policy.

Industrié 4.0: Contributions to Improving Business Effectiveness and Efficiency-Based Capabilities

The benefits offered by Industrié 4.0 digital connectivity include:

Improvements in Customer Relationship Management
    Digital systems manage customer orders by connecting a customer relationship management (CRM) capability system (that stores customers order details product services, frequencies, delivery requirements and payment and credit facilities histories) to an enterprise resource planning (ERP) system that identifies order fulfilment capabilities, forecasts likely future orders, provides process updates throughout the process and allows online collaboration during the planning phase.

Increases the Application of Automation
    Digitizing, integrating, and automating manually administrated processes (such as stand-alone excel systems) processes can be accelerated, accuracy improved and maintained, and productivity improved as staff are allocated to non-routine tasks.

Optimize Manufacturing Operating Schedules
    Given digitized control over materials ordered and in inventory it is possible to optimize the production schedule, speed-up processes and reduce unnecessary costs. Digital planning systems ensure items are ordered and delivered in the right order, in the correct number, in time for processing, by the relevant number of staff, thereby improving "total factor productivity"; materials, labor and, asset utilization, thereby enhancing an efficient operations capability.

Data Capture, Analysis, and Improved Decision Making
    Connectivity is a major benefit offered by Industrié 4.0. Using scanners and sensors to provide connectivity data linkages improves capability responses for example; machine2machine (M2M), machine2operator (M2O), by digital thread (equipment performance data transfer from equipment to user organization and equipment manufacturer, or to digital twin, provide real-time performance data that can be used to improve operating performance capability, by providing notification of maintenance requirements, and provide data for product improvement and development).

Traceability Details

It is becoming an increasingly important capability for organizations (particularly value chain networks) to have the capability to trace the origins of materials used in production in detail and rapidly to ensure compliance with input and production process standards. Food manufactures may need to prove provenance; aerospace manufacturers may be asked to demonstrate tolerance specification levels, and chemical processors required to meet specified levels of manufacturing processes. Regulators are requiring value chains networks to maintain "documentation management tools" to demonstrate compliance increasingly being provided by Block Chain technology.

Enhanced Security

As manufacturing and distribution organization embrace Industrié 4.0's facilities companies cannot ignore the threat of cyber espionage and cybercrime. Reduced vulnerability is an essential capability for businesses to ensure the security of digital linkages between all value chain network stakeholders is maximized. The adoption of Blockchain has added both traceability and security.

As organizations gain confidence in using these features, they are becoming essential rather than desirable and as such become "required support capabilities."

The response capability approach can reinforce a market (and value chain) positioning as well opening up alternative market opportunities. This suggests that response capabilities are not only useful for creating new innovative solutions but can also be used to search for alternatives around a specific capability. For example, preservation may be one response for which there are constraints such as the reuse of resources, requiring product-services to be designed such that a fixed amount of the retired product will be remanufactured and form the basis of a second generation of the item. Similarly, the response capability can set prerequisites; for example, organizational policy, not strategy, can establish ground rules concerning levels of returns, such as a margin of the amount by which an EVA return should exceed the annualized weighted cost of capital or perhaps the amount of increase in capital intensity the organization is prepared to undertake as further investment. Non-financial constraints may also be policy issues, such as environmental sustainability and/or environmental damage the organization is prepared to accept from any new processes or activities it undertakes. Given this is an important consideration for the individual organization and the network we can appraise any relevant opportunity. A third and possibly the most important benefit concerns the "mutuality"; Fig. 3.4 explains that response capabilities should be two-way; they are far more effective if they create value advantage for both the vendor and the customer. The recent transition in capital good markets of the PRODUCT-service becoming a product-SERVICE is an example. It has benefited the supply response activity of the value chain by focusing (and rationalizing) capital expenditure. *Serviceability* is a much broader response than *servitization;* capital equipment companies can either own or leverage serviceability; either way capital equipment required for implementing serviceability facilities is used more efficiently (see the Small Robot case in which the company suggests that agricultural equipment in arable farming is typically used for 4/6 weeks in a year). Figure 3.4 suggests that vendor/customer activities and processes would benefit from collaborative capital expenditure planning.

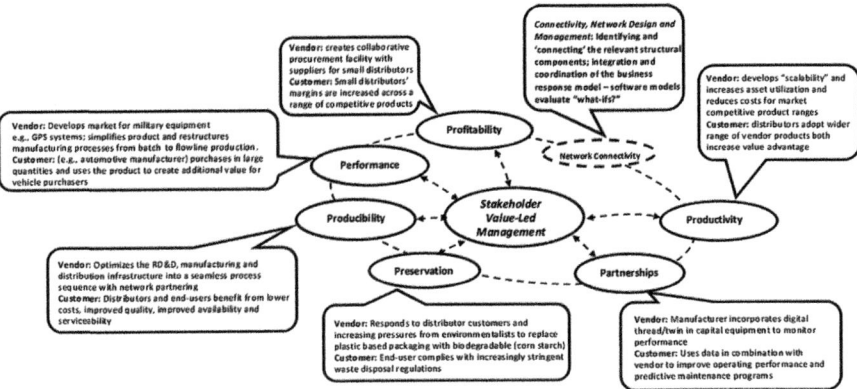

Fig. 3.4 Response capabilities function "two-ways" creating value advantage benefits for vendors and customers

Convergence

The convergence of Industrié 4.0, Value Chain 2.0, and the stakeholder value management perspective requires "thinking" networks capable of integration, agile responses, and resetting the benchmarks. Required *response capabilities* are *performance* (stakeholder-wide, performance, fit4purpose, etc.), *profitability* (financial viability, economic profit), *productivity* (total factor productivity, optimal utilization of capital, labor, materials, and service inputs), *partnerships* (expertise and skills of value-generating network members), *producibility* (the cost-effective and efficient network operating infrastructure that creates, produces, and delivers value), and *preservation* (a socially responsible use of environmental resources used in creating stakeholder value).

This chapter has considered a number of important changes and challenges confronting organizations. It has identified Industrié 4.0, Value Chain 2.0, and stakeholder value management as being "strong game changers" currently and, increasingly, in the future.

The organization is connected and functioning within a value chain network. It is essentially a "customer-serviceability" response capability, focused in its approach to selecting and responding to opportunity and in adjusting its market and value chain positioning by identifying the relevant capabilities for success. Typically, these are available as leveraged response capability resources within the networked structure within which it is involved.

> *Performance:* Integrating value creating activities from RD&D through to remanufacturing offers network stakeholders an opportunity to optimize the use of their resources in a collaborative activity to create optimal stakeholder value.

Profitability: Each of the network partners expects to be able to maintain a level of profitability that will meet strategic and operational investment objectives if their contributions to the network success are to be long-term.

Productivity: Network structures enable partners to invest in relevant capabilities, capacities, and processes to participate in the value creation, production, and deliver the network's value proposition.

Partnerships: Network success is dependent upon integration and collaboration; these are core response capabilities for strategic effectiveness and operational efficiency.

Producibility: The fusion of value creation *(value engineering),* production, delivery and serviceability *(value delivery)* requires shared, agile collaboration and communication response capabilities that meet the changes and challenges of the business environment emerging as the industrial convergence develops opportunity.

Preservation: Changes and challenges have become a given in a developing and demanding business environment that is placing demands upon both business organizations and the environmental resources they use. *Sustainability* has dual consideration for decision making; the longevity of the business *and* of the environment and its resources.

Network Connectivity: Network Design and Management; A core response capability. Identifying and 'connecting' the relevant structural components; integration and coordination of the business response model - software models that evaluate "what-ifs?" using digital twinning.

Relevant financial structures are important; the current move from Capex structures to Opex structures (by becoming output focused) suggests a view that intangible assets are gaining in importance as they are customer-facing. The Opex structure would appear to offer stakeholders realistic returns. An increasing number of organizations are realizing the benefits of measuring economic profitability using EVA as a metric.

Activities suggest the increasing importance of the move towards "connected" activities. The growth of data management and analysis capabilities, the capturing of data from real time activities decision responses, together with the transparency and sharing of information have contributed towards an enhancement of effective (strategic) management, and efficient (operational) management.

References

Barney, J. (2001). Resource based theories of competitive advantage. *Journal of Management.* Sagepublications.com.

Christopher, M. (1998). *Logistics and supply chain management.* London: Financial Times, Prentice-Hall.

Hamel, G., & Prahalad, C. K. (1994). *The core competences of the corporation.* Boston: Harvard Business School Press.

Kay, J., McKiernan, P., & Faulkner, D. (2003). The history of strategy and some thoughts about the future. In D. Faulkner & A. Campbell (Eds.), *The Oxford handbook of strategy. Volume one: Strategy overview and competitive strategy. Overview and competitive strategy. (2003).* Oxford: OUP.

Kay, J. (1993). *Foundation of corporate success.* Oxford: Oxford University Press.

Leinwand, P., Mainadi, C., & Kleiner, A. (2016). *Strategy that works: How winning companies close the strategy-to-execution gap.* Boston: Harvard Business Review Press.

Leonard-Barton, D. (1992). Core capabilities and core rigidities: A paradox in managing new product development. *Strategic Management Journal, 13*(S1), 111–125.

Penrose, E. T. (1959). *The theory of the growth of the firm.* New York: John Wiley.

Teece, D. & Pisano, G. (2003). Dynamic capabilities of firms in Handbook on knowledge management (INFOSYS, volume 2). New York: Springer.

Ulrich, D. G., & Lake, D. (1990). *Organizational capability: Competing from the inside out.* UK: Wiley.

Part II
Introduction

Part II comprises six chapters; each chapter considers a response capability in the process of doing, so it also offers a review of current practice and application of management thinking and application within the disciplines outlined as interacting capabilities seen as requirements for success in an increasingly changing business environment.

Chapter 4 explores performance. Being "value competitive" now requires a stakeholder focus, implemented by an analytical and evaluative approach to both strategic and operational opportunities. For both, irrespective of the planning horizon being explored, a stakeholder focus offers management the facility to evaluate the capabilities required of the organization (and its network partnership) to undertake identified opportunities and their impact on current structures and activities and to explore the creation of an optimal strategy and structure format (infrastructure) that will optimize stakeholder-led value management. Capability response focuses on the ability to create inter-organizational performance output. It moves further away from the traditional silo approach to consider the contribution of disciplines. The disciplines identified are *performance* (strategic and operational marketing, finance, and operations), *profitability* (economic profit, economic value added (EVA), asset management and gearing financial and operating, profit pool analysis as a method of reviewing past and future investment activities, and risk assessment), *productivity efficiency management* (the problems of current measurement methodology and the use of EVA as an alternative), *partnerships management* (the increasing relevance of networks (Value Chain Networks 2.0) increased integration, coordination, collaboration, and transparency brought about by the richness and reach of data availability and analysis), *producibility* (the concept identified by the creation of production infrastructures that link the activities required to create an end-to-end "connected" series of activities from RD&D, manufacturing, distribution, consumption, and, increasingly, remanufacturing and repurposing), and *preservation* (the notion that sustainability concerns organizational longevity providing stakeholder satisfaction, as well as longevity of environmental resources and limitation of damage caused by production process pollution). The chapters in part two consider these topics and how they each contribute to the overarching objective of creating stakeholder value.

The sole purpose of a *performance management system* is to assess and ensure that its value creation, production, and delivery system (the value chain network) partners conduct roles and tasks for which they have to undertake in an effective and satisfactory manner, which is contributing to the business objectives of the value chain network. Delivering optimal stakeholder satisfaction, performance management requires some explanation. At a transaction level, it is a mutual, bilateral process in as much as it requires both customer and vendor agreement on expectations and satisfaction for it to work. This is essential because if this does not occur, there can be no stakeholder satisfaction. Planned and managed "performance management" cognizant of the primary stakeholders' *expectations* and responses and of the secondary stakeholders' *interests and concerns* is the basis of a successful value proposition.

Performance management is a response to a customer organizations value drivers. The value drivers in any business depend on the specific setting, competition, and the market structure. Their time perspective is clearly short-term given that they are factors that "drive present value" and are levers of present value. They are mutual; supplier organizations also have performance drivers. And creating value creates costs; hence the performance of both customers and suppliers is linked; it is the concept of the value chain (integration and collaboration) that manages performance. It is not unknown for competitors to collaborate within a value chain to overcome competition from within the industry/market. Tesco (UK) and Carrefour (France) announced a collaborative project to purchase their "own brand" products and other items (for 3 years) in a competitive response to Aldi and Lidl.

Value builders are long-term and may have no relationship with the current business model. They comprise *experience capability characteristics* (skill sets, expertise) and *positional capability characteristics* (features/characteristics the organization has (or is) investing in that have a long-term payoff). The ability to capture value in a dynamic market environment requires management to provide "customer aligned solutions," developing adaptive organizational structures, creating network modularity and network orchestration, and developing value chain loyalty relationships that encourage increased comprehensive customer cooperation and commitment. An overriding factor concerns the ability of management to identify and analyze macro-developments in the business environment and their implications for the future of competitive capability. Strategic value builder analysis requires a more open approach than that required for operational performance value drivers. There are two considerations: the long-term strategic intentions of the customer organization and those of supplier organizations. It is for these reasons that scenario analysis has an important role. While the customer and supplier organizations may have congruent interests in the short-term, there may be reasons why these may diverge over time and require both to consider the longevity of the partnership.

The topic in Chap. 5 concerns profitability. However, profitability has many interpretations, often competing with other metrics as a means of measuring corporate performance. The DuPont ratio spread is explored in this chapter as useful in identifying meaningful interpretations across a range of interests, strategic effectiveness, investment strategy, operating efficiency, and shareholder returns. However, there are differing views on what comprises "profit." Kay (1993) argued that while organizations appraise success by size of sales, or share price performance, market share, return on a measure of investment, or perhaps sales or growth rate, a realistic measure of corporate success is *added value, a derivative of economic profit*. To be of use as a performance metric, added value requires to be quantified. Operating margins are one indication of value that is added at each stage of production. However, this has problems. One problem concerns the charge made for the use of capital in the production process. Here the concept of economic profit can be used. Economic Value Added (EVA) is a financial management technique developed by Stern and Stewart (www.sternstewart.com), a financial consultancy that built upon Kay's conceptual work.

Other relevant topics are introduced and discussed in the context of their importance of value creation.

Reference

Kay, J. (1993). *Foundation of corporate success*. Oxford: Oxford University Press.

Chapter 4
Performance Management: Value Drivers and Strategic Value Builders

Abstract "Performance management" has both short-term and long-term perspectives. Short-term performance management comprises modifications to a product service and possibly to the activities and processes that are responsible for its "production." It is the short-term aspects that are most immediately impacted by Industrié 4.0 and Value Chain Network 2.0. Understanding the importance of "value" to customers and other stakeholders helps strengthen relationships among stakeholders, customers, suppliers, shareholders and investors, and organizations with indirect interest and influence, as these "value-based" relationships are the links an organization needs if it is to develop a strong competitive position. To do so, it clearly needs to identify the *operational performance value drivers* (and in the longer term, the *value builders*) that are important to the end-user customer to structure a value delivery system that reflects these *and* the objectives of the other value chain participants. Within the context of a value chain network, operational performance value drivers assume short-term significances. One is clearly that of the role of the process of adding *relevant value* for customers and its ability to differentiate the value offer (via its value proposition) such that it creates value advantage for both the customer and the supplier organization. The second is that like their customers, their suppliers also have operational performance value drivers, and *creating value creates costs* for supplier organizations, thereby raising questions on the impact on the value *and* cost drivers of the supply/vendor organization. The third questions the impact on the value *and* cost drivers of the supply/vendor organization network. Accordingly, an analysis of the impact of enhancing a customer value driver should be accompanied with an evaluation of its impact on the supplier's own operational performance value drivers *and on* the impact of the supplier's cost of focusing on the customer's value driver(s). Clearly, these two issues are linked as a supplier's operational performance value drivers include supplier financial performance, and it follows that unless the marginal revenue generated by enhancing a value driver exceeds its marginal cost, there would be little point in pursuing the proposal.

Keywords Performance · Performance management · Operational performance value drivers · Strategic performance value builders

© Springer Nature Switzerland AG 2020
D. Walters, D. Helman, *Strategic Capability Response Analysis*,
https://doi.org/10.1007/978-3-030-22944-3_4

Introduction

Being "Value Competitive" now requires a stakeholder focus, implemented by an analytical and evaluative approach to both strategic and operational opportunities. For both, irrespective of the planning horizon being explored, a stakeholder focus offers management the facility to evaluate the capabilities of the organization (and its network partnership) required to undertake identified opportunities, their impact on current structures and activities, and to explore the creation of an optimal strategy and structure format (infrastructure) that will optimize stakeholder-led value management. The digital connectivity feature of Industrié 4.0 has "enabled" management to review its entire role in an industry value chain network. The large number of divestment and investment activities is evidence of changes that are occurring as network partners adjust their value chain network positioning. The increased integration and coordination of Value Chain 2.0 has brought new approaches to interorganizational collaboration. Notable examples are the large automobile manufacturers (with their acquisition of software companies and of value sharing service-related companies) and large industrial equipment companies who have undertaken similar activities but have extended the value chain positions by focusing on using the data generated by their hardware products to include customer performance management in their product portfolios. The corporate awareness of the wider responsibilities of organizations has widened the focus of organizational performance to include primary and secondary stakeholder interests. The sole purpose of a performance management system is to assess and ensure that a system's (value chain network) partners are capable of conducting the roles and tasks they have agreed to undertake in an effective and satisfactory manner, that their performance is contributing to the overall business objectives of the value chain network, and that they have proven and continue to prove that they are fit4purpose.

"Performance Management" requires some explanation. It is mutually bilateral in as much as it requires both customer and vendor agreement on expectations and satisfaction for it to work. This becomes clearer when we consider a component-based definition of "performance management":

> The bilateral expectations of "performance management" are based upon a positive impact on the generation of added value for both customer and customers' customers either by increases in revenues and/or productivity efficiencies.
>
> Fit4purpose identifies positive bilateral impact on processes and activities as well as on product-service outputs.
>
> Shared data management output identifies bilateral infrastructure operating and strategic capability improvement potential.
>
> Planned and managed "performance management" optimizes the organization's role in the producibility infrastructure.
>
> Planned and managed "performance management" cognizant of the *primary stakeholders' expectations and responses* and of the *secondary stakeholders' interests and concerns*, is the basis of a successful value proposition.

"Performance management" has both short-term and long-term perspectives. Short-term performance management comprises modifications to a product service and possibly to the activities and processes that are responsible for its "production." It is the short-term aspects that are most immediately impacted by Industrié 4.0 and Value Chain 2.0.

Many of the long-term perspectives will be based upon developments of the ongoing digitization of product-services, of processes and activities that continue to offer a framework for developing a perspective on "business futures" and their shape and the response capabilities that maybe required for success, as suggested by Phelps (2004).

Understanding the importance of "value" to customers and other stakeholders helps strengthen relationships among stakeholders, customers, suppliers, shareholders and investors, and organizations with indirect interest and influence, as these "value-based" relationships are the links an organization needs if it is to develop a strong competitive position. To do so, it clearly needs to identify the *operational performance value drivers* (and in the longer term, the *value builders*) that are important to the end-user customer to structure a value delivery system that reflects these *and* the objectives of the other value chain participants. Using Slywotzky and Morrison's (1997) "customer-centric" approach to the value chain network the "things that are so important to customers" are the customers *operational performance value drivers*, and the important operational performance value drivers are those adding *significant* value to customers (these are likely to demonstrate a Pareto (80/20) profile).

Within the context of a value chain network, operational performance value drivers assume short-term significances. One is clearly that of the role of the process of adding *relevant value* for customers and its ability to differentiate the value offer (via its value proposition) such that it creates value advantage for both the customer and the supplier organization. The second is that like their customers, their suppliers also have operational performance value drivers, and *creating value creates costs* for supplier organizations, thereby raising questions on the impact on the value *and* cost drivers of the supply/vendor organization. The third questions the impact on the value *and* cost drivers of the supply/vendor organization network. Accordingly, an analysis of the impact of enhancing a customer value driver should be accompanied with an evaluation of its impact on the supplier's own operational performance value drivers *and on* the impact of the supplier's cost of focusing on the customer's value driver(s). Clearly, these two issues are linked as a supplier's operational performance value drivers include supplier financial performance, and it follows that unless the marginal revenue generated by enhancing a value driver exceeds its marginal cost, there would be little point in pursuing the proposal.

Operational Performance Value Drivers

Performance value drivers are short- to medium-response actions that are responses to the market based on existing capabilities. These might build competitive value necessities, correct competitive value disadvantages, and create competitive value advantage largely using the existing business model. Efficiently managing working capital is a good example of a short-term value driver, as is customer performance management.

The 1980s–1990s introduced the concept of customer centricity (Cleland and Bruno 1996; Slywotzky and Morrison 1997), with Slywotzky and Morrison suggesting:

> The value of any product is the result of its ability to meet a customer's priorities. Customer priorities are simply the things that are so important to customers that they will pay a premium for them or, when they can't get them, they will switch supplier.

The "things that are so important to customers" are the *value drivers*, and the important value drivers are those adding *significant* value to customers. Within the context of the value chain, value drivers assume a twofold significance. One is clearly that of adding value for customers; the other is the ability to differentiate the value offer such that it creates competitive advantage. Four questions emerge:

- What is the combination of value drivers required by the target customer group? What is the customer groups' order of priority?
- What are the implications for differentiation decisions? Are there opportunities for long-term competitive advantage?
- What are the implications for cost structures?
- Are there opportunities for trade-offs to occur between value chain partners that may result in *increased* customer value (and stakeholder value) or *decreases* in the value system costs or the costs of the target customer group?

Walters and Rainbird (2012) suggested a portfolio of value drivers that applied to both the customer *and* the vendor network: asset management, time management, cost management, and performance management comprising a range of characteristics (e.g., financial, marketing, customization, quality, service, agility response, customer choice, convenience), risk management, and information management.

Industrié 4.0 and Value Chain 2.0 and to some extent the broader notion of accountability expressed by the acceptance of stakeholder value management has expanded customer expectations. In Fig. 4.1, a post "industrial renaissance" perspective on operational performance value drivers reflects the changes occurring in the generic set of value drivers.

Figure 4.2 extends the discussion to the capability responses required of network organizations and is the result of ongoing observations of the activities of organizations across a range of product-service activities and a range of organizational sizes. There are a range of considerations that contribute to stakeholder satisfaction; furthermore they each have influence on the overall objective of an optimal solution. Figure 4.2 identifies a generic set of "drivers"; clearly each transaction participant will have quantitative and qualitative objectives, and these will need to be matched with those of the responding network. Creating value creates costs, and the responding network will consider the implications of offering a value proposition that is a balanced offer to all stakeholders. The network's management

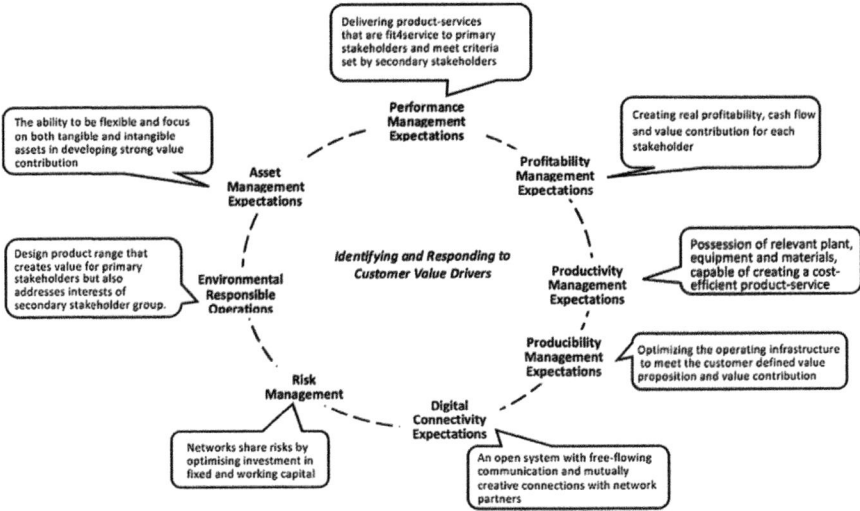

Fig. 4.1 A post "convergence" perspective on operational performance value drivers

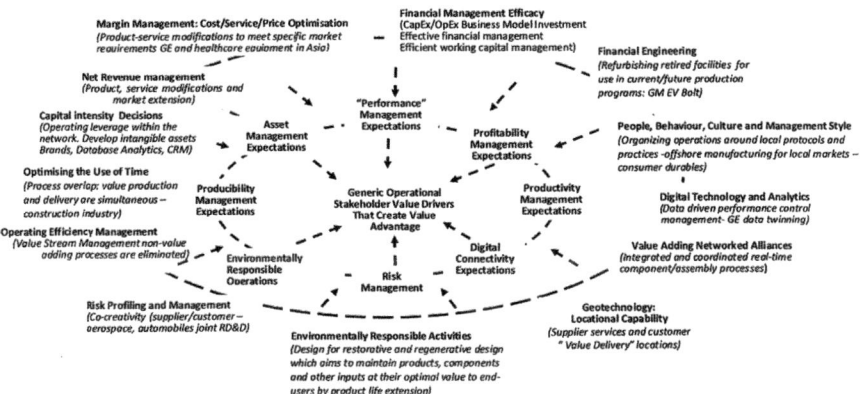

Fig. 4.2 Creating capability response value drivers

should take the secondary stakeholders' interests in mind when developing a value proposition; they may be offering advice that could be important and that has been overlooked. This group is expanding in both the number of stakeholders and in the number of specific interests; this clearly is due to the impact of the amount, availability, and transparency of information now available. Developing a feasible (workable) and viable (profitable) value proposition is a complex task. However, as the success of the large industrial organizational networks suggests it is complex and time-consuming but can bring success.

Identifying Performance Value Drivers

Identifying value drivers begins by asking "What drives value in your business? Who are the competitors? What are the characteristics of the market?" The value drivers in any business depend on the specific setting, competition, and the market structure. Their time perspective is clearly short term given that they are factors that "drive present value" and as levers of present value. Focus on adjustments to the value drivers results in short-term improvements in performance. Performance value drivers include adjustments and operational implementation characteristics such as product mix, manufacturing and distribution capacity, employee motivation, supply chain configuration, generating strong positive cash flow, excellence in customer service, etc. Phelps (2004) argues there are no generic value drivers; they can be as diverse as brand image for one organization and employee recruitment policies for another: what is common to all organizations is that value drivers create short-term performance improvements.

Figure 4.2 extends the notion of generic value drivers further by identifying their basic characteristics. The earlier comments made concerning the application to specific organizational situations are given emphasis in this section, i.e., they differ in specific content application and in importance when considered by networks. Also emphasized is that it is important to remember that both customers and vendor response networks have performance value drivers.

A comprehensive list of support characteristics has been developed by researching the activities of a number of international organizations; they range from General Electric, Siemens, IKEA, and Rolls Royce Engines to specialist small-medium enterprises (SSMEs) offering focused product services. These support characteristics are the result of the recent capabilities offered by the industrial renaissance coalition of Industrié 4.0, Value Chain Networks, and the broader concept of accountability, stakeholder value management. Each of the characteristics is supported by comments and examples:

1. *Net Revenue Management and Market Response*

A customer-focused approach identifying changes in customers, products, and processes and creating innovative responses to changes in customer expectations and how they may be satisfied. It is arguable that investment of cash and time in this activity leads to competitive value advantage. The connectivity provided by digital products and operations facilitates customer centricity without imposing time and cost penalties, product platforms, and standardized components and processes make a significant contribution.

(By focusing on a specific customer segment and working on collaborative projects (component and OEM companies), revenues are consolidated and costs reduced. Adopting platform concepts for product-service development, using partner expertise and facilities, and designing service processes into products lock in the full added value. ("If its broken, you can't fix it". 2017).

2. *The Capex Business Model Versus the Opex Business Model: Financial Management Efficacy*

The connectivity facility of Industrié 4.0, digital threads, data twins, and "On-Demand Manufacturing" is resulting in a change of managements' views concerning the financial structure of individual organizations and the network partnership structures. The online/real-time facilities of connectivity are having an impact on the location of manufacturing processes within the value chain. In automotive manufacturing, component and module building is becoming increasingly capital intensive as the nature of the assembler activities and investment is focused on creating options that differentiate product service within the range and from competitive product-service offers and in expanding the product range offer by adding "service features" in car features that offer external connectivity and RD&D and investment into alternative energy forms that may replace fossil-based energy. Nevertheless, fixed capital effectiveness and working capital efficiency remain major considerations; it is the nature of their structure that is changing:

Fixed capital effectiveness
    Ensure network core resources are 'deployed' via value creation, creative logistics and supply chain management to contribute to stakeholder wealth creation (for both end-user customers and network partners). Alternative delivery systems should be evaluated against the need to deliver end-user customer value *and* the impact on partner fixed assets (facilities) as well as working capital (inventory levels). New projects are 'designed' such that they may by financed by leveraging existing assets (owned and those of partners) that are operating with the benefits of scale economies, thereby sharing fixed costs and facilitating a competitive market entry price.

(Product-service design is evolving around shared platforms (automotive industry practice). This has an impact on fixed costs as they are shared with competitors)

Working capital efficiency
    Efficient working capital management through effective and efficient supplier and customer relationship management results in service characteristics (availability, on time deliveries, flexible responses, etc.) that make for prompt collection of payments and, therefore fiscal efficiency.

(Inventory holding costs are being contained by developing product hardware around platforms and standardized service parts) Adopting platform concepts for product-service development, using partner expertise and facilities, and designing service processes into products locks in the full added value. Additive manufacturing is being used as an alternative to field inventory for service parts availability)

Financial engineering
    Recycling capital, refurbishing retired facilities for use in current/future production programs.

(GM are gaining value advantage by manufacturing the EV Bolt in manufacturing facilities that were idle, therefore avoiding the costs of building a dedicated facility).

3. *Operating Efficiency Management*

Cost-efficiency management, or value stream analysis, identifies waste within value chain networks. Creating value creates costs, and the purpose of operating-efficiency management is to ensure the value proposition/value delivery profile is optimal, i.e., the market offer is made with a lean capability/resources mix that optimizes the stakeholder value delivery.

(As part of the process of developing a value proposition benefits (value added)/total cost of ownership (value delivery costs) are identified. Robins has selected a niche segment in the ladies' fashion shoe market in which quality, style and availability (time) are important performance characteristics and has responded by retaining domestic manufacturing in Australia because an important value driver is the ability to meet a two-week lead time as opposed to the six months that is typical from off-shore suppliers).

4. *Margin Management: Cost/Service/Price Optimization*

By reducing product-service complexity and increasing product simplification (standardizing components, platform processes), using lean product design and assembly and distribution (optimal operational costs and the network quantity sizes) and intraorganizational information systems, costs of both fixed asset and current asset costs are contained.

("Design for Use" considers the entire process of RD&D through to the possibility of recycling/remanufacturing as seamless. As a result, the problems of access to serviceability are met at the design stage).

5. *Optimizing the Use of Time*

Ensure a relevant "time-to-customer" response by managing order cycle length and operating cycles, resulting in acceptable cash2cash cycle and market product life cycles. Optimize *horizontal (process) time where* value adding (or is wasted) occurs. Eliminate/reduce *vertical (no-activities) time* where no value is added.

(Automotive assembly plants have used JIT component delivery to minimize the storage and handling costs of inventory management in the manufacturing process)

6. *Geotechnology*

The ability to respond to customers (directly or indirectly) within specified time spans and across specified responsibilities and actions. Connectivity offers increasingly cost-efficient solutions for servicing distant demand. Prior to the developments in ICT, and then subsequently connectivity, distant demand was serviced by several methods. The problem of compressing time was met by rapid dedicated transportation, company-owned field service distribution centers, franchised sales and service agents, and/or co-opetition shared supply chain response facilities with competitors. An attempt at responding to international competition is the purpose of the recently announced collaboration between Tesco (UK) and Carrefour (France) to purchase their "own brand" products and other items (for 3 years) in a competitive response to Aldi and Lidl (Agnew and Eley 2018). Each of these alternatives was expensive and not always an effective solution.

Ordnance Survey (OS) founded back in 1791 was made by a UK government-owned company in 2015. The organization makes most of its £150 m yearly revenue from the private companies who use its data – such as Garmin's "sat nav" equipment – and through big public-sector agreements providing data for bus routes, planning, and flood prevention. A third of its products can be used for free. It is the information on roads and rivers, which provides the core data inputs for so many other companies, including Google Maps. OS decided to open the *Geovation* hub in London in 2015. The company felt it was "behind the curve" on how its mapping data was being used and that it wasn't seeing the benefits of giving the information away for free. Uber, Deliveroo, Airbnb, the development of driverless cars, and drone deliveries – all of these companies need location data to power their software. OS sees its long-term future in embracing new technologies such as providing data for smart cities, the rollout of 5G, and driverless cars. Ordnance Survey technology is also being used in the development of driverless tractors (Prescott 2017).

(Digitization offers some interesting solutions. Caterpillar's remote diagnostics provided the company with the capability of matching response with demand in order cycle response times that met an identified customer requirement. Boeing and Rolls Royce considered their customers' approach to operations by considering airframes and engines as assets that should be managed as tangible assets in a financial context, just as other investments. Both organizations have service subsidiaries that align their services around the routes their airline customers fly and locate relevant service-parts kits to meet both regular and emergency service requirements).

7. *Digital Technology and Analytics: Connected Operations*

The characteristics of business operations activities are undergoing constant change and challenge. The capacity issues that once dominated operations managers' activities are being superseded by decisions on capabilities as customers' expectations change to meet the challenges of customer dynamics – probably the most important performance driver.

The development of sensors and digital applications (digital threads and twins) enables continuous monitoring of engines and their components, thereby being able to respond to a service problem "as it is occurring." GE introduced a new method to monitor their engines once they are in use and attempt to predict how and when they will need repair. The first part of the new system is to create what they call a "digital twin" of every engine they build. During the design and manufacturing phase of the engine, engineers compile thousands of data points specific to each engine, which they use to build a digital model. This allows them to know exactly how hot that engine should be in each of its modules, what the pressure should be, and how fast the airflow should be moving.

(Industrial and consumer durable product manufacturers are using sensors to monitor their customers' use behavior of their products to analyze performance productivity and offer advice for improvements. GE and Alitalia are a recent example. Breville recently appointed a CEO (previously with Samsung) to develop connected kitchen appliances).

Industrié 4.0 has introduced revolutionary thinking concerning the use of information within the organization. Data management and decision-making are "connected" in real time and can delegate decision-making to mobile machines that are capable of being managed from a distance, thereby changing the nature of management at all levels.

(General Electric launched Predix and Siemens TeamCenter, as a data management and analysis platform service for customers to explore "connectivity" opportunities).

8. *Value-Adding Networked Alliances*

A structured business model that focuses core resources (assets, processes, capabilities, capacities and availabilities) and on-response processes that together with those of partner organizations enhance the benefits accruing from an integrated business system (network) that is market-proactive, flexible, agile and capable of a cost-effective response to opportunity.

(Value chain networks have developed as cost-effective holistic organizations that operate based on integrating and coordinating value delivery to end-user customers. Currently the automotive industry is an example. Healthcare is becoming a holistic activity with centralized patient monitoring working with local healthcare practitioners. Aerospace is also an example, the role of servitization, the focus on product-SERVICE outputs and maintenance are also an example of the expanding influence of network approaches).

9. *Risk Profiling and Management*

Assessing individual project risks by using net present value (NPV) analysis with a β (Beta)-adjusted risk estimate based upon the "returns spread" or a similar treatment of aggregated network partner EVA (economic value added) expectations – see below.

(Risk analysis is undertaken at two basic levels. FMCG companies consider the return/risk profiles involved when offering "variety" across product categories. Industrial product/ project organizations use industry specific risk adjustment data when assessing unfamiliar product/project opportunities. Increasingly geopolitical risk factors are applied, given the levels of terrorism that are dominant in many markets).

10. *People, Behavior, Culture, and Management Style*

People are often an organization's most significant asset. Often the organization requires new thinking and seeks this outside the company and its industry. "Imported" staff must accept "workplace" culture and adapt to alternative methods and attempt to introduce the changes for which they have been recruited with senior managements' guidance. Workplace culture is an important influence on business success. A recent contribution (Vuori and Quy 2016) suggests Nokia's problems were due, in part, to a "culture of fear"; employees failed to report problems because of the reactions of senior management. Equitable remuneration and benefits policies monitored for currency.

(Gender balance, equal opportunity, and racial discrimination feature strongly in the workplace balance debates. General Electric and Panasonic developed "localized" marketing/ manufacturing operations to meet use and cost parameters of healthcare equipment and consumer durable products).

and

11. *Environmental Responsibility*

Increasingly preserving the environment is becoming of interest. Climate change, global warming, resources recycling, and remanufacturing (value renewal) are prominent stakeholder concerns. The Global Reporting Initiative, Sustainable Development Goals, and Task Force Financial Disclosure are efforts gaining industry support.

> *(The Ellen MacArthur Foundation has been very active in gaining support for the Circular Economy/Value Chain and has gained support from global businesses. A recent project (June 2018) focusing on apparel production and marketing was launched with significant support of the industry).*

Creating customer and organizational value advantage has time considerations. Whereas the importance of operational performance value drivers is their facility to assist in identifying which of them has real impact on creating value for the customer organization (and the network) *in the short-term*; strategic value builders' *impact on future value* and typically there is no 'hard' data available with which to analyze the potential impact of each value builder.

Longer-term actions to create new response capabilities that can provide sustainable value advantage are *strategic value builders*. These generally require some sort of investment in organizational and network relationship changes, for example, capital investment or building (or transferring) intellectual property. Not all business models are bespoke. Indeed, the whole idea of franchising (networks) is to buy one off-the-shelf where somebody else has already worked out what the key operational performance value drivers and strategic value builders are and provides a "how to" kit. Most firms, however, construct their business models over time, reacting to changing market opportunities and resource constraints – collaborating with network partners if it is considered to be financially advantageous to both organizations.

Strategic value builders help build *long-term* future value. They provide an organization with the ability to plan to take advantage of opportunities as they arise and help to avoid threats and risks. For this to be effective strategic value builders are built on positional characteristics (a value chain network positioning strategy), investment levels, and funding, partnerships and strengthening external and internal relationships), and expanding (or at least maintaining) stakeholder value; and upon *capability characteristics* (quality of management, innovative expertise, flexible response processes).

Strategic Performance Value Builders

Value Builders are long term and may have no relationship with the current business model. They comprise *Capability Characteristics* (skill sets, expertise) and *Positional Characteristics* (features/characteristics the organization has (or is) investing in that have a long-term pay-off).

Creating value in the long run differs from the short run not only in the context of time but in relevance to the future business environment. Value builders also differ in that there is an amount of risk involved, and the available information for decision-making is limited. Furthermore, value builders (especially capabilities) are intangible in their nature and as such very difficult to measure. Phelps (2004) made a point with comments concerning matching current managerial expertise with future opportunities, and the extent to which value can be added will depend on the behavior of not only one organization. Phelps suggests two categories of value builders: *positional led* and *capability led*.

A *positional value builder* suggests that for a number of organizations, they are future options; the development of digital-led manufacturing offers potentially higher profitable investment by saving time (product development and testing) and in the manufacturing of "one-off" components for "one-off" products. *Access to emerging distribution networks* is another example; the development of FMCG home shopping delivers both failures and successes. For those companies without established distribution (i.e., conventional retail outlets with national coverage), a high rate of failure was observed due to the investment required in physical distribution facilities and systems. The concept was new, and many problems existed (e.g., redeliveries if customers were not available to receive purchases ordered); however, the national multiples (e.g., Tesco, Sainsbury in the UK, Woolworths and Coles in Australia) each had an efficient supply chain to leverage when they introduced their services, and this subsidized the very high initial costs when the service was introduced. It is of interest to note the success of Ocado (UK) in this respect as they had no existing (traditional) outlets or identifiable brand to leverage but used communications technology to develop an online value proposition that met with the emerging change in consumer expectations that rated time as an important life style factor. The Amazon acquisition of Whole Foods is another example; here we see an organization acquiring in a market segment within which it expects significant growth.

Capability value builders are more of a challenge; they are intangible resources and as such are key components of any business. Phelps (2004) argued that *structure and organizational culture, perspectives on customer trends*, and *awareness of and willingness to accept change* are all examples of capability value builders. *MinuteClinic* (Chap. 1; industry value drivers and enablers) is an example of how these characteristics combined to create a very successful auxiliary healthcare business. An analysis of what made fitting just vehicle exhausts and tires successful as a niche business in a very large market was convenience and time and a focus on the most frequent elements of vehicle servicing. When applied to healthcare, it met with huge success. Champy (2008) summarizes the success of *MinuteClinic* as being based upon an understanding of customer/patient expectations and the application of philosophies and practices from outside of the industry. Applications of customer service techniques from retailing and the adoption of operating processes from quick service organizations have given *MinuteClinic* a strong position in the healthcare value chain network.

The ability to capture value in a dynamic market environment requires management to provide "customer-aligned solutions," developing adaptive organizational structures, creating network modularity and network orchestration, and developing value chain loyalty relationships that encourage increased comprehensive customer cooperation and commitment. An overriding factor concerns the ability of management to identify and analyze macro-developments in the business environment and their implications for the future of competitive capability. An understanding of the development of industry dynamics is helpful. Identifying the "dynamics" for which little doubt concerning their continued impact exists provides a basis for capability planning. However, there may be indications of emerging dynamics that may have an impact on potential opportunities, some of which exist, and for others, they are likely to emerge in the future. These have a strong influence on the development of the strategic value builders relevant for organizational and network partnerships and how these are structured. A novel approach to viewing the organizational perspective of strategic value builders was suggested by Olvé et al. (1997). The authors consider the capability and positional characteristics assets and liabilities and create a "Balance Sheet" of an organization's current and/or required future strategic value builders identifying the organization's direct ownership and indirect access to them.

The authors argue that the traditional way of evaluating a company is to analyze its balance sheet, which after adjustments is one way of representing the value of the company. Using the balance sheet as a model, they develop a competence balance sheet. *The "assets" are the capabilities/competencies required for success and the "liabilities" show how they are financed or who owns and provides them.* Figure 4.3 suggests "tangible and intangible assets" and "liabilities." The argument continues by suggesting the company determines what capabilities are needed to compete and then to decide on how they should be "financed." A number of factors may influence the resulting decision. Clearly, the risk/return consideration is important as is the expected life span of capability, particularly so for technological capabilities. Other concerns likely to be important are the impact/urgency relationship (i.e., the relative influence of specific value builders on performance *and* the time span over which they could make an impact) on the strategic positioning, performance, and resultant competitive advantage sheet.

It follows that strategic corporate planning is following a collaborative approach to productivity (the optimization of resources – broadly defined as assets, capabilities, capacities, collaborative activities, and business process developments by a networked organization). The current focus on energy sources and its impact on the automotive industry is an example of "collaborative" planning for strategic productivity, and the ongoing activities within the industry suggest that *strategic collaborative productivity* best describes what is happening. In the following examples, it is apparent that individual organizations are identifying partners with resources that are required if long-term opportunities are to be pursued. A collaborative (network based) approach will ensure the greater productivity of resources applied.

The added value in the context of strategic value performance builders is probably best investigated by exploring alternatives available to an organization by comparing inputs and outputs in much the same way as would be used for operational

"Assets" Tangible & Intangible	*"Liabilities"*
(Exclusive Distinctive Capabilities)	*(Externally Owned Reproducible Capabilities)*
Revenue and Margin Growth Management	**Inter-Organizational "Production" Facilities and Networks**
• Data access and analytics expertise	• "Seamless", integrated, collaborative and coordinated
• Flexible structure: market *and* value chain positioning changed	process of product-service-development ,production,
to build significant income streams	delivery, & service
• Long-term performance gives added emphasis	• "Access" to 'exclusive' inputs (rare earth inputs)
to corporate/shareholder value	• "Access" to non-specialist facilities, equipment & processes
• Risk management expertise	• Capacity and quality management expertise
	• Real time data-base product/service performance
Innovation, Creativity, Product-Service & Process Research and Design	delivery and maintenance expertise
• Market/customer trend knowledge	
• Strategic customer focus	**Specialist Assets, Processes & Capabilities**
• Product-Service-Market design and development expertise	*(Partner Owned: Exclusive Capabilities)*
• Operational processes design	• Patents and brands (e. g. Intel)
	• Specialist processes and services. (e.g., design
Integrated Value Creating Networks	and development)
• Integrated -Supplier/Customer/Product-Service Databases	• "Access" to specialist facilities, equipment and processes
• Integrated Network Interactions	• Service response management networks
• Collaborative based design and development	• Product/service performance delivery
• Market connectivity	and maintenance
• Patents and licences management expertise	
• Specialist SMEs processes and services	**Complementor/Facilitator Networks**
	(Externally Generated "Market Opportunities"
Financial & Investment Management Expertise	• Access to parallel markets
• Familiarity with sources of funds	• Access to additional markets/market segments
• Access to relevant sources of funds at	• Access to new markets
acceptable terms, amounts, and interest rates	
• Ability to raise capital in international markets	
• Asset management expertise	
• Investment cycle planning (Technology-led demand cycles,	
SBU structures, coordinated management	
• Risk profiling, assessment, and management	

Fig. 4.3 The Virtual enterprise "capabilities balance sheet." (Adapted from: Olvé et al. (1997). Performance Drivers. Chichester, Wiley)

performance value drivers in the short term, that is, by evaluating the resource requirements, the resource owners' expectations for returns, and the optimal output acceptable to network members. Given the long-term nature of the projects being considered, the initial appraisal should be one using NPV methods such as CFROI (cash flow returns on investment). Another consideration is the type of relationships involved. Operational performance value drivers are short term in their nature and operation; the *operational partnering* involved is typically cost-led and robust for as long as the advantage prevails. However, strategic value performance builders are, by definition, long term and involve *strategic partnering* in which individual organizations create networks of organizations, each of which contributes an essential core resource, in order to create a value chain network that can compete with other networks successfully. The recent economic and financial problems have brought about a fiscal conservatism and a preference for minimizing investment. Large organizations are increasingly turning to "capital recycling" whereby they are selling assets (businesses or divisions) of activities that no longer return a substantial proportion of the organization's income: *transformational partnering* offers organizations the opportunity to change both strategy and structure in a cost-effective manner.

Creating Strategic Stakeholder Performance

Organizations operating in value chain networks have the facility to undertake projects that might otherwise be beyond their resources. Clearly there is an obligation to meet their own stakeholder needs in both the short-term and long-term. Strategic stakeholder performance is based upon a number of criteria such as: revenue and margin growth management, innovation, creativity, product-service and process research and design, integrated value creating networks, financial and investment management expertise, inter-organizational "value production" facilities and networks, specialist assets, processes, and capabilities: strategic stakeholder performance topics and examples include:

1. *Review Revenue and Margin Growth Management from Related Opportunity*

Core resources (assets, capabilities capacities, collaborative activities, and business processes) and partners are likely to change over the planning horizons that need to be considered – possibly for 10–20 years ahead. Revenues are very likely to be generated from markets that currently use concepts and ideas requiring industry clusters and delivery infrastructures that differ from current markets for them to be successful. Margin expectations may change as industries and market boundaries change. Organizations are considering revised market *and* value chain positioning if they are to build significant income streams. Long-term performance gives added emphasis to corporate/shareholder value – a number of measures are used; market valuation (market capitalization/asset value, and free/anticipated free cash flow).

(The context of value will have different characteristics; IBM and Dell now see themselves as solutions providers rather than as organizations that manufacture computer hardware in a highly competitive market. Automobile manufacturing is redefining 'mobility' requirements faced with sharing rather than owning vehicles and other assets. Future product-services may develop packages with value delivery distributed across the product-service and the delivery infrastructure; this appears to be the emerging direction for electric vehicles. Ownership/access to new core resources will be important issues to be decided well ahead of market development).

2. *Innovation, Creativity, and Product-Service Process Research and Design*

It is almost certain that product-service complexity and increasing product simplification will continue to be significant features as cost management maintains prominence in contributing to value-led productivity; increasingly PRODUCT-service-based "products" are becoming product-SERVICE based. Digital developments will continue to have prominence in the long term. The ability to use additive technology to produce models rapidly in the product development process is already reducing NPD cycle times and costs significantly; the application of this technology is expanding rapidly as input material types increase, from plastics to metals (steel, aluminum, "super alloys" such as Iconel") and to wood ceramics, glass, paper, graphene, and living cells, etc. Healthcare is being restructured with digital applications *together with* medical treatment developments that use technology from other industry sectors *and processes*. Further developments in collaborative activities will enhance data generation and analytics, transparency in inter-organizational information systems, and the costs of both fixed asset and current asset costs can be contained. Manufacturing process and product design are merging as digital manufacturing (MPM, manufacturing-process management) through software-driven product life cycle management is leading to potential benefits: decreased product-development cycles, faster time-to-market, reduced manufacturing costs, support for lean manufacturing and agility initiatives, enabling integrated design initiatives, e.g., design-for-manufacturability, design-for-assembly (*IKEA* is a leading organization in this area), improved product quality, and enhanced product-knowledge dissemination. Shared product and process platforms (eco-business systems, eBay, PayPal, Li, and Fung) have made a large impact on operating costs, and the impact of Industrié 4.0 on data collection and analysis is forecast to have an impact on operating costs. The data availability and analytics capabilities of Industrié 4.0 will ensure cost management is likely to become a focus of organizational product service at an earlier stage (RD&D) and continue to expand productivity management throughout the network as both vendor and customer partners fixed and current assets' utilization is increased. Centralized procurement is also contributing.

(PepsiCo has implemented a procurement strategy for hedging forward commodity purchases; these amount to US$18 billion annually. Prices are effectively fixed for nine months thereby avoiding price increases (and not benefiting from unlikely decreases); local regional managers cannot deviate from the planned costs without very good reason. The Company CPO argues this gives the organization the ability to forecast costs accurately whereas forecasting the costs of inputs is becoming very difficult. Proctor & Gamble consider hedging to be an operational solution, and therefore short-term; the CPO argues they are also expensive. P&G adopt a longer-term strategic solution; "innovative substitution". For example, the new packaging for its Pantene hair products uses biodegradable corn-starch rather than petro-chemical resins. They argue such changes pass unnoticed by customer stakeholders but are applauded by shareholder stake holders – particularly when they can reduce the cost of US$2 billion incremental commodity costs by 50 per cent!

3. *Integrated Collaborative Value Creating Networks*

A shift in the emphasis of order winning criteria is expected – the success of the GE and Panasonic "15/50" concept is introduced into developed markets following their success in the emerging markets. Both companies are suggesting that the substantially lower purchase prices will have strong purchasing incentives for consumers. Opinion based upon the extended impact of the 2008–2009 recession suggests these strategies would appeal to developed market consumers. The emphasis on the product-service requirements of the emerging markets has seen shifts of emphasis (and location) of RD&D and operations into the emerging markets.

(Within industrial markets a trend that is noticeable is the efforts being made by SSMEs (specialist small/medium enterprises) to focus on collaborative opportunities with OEMs, for example, Quickstep an Australian company specializing in metal finishes for the aviation industry).

4. *Financial and Investment Management Expertise*

Long-term strategies and organizational (and network) structures will change to adapt to changing opportunities. What is currently emerging is a suggestion that "capital recycling" will be used by many organizations as they move out of markets that are declining (or possibly offer decreasing returns) into growth industries and markets requiring investment. Reinvesting in "idle" assets to manufacture new product-services and together with a focus on intangible asset management (brands, data analytics, and customer performance management).

(GE, Panasonic, Lego, Dell, and IKEA have all made changes in their strategic directions and the organization structures needed to make them effective. Capital recycling, financial engineering, and leveraging partner strategic assets are components of financial effectiveness).

Planning working capital management through product-service design (platforms, standard service replacement parts) and effective and efficient supplier and customer relationship management will remain high-priority activities. Planned increased collaboration (co-productivity and co-opetition) fueled by yet more efficient digital data analytics systems will improve the long-term efficiencies of customer service characteristics and customer performance management. Demand chain analysis combined with response management responses that are likely to include more use of advanced manufacturing (additive manufacturing), VMI (vendor managed inventory), and JIT (2) (supplier company personnel working in customer organizations with procurement and manufacturing management to ensure supply continuity) will impact upon inter-organizational working capital characteristics.

(Demand chain analysis and response management can be expected to become more closely coordinated as relationship management promotes closer collaboration through additive manufacturing to meet on demand for high-cost service parts, VMI, and CPFR programs).

Long-term cash flow requirements change as organizations buy and sell assets to reposition within new growth markets and in new value chain positions. *Strategic cash flow*: changes in long-term working capital (inventory requirements, receivables, and payables), long-term fixed assets (tangible and intangible assets), and entry and exit costs (processes, capabilities, technology, and relationship building.) *Transformational Cash Flow*: investment (equity or debt) is required to make changes to the long-term organization by developing inter-organizational response cash flow management.

(Actions by large organizations (General Electric and Amazon to use cash generated from operations to expand growth) may become a wider spread activity).

5. *Inter-organizational "Production" Facilities and Networks*

Market/industry leadership that integrates and coordinates "seamless" value engineering and delivery. Localized RD&D that relates to local "needs" and capabilities. GE's distributed power business designs and produces engines and power equipment for industrial power generation and gas compression at or near the point of use.

(GE's new, "multi-modal 'Brilliant Factory'" concept. The manufacturer is adopting a range of data gathering and networking technologies to equip its operations for real-time reporting and supply-chain updates, in line with Industrial Internet of Things concepts to improve process efficiency, product quality, and plant profitability.)

6. *Specialist Assets, Processes, and Capabilities*

Global suppliers for the automotive industry, are delivering superior value to customers through innovative products and processes; offering complete vehicle engineering and contract manufacturing expertise, as well as product capabilities which include body, chassis, exterior, seating, powertrain, electronic, active driver assistance, vision, closure and roof systems.

(For example, Magna Steyer), offer global automotive suppliers operating 312 manufacturing operations and 98 product development, engineering and sales centers in 29 countries. Magna specialize in manufacturing low-volume vehicles for global vehicle manufactures whose plants are built to operate at high-volume output to meet economies of scale cost targets).

Strategic value builder analysis requires a more open approach than that required for operational performance value drivers. There are two considerations: the long-term strategic intentions of the customer organization and those of supplier organizations. It is for these reasons that scenario analysis has an important role. While the customer and supplier organizations may have congruent interests in the short-term, there may be reasons why these may diverge over time and require both to consider the longevity of the partnership. For example, the automotive industry is undergoing major changes in strategic direction (developing electric vehicles and alliances with software-based companies such as Uber and Lyft) as the doubts concerning fossil fuel life expectancy raise (further) doubts concerning the design of future vehicles, as well as having doubts concerning attitudes towards vehicle ownership. We have already seen the emergence of new alliances between vehicle manufacturers to share R&D development programs and costs in an attempt to minimize time-to-market schedules; oil companies are diverting R&D expenditures into battery technology and into sourcing lithium for battery inputs. Customer and supplier organizations alike are both considering *positional* and *capabilities characteristics profiles* required for competitive positioning in their future markets. Suppliers will also be reviewing the implications of potential changes on the performance metrics as well as their market and value chain network positioning.

Major specialist suppliers in the industry that have exclusive positioning in the value chain network are likely to face significant changes. Typically, these organizations have expertise at managing scale economies in relatively low production volumes, whereas a large vehicle manufacturer would consider 300,000 vehicles per year an optimal, economic, level of production. Organizations such as Magna Steyr engineers develop and assemble automobiles for other companies on a contractual basis. In 2002, it absorbed Daimler AG's *Eurostar* vehicle assembly facility. The company's aggregate vehicle assembly capacity reached 200,000 vehicles a year. It is the largest contract manufacturer for automobiles worldwide. Magna Steyr developed Mercedes-Benz's "4Matic"four-wheel drive (4wd) system and assembles all E-Class 4Matic models. The company also did substantial development on the BMW X3 and manufactures all X3s. CargoWise, logistics software company, is aiming to create an international "container chain" that links connecting manufacturers, freight companies, customs, warehouses, trucking operators, and last mile transporters. The growth had been driven by its CargoWise One platform, not acquisitions, and the business was "accelerating the flywheels" of its global footprint. "We've never bought companies to buy growth. We've bought them to build something vast that's needed in the supply chain," Richard White, CEO. In the last year, WiseTech has made 22 acquisitions, the majority of which have been for geographic expansion, but 7 have been for technological capability or product expansion, fitting with its vision to create a software platform for the whole shipping and delivery ecosystem.

The Importance of Industry Dynamics

Industry dynamics are very likely to influence the selection and subsequent performance of value builders. The current advances being made in robotics, artificial intelligence and algorithm-based decision models suggest a future in which many of the existing "barriers" across industry sectors are likely to disappear. The interest of the automotive manufacturers include software development and is an example of the shift in strategic developments beyond mobility and towards information services. PwC (2017) undertook "futures studies" in the automotive industry and in healthcare. See Fig. 4.4. The findings (that will be explored in detail in a later chapter) support the view that value builders are very likely to be influenced by industry dynamics; as such they will play a major role in understanding the development of customer's strategic value builders as an essential input into identifying the future directions of customer relationship management. Typically, these will be based on the scenario analysis and as a result of discussions with the customer organizations. The output from these activities will be of major significance to both customer and supplier organizations, as based upon the outcome will be the need to consider strategy, structure and partnerships, and future investment programs.

Initially, responses to customer strategic value performance builders will be largely speculative because future network roles and tasks will require investigation to identify likely investment costs and their alternative structures. However, in Fig. 4.4, we make tentative proposals on what the responses may appear to be given current trends and extrapolating these into the future for the automotive industry. The point was made earlier that industry dynamics (see Chap. 2) will continue to develop and may converge. The healthcare example of a "mass-production" philosophy being applied to heart surgery is an example of a process management technique being imported into a different industry. Increasingly, the role of relationship management is making such migration a possibility, and the role of inter-organizational interactions

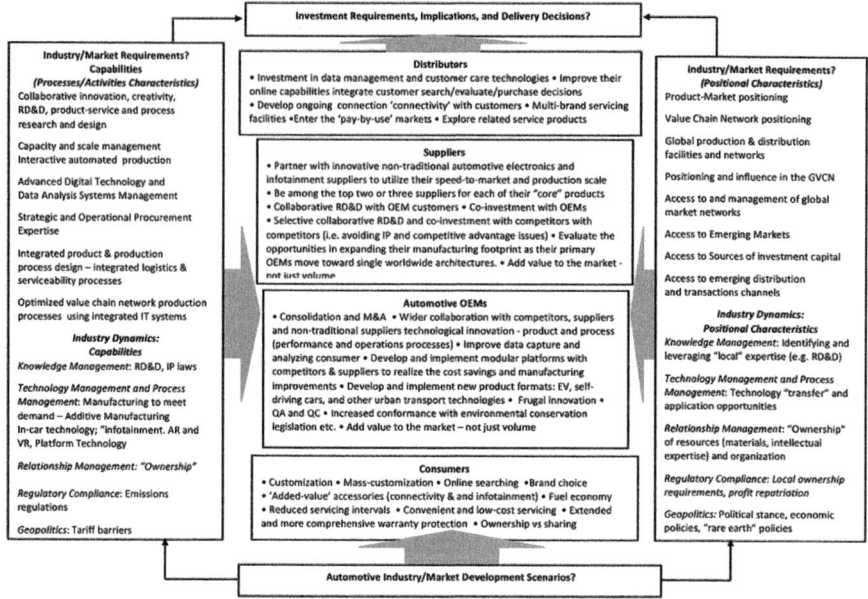

Fig. 4.4 Combining value builders and industry dynamics to identify future capability response requirements. (Adapted from: Strategy + Business PwC, 2017)

is increasing the trust and transparency of intraorganizational collaboration. This will be necessary if the responses suggested in Fig. 4.4 are to be pursued and strategic effectiveness and operational efficiency encouraged.

Volkswagen AG has mapped out details of its transformation from a mass manufacturer of cars to a provider of transportation services by unveiling a car-sharing service and promising digital acquisitions as part of a 3.5 billion-euro (US $4 billion) push into next-generation automobiles. The "We Share" label will compete with the DriveNow and Car2Go offerings of Daimler AG and BMW AG. Volkswagen plans to start technology partnerships this year, with purchases likely before the end of 2018. Juergen Stackmann, head of sales at the VW namesake brand, said in a Bloomberg Television interview in the German capital. Volkswagen says last generation of combustion engines to be launched in 2026, (Comments made by Michael Jost Strategy Director, Volkswagen AG,) Business News Reuters, (2018).

"a device and software company." "To deal with this development, we need to reinvent the automobile."

Industry dynamics and value builders will have a powerful influence on value engineering (the development of future "value" and "value delivery options" will involve value chain networks in forward thinking and entail consideration of a range of possible events. Clearly, the further (in time) projections are made the more difficult it becomes to forecast both revenues and input costs. In "Manufacturing's next act" (2015), McKinsey projected future value builder characteristics based upon the Industrié 4.0 developments that were either ongoing or appeared to be highly

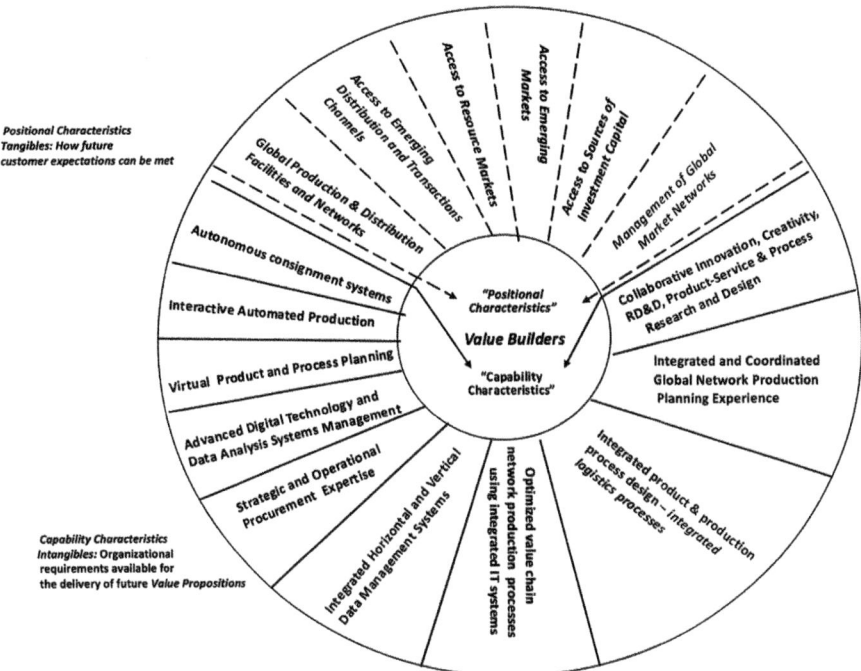

Fig. 4.5 Developing response facilitators to value builders: The impact of Industry 4.0. (Adapted from: Baur and Wee (2015))

probable. Figure 4.5 is based on suggestions made in the McKinsey document but has been structured around capability and positional characteristics. We shall be considering ways and means of tackling this issue in a later chapter, identifying some examples to demonstrate the efficacy of the process.

It is essential that performance value drivers and strategic value builders should identify with stakeholder expectations. Value chain networks are built upon the understanding and management belief in collaborative planning and forecasting and management of inter-organizational response to market opportunities with shared resources and responsibilities. This raises the important topic of performance management and measurement. The starting point for researching performance expectations is to identify the stakeholders who will impact on the project and/or be impacted by it. Figure 4.6 identifies the current stakeholder value builder portfolio options confronting organizations. There are questions that need to be asked concerning the entities and their expectations (and indeed their very existence): as we move forward in time, both may change, the entities and their expectations. Stakeholders and their expectations differ by industry/market type, for example, in defense, market governments dominate stakeholder groups as military expenditures are part of a defense strategy with specific tasks within the overall strategy. In FMCG markets, consumer/customer perspectives have a dominant role.

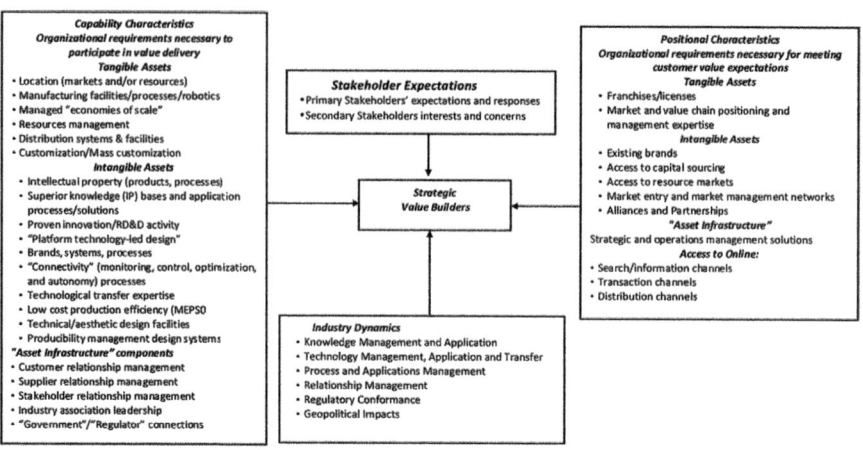

Fig. 4.6 Generic stakeholder performance value builder options

Convergence

It follows that once a range of market opportunities are identified, a qualitative approach to questioning the implications of the stakeholders' perspectives should be undertaken in order that any significantly large group with significant interests can be identified and explored very closely for indications that their interests may have a serious impact (i.e., positive or negative) on the outcome of manufacturing and marketing aspects of the opportunity. From such analysis, response capabilities emerge. Figure 4.7 demonstrates how the model contributes to identifying, evaluating, and aiding decision-making on appropriate responses by raising "what if?" questions concerning implications and concerns of decisions within the *performance* response capability activity center.

Performance: Maximize the "value added" by a network to its "stakeholders" – by managing and supporting suppliers and customers and working with other network partners (physical distribution and serviceability suppliers) to improve operational performance value drivers and their strategic value builders to create unique or exclusive "Value Advantage" for ALL Stakeholders.

Issues and Concerns for the Response Capability Activity Centers are:

Profitability: There are essential responsibilities within the profitability response capability area; managing the efficacy of organizational investment in tangible and intangible assets identifying; sources, availability, and managing the cost of capital so that the network may achieve an optimal cost of capital (for example, an offshore partner may have access to funds at lower cost than elsewhere, this would suggest the activities within the network be structured to meet this opportunity if can meet investment amount, cost and time span). Adopting and applying realistic measures of *economic profit performance* i.e., economic profit, economic return on investment (EROI) and EVA, and using EVA based a value contribution-based metric to monitor intra and inter-organization performance,

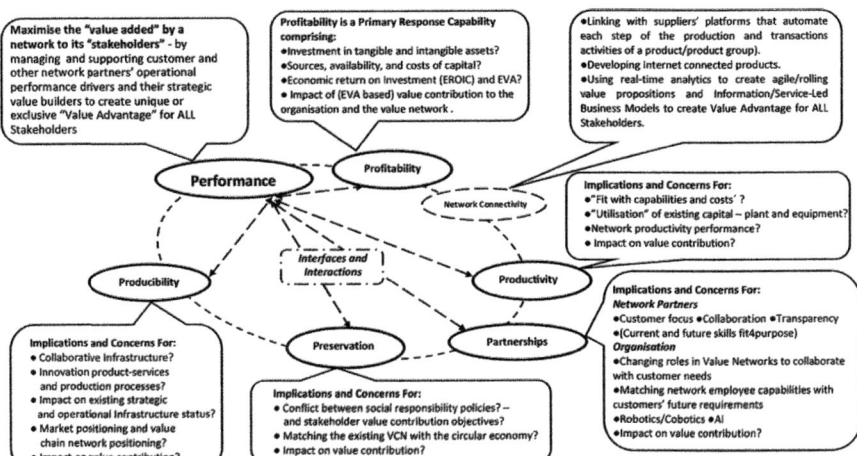

Fig. 4.7 Interfaces and interactions strategic and operational operations – performance management

Productivity: Measurement of productivity is currently a controversial topic of debate; clearly a standard, transparent measurement offers advantages when the impact of expansion of the network's activities, and therefore, capabilities is considered. There is evidence of successful use of EVA as a business model metric (Volkswagen) using a derivative; value contribution and as an incentive (Eli Lilley, Whirlpool, and Halliburton). Productivity measures important aspects of network stakeholder value such as; utilization of existing capital – plant and equipment, and of network productivity performance.

Partnerships: Network strategy and structure decisions depend upon current and future response capabilities requirements. While they may differ over time they have "basic" criteria such as: for *Network Partners, a* customer focus, agreement on collaboration, transparency, and to meet current (and future) skills requirements, and product-service fit4purpose criteria. *Individual Organizations* need to monitor and adapt to change, when and where necessary, to maintain a competitive profile/value proposition and to maintain a satisfactory stakeholder value contribution.

Producibility: Essentially producibility is a network infrastructure designed to meet specific customer-based outputs. Change in any of the following may change the optimal network resource combination and unless their impact is assessed, and changes made if they are required, the output will be less than optimal. These are value proposition, market positioning, product complexity, serviceability complexity, value engineering (integrated design), integrated network design for value delivery, network connectivity and collaboration, and value contribution equity.

Preservation: Sustainability is corporate, network, and business environment concerns. Policies adopted will have an impact on performance, for example, a change in a major material input, change in labor regulations (payment amounts working hours, etc.) A significant change from a linear to a circular value chain network structure will have implications for performance value delivery.

Network Connectivity: Digitized connectivity is a primary response capability as it provides linkages with suppliers' platforms that automate each step of the production and transactions activities and with customers, connectivity and inter-organizational liaison for developing internet connected products. Access to real-time analytics will create agile/rolling value propositions and information/service-led business models to create value advantage for all stakeholders.

References

Agnew, H., & Eley, J. (2018, July 3). *Tesco and Carrefour strike supply deal to take on big German rivals*. London: Financial Times.

Baur, C. & Wee, D. (2015), Manufacturing's next act, *McKinsey Quarterly*, June.

Champy, J. (2008). *Outsmart*. New Jersey: Pearson Education/FT Press.

Cleland, A. S., & Bruno, A. V. (1996). *The market value process: Bridging customer & shareholder value*. San Francisco: Jossey-Bass.

Olvé, N., Roy, J., & Wetter, M. (1997). *Performance drivers*. Chichester, UK: Wiley.

Phelps, R. (2004). *Smart business metrics*. Harlow: Pearson Education.

Prescott, K. (2017, October 11). How maps are powering the tech revolution. *BBC News*.

Slywotzky, A. J., & Morrison, D. J. (1997). *The profit zone*. New York: Wiley.

Strategy+Business PwC (2017). Automotive and Healthcare Industry Futures Studies. November.

Volkswagen says last generation of combustion engines to be launched in 2026 (2018, December 4) *Business News. Reuters*.

Vuori, T., & Quy, N. H. (2016). Distributed attention and shared emotions in the innovation process: How Nokia lost the smartphone battle. *Administrative Science Quarterly, 61*(1), 1–43.

Walters, D., & Rainbird, M. (2012). *Managing in the value chain network*. Sydney, Australia: Prestige Books.

Chapter 5
Profitability: Interpretations and Considerations

Abstract Profitability has many interpretations, often competing with other metrics as a means of measuring corporate performance. The DuPont ratio spread is explored in this chapter as useful in identifying meaningful interpretations across a range of interests: strategic effectiveness, investment strategy, operating efficiency, and shareholder returns. However, there are differing views on what comprises "profit." It has been argued that while organizations appraise success by size of sales, or share price performance, market share, return on a measure of investment, or perhaps sales or growth rate, a realistic measure of corporate success is *value added, a derivative of economic profit*. To be of use as a performance metric, added value requires to be quantified. Operating margins are one indication of value that is added at each stage of production. However, this has problems. One problem concerns the charge made for the use of capital in the production process. Here the concept of economic profit can be used. Economic value added (EVA) is a financial management technique developed by Stern Stewart Inc, a financial consultancy that built upon earlier conceptual work. Other topics having an influence on profitability include cost driver analysis, total cost of ownership, and the importance of cross business network synergies. This chapter uses relevant finance and investment applications together with Industrié 4.0 and Value Chain Network 2.0 characteristics to demonstrate how finance and investment management can deliver successful stakeholder value management and introduces the notion of *value contribution. Volkswagen* uses value contribution as "a key measure of operating efficiency" and is calculated based on the operating result after tax and the opportunity cost of invested capital. The operating result shows the economic performance of the organization.

Introduction

Profitability has many interpretations, often competing with other metrics as a means of measuring corporate performance. The DuPont ratio spread is explored in this chapter as useful in identifying meaningful interpretations across a range of

© Springer Nature Switzerland AG 2020
D. Walters, D. Helman, *Strategic Capability Response Analysis*,
https://doi.org/10.1007/978-3-030-22944-3_5

interests: strategic effectiveness, investment strategy, operating efficiency, and shareholder returns. However, there are differing views on what comprises "profit." Kay (1993) argued that while organizations appraise success by size of sales, or share price performance, market share, return on a measure of investment, or perhaps sales or growth rate, a realistic measure of corporate success is *added value, a derivative of economic profit*. To be of use as a performance metric, added value requires to be quantified. Operating margins are one indication of value that is added at each stage of production. However, this has problems. One problem concerns the charge made for the use of capital in the production process. Here the concept of economic profit can be used. Economic value added (EVA) is a financial management technique developed by Stern Stewart Inc (www.sternstewart.com), a financial consultancy that built upon Kay's conceptual work. Other topics having an influence on profitability include cost driver analysis, total cost of ownership, and the importance of cross business network synergies. Walters and Rainbird (2007) discussed the application of EVA in value chain network.

The purpose of any business is to convert its resources profitably into a product-service that creates value for another business or an end-user. Accounting and financial management activities in organizations have seen customer satisfaction (and the deployment of resources) as a component in the process of generating wealth for the business for the shareholders. This emphasis on generating shareholder value has been challenged (see Freeman (1984)). However, many organizations now see the role of the CEO as defining the business in terms of *a broader purpose* – that is one that benefits a larger set of *stakeholders* (customers, employees, suppliers, and future generations) than would otherwise gain from simply increased profits and shareholder value and indeed shared value. It follows that the *broader purpose* approach requires strong data support; Industrié 4.0 makes this possible.

This chapter uses relevant finance and investment applications together with Industrié 4.0 and Value Chain Network 2.0 characteristics to demonstrate how finance and investment management can deliver successful stakeholder value management. It introduces the notion of *value contribution*; As mentioned above, *Volkswagen* uses value contribution as a key measure of operating efficiency and is calculated based on the operating result after tax and the opportunity cost of invested capital. The operating result shows the economic performance of the organization.

The primary role of the accounting and finance activity is one of stewardship; the protection of shareholders investment is a legal requirement, and as such finance and investment management has this as a primary role. However, within the broader remit of responsibility to the stakeholders, the activity's stewardship extends to managing the commitments to customers, suppliers, distributors, community, and government to manage the organization's cash flows such that contractual payments (including taxation and other statutory fiscal requirements) both inbound and outbound are met within the terms of these contracts.

The financial and investment management activity should ensure that the implications of investment concerning *productivity* are understood and applied. Some large organizations have introduced procurement strategies that contribute to increased productivity; for example, *PepsiCo* has implemented a procurement strategy for hedging forward commodity purchases; these amount to US$18 billion annually. *P&G* adopted a longer-term strategic solution; "innovative substitution" the packaging for its Pantene hair products uses biodegradable corn-starch rather than petrochemical resins. The CEO added that such changes pass unnoticed by customer stakeholders but are applauded by shareholder stakeholders – particularly when they can reduce the cost of US$2 billion incremental commodity costs by 50% (see Quirk (2011)! Network collaboration results in *producibility* (enhanced effectiveness and efficiency of production), an integrated process whereby the cost-effective design of the production/service/consumption system can be designed and implemented. Finance and investment management should ensure the financial efficacy of the product-service investment and working capital requirements and budget accordingly. Successful companies such as Dyson and Haier are aware of the impact of changing operations technology on the workforce and develop people and workplace skills to keep pace. A finance and investment management role should ensure these activities are adequately funded and should monitor their program effectiveness. Finance and investment management, need to identify and evaluate network investment requirements and the relevant optimal profitability capabilities (i.e., effective and efficient investment and liquidity management of network partners) and encourage the coordination of operating and cash2cash cycles, as the activity is becoming involved in industry and competitive activities; monitoring both profit and productivity pools, providing an innovative method of monitoring the business model, as well as the capability requirements of continued successful market engagement.

Neely et al. (2002), *The Performance Prism,* contributed to performance planning and management by suggesting that the "demands" made on organizations are in fact two-way. While the stakeholders seek to improve their "lot" through more "profitable" relationships (looking for closer and longer relationships, not simply increased profitability and productivity), organizations are beginning to program and structure their expectations of stakeholders. The authors identified typical contributions expected of stakeholders:

- *Investors:* capital for growth, assume more risk and long-term support
- *Customers:* profitable and long-term loyalty, feedback
- *Intermediaries:* planning and forecasting, inventory management
- *Employees:* flexibility, multi-skilling, antisocial hours, loyalty
- *Suppliers:* increased customization, total solutions, integration
- *Regulators:* cross-border consistency, advice, involvement, grants, and aid
- *Communities:* skilled employment pools and education
- *Pressure groups:* closer cooperation, shared research
- *Alliance partners*: co-development, co-productivity, shared information, and shared costs
- And we would suggest adding:
- *Governments: IP protection, technology transfer, relevant local skills, profits repatriation, and no unnecessary interference*

Clearly the interrelationships between stakeholder and organization are becoming structured as corporate boundaries reach beyond their legal entities and become inter-organizational. Neely and his colleagues concluded that:

> Indeed, the very concept of stakeholder value itself should perhaps be quantified in terms of the strength of the interrelationship.

This is an important consideration concerning the topic of relationships, their characteristics, and their management and is the topic of the next chapter.

DuPont and Managing the Business: Decisions and Performance Measurement

Some years ago, the DuPont Company developed an approach to planning and control which was designed to monitor divisional performance. This approach has subsequently been used by numerous companies. Its attraction is that it identifies the *activity ratios* which measure how effectively and efficiently an organization (or its strategic business units) employs the resources it controls with the profit margins on sales and how these ratios interact and determine the profitability of the assets. Figure 5.1 outlines the nature of the ratio system.

The lower half of Fig. 5.1 develops a net asset turnover ratio, an asset productivity measure. The upper part of the figure develops an operating profit (margin percentage) on sales, a measure of its *financial operating efficiency*. When the net asset turnover ratio is multiplied by the operating margin, the result is a return on net assets/net worth – its strategic capital management *effectiveness* is identified. Figure 5.1 illustrates an example of the model – the *strategic profit mode;* this

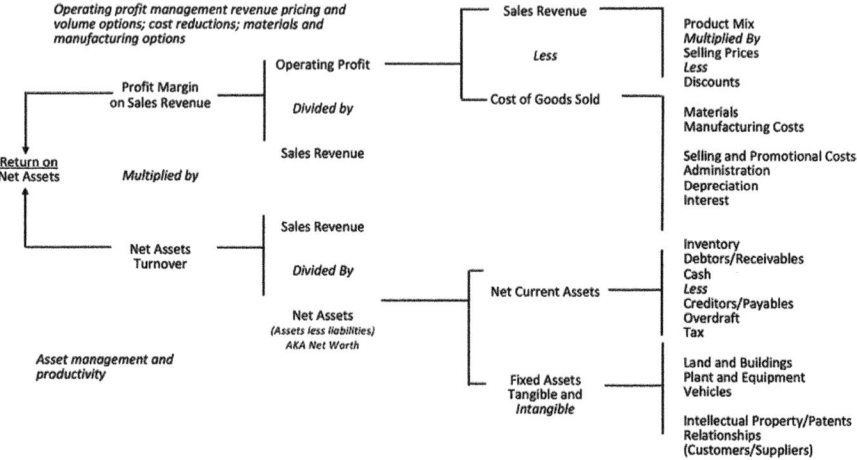

Fig. 5.1 DuPont system of financial analysis

application allows us to relate management activity components quantitatively and helps with managerial decisions in four ways:

It identifies one of the primary objectives of the business: to optimize returns to the shareholders.

It identifies the growth and profit paths available to the organization (improve the margins earned, increase asset productivity, gearing).

It highlights the principal areas of decision-making: asset management, margin management, and financial management.

It provides a useful model for appraising the strategic, marketing, and financial options by providing the facility to ask what if questions; for example, if the organizational network expanded into an adjacent, but related, market sector by acquiring a specialist activity (strategy), what would be the cost alternatives and implications for both marketing and financial activities and their costs?

The DuPont model offers an interesting and useful approach to working the interfaces of operations, finance, and marketing by combining operating profit management: revenue pricing and volume options, cost reductions, materials and manufacturing options, and asset management and productivity. Figure 5.2 follows through on some simple cross multiplication to demonstrate how operations, finance, and marketing contribute to creating value-added, or a quantitative measure, profitability, i.e., "profit"/equity; clearly this is a critical metric; without

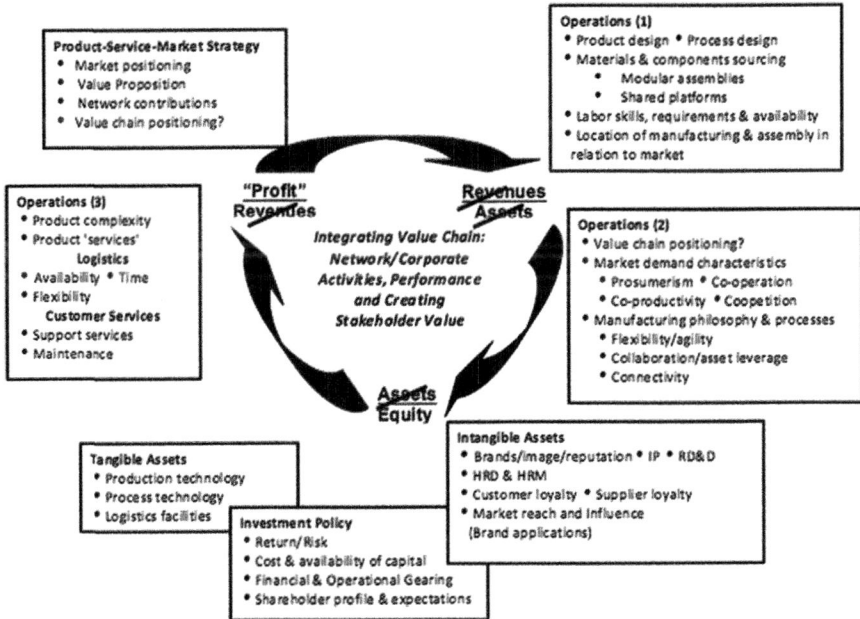

Fig. 5.2 Investment and finance work together as integrated functions

ongoing profitability, an organization loses the confidence of ALL stakeholders, not only its investors. Figure 5.2 is important because it demonstrates how the decision-making areas interface and interact, for example, *product-service-market strategy* is a response to a market opportunity and is responsible for developing a feasible (customer acceptable) and economically viable (producible) response, i.e., a value proposition, with significant decisions concerning the market position of the offer and where it envisages the optimal positioning of the organization in the value chain network. This is a process of value engineering, a process which requires a cost-effective coordination of relevant resources to ensure not only customer satisfaction but also satisfactory outcomes for ALL the stakeholders. *Operations* has a major role in identifying the resource capability alternatives and manufacturing options. The range of options includes where, how much, and who will be involved in the overall production activity. *Finance and investment* management will be required to assess the financial implications of each of the operational options, following with how best the project is to be funded.

This takes us back to "profit"; the emphasis indicates the need to identify the most realistic measure of "profit" to be used. Profitability performance is influenced by the size and nature of the investment required (this is discussed below), but just as important is what or which measure of "profit" is used and how this contributes to creating stakeholder value.

Economic Profit: Economic Value Added and the Impact on Stakeholder Value Management and Value Contribution

There are both qualitative and quantitative aspects of value contribution to be considered. Kay (1993), for example, takes a quantitative view, by considering the notion of added value as a measure of corporate performance as "the key measure of corporate success" and defined this as:

> Added value is the difference between the (comprehensively accounted) value of a firm's output and the (comprehensively accounted) cost of the firm's inputs. In this specific sense, adding value is both proper motivation of corporate activity and the measure of its achievement.

Kay (1993) calculates added value by subtracting from the market value of an organization's output the cost of its inputs:

> *Revenues*
> *Less (wages and salaries, materials, capital costs)*
> *Equals*
> *Added Value*

Added value in this context includes depreciation of capital assets and also provides for a "reasonable" return on invested capital. Calculated this way added value is *less than* operating profit, the difference between the value of the output and the value of materials and labor inputs and capital costs. Kay's (1993) measure of competitive advantage is the ratio of added value to the organization's gross or net output:

Competitive advantage or

$$\text{competitive value advantage} = \frac{\text{Revenues} - (\text{Wages} + \text{Salaries} + \text{Materials} + \text{Capital costs})}{\text{Wages} + \text{Salaries} + \text{Materials} + \text{Capital costs}}$$

These are viewed as comparisons of added value to either gross or net output. Kay's added value has similarity with the concept of a producer's surplus and has the additional benefit of being able to be calculated from accounting data. However, it should be said that inaccuracies are very likely if direct comparisons between organizations are made on a one-off basis. Local accounting practices differ, and accounting statements therefore are not strictly comparable due to differing procedures and practices. But longitudinal comparisons are worthwhile, particularly over periods of 3–5 years when input/output and added value/output ratios may be compared.

Kay's (1993) model may be illustrated, and Fig. 5.3a depicts two possible outcomes.

Figure 5.3a shows a successful organization making a positive added value (in Kay's terms, it must be remembered that added value > operating profit because it includes an economic return on capital; therefore it *may not* be profitable). In Fig. 5.3b the organization represented has difficulties; wages and salaries and capital costs are greater than revenues. Figure 5.3c suggests the added value structure of a value chain. At the raw materials (primary production), stage materials costs are a low proportion of total input costs, but as value is added by successive processes, materials costs become significantly more important as do the capital costs of inventory financing.

A potential source of error is confusion of the terms *added value* and *value added*. Value added is the principle basis of expenditure taxation in a number of countries. Value added is, therefore, not the same thing as added value; the former is the difference between the value of the material and semi-produced inputs which a firm buys in and the value of the output which it sells. It is therefore equal to the firm's net output. Added value is only a part of value added, as Fig. 5.3c makes clear. Often the terms are used interchangeably, and this is incorrect and confusing. In this book our interest is in both value added (the increase effect of economic efficiency) and added value (the impact of a change in product or process that can have a positive impact on a customer that increases the customer's value advantage in the market).

It follows that when a firm is generating *economic profit* (considered here to be operating profit less a charge for the use of capital involved in the production of the operating profit), it is creating and appropriating added value. The value of the sales generated is more than sufficient to cover the cost of all the resources used by the collective. This surplus represents an addition to the value of all the resources tied up in the networked coalition.

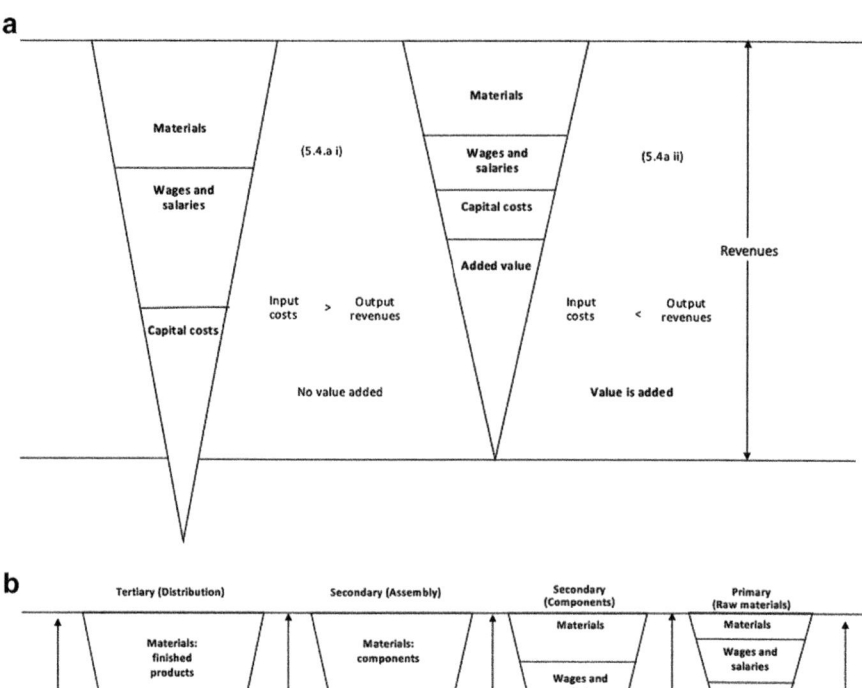

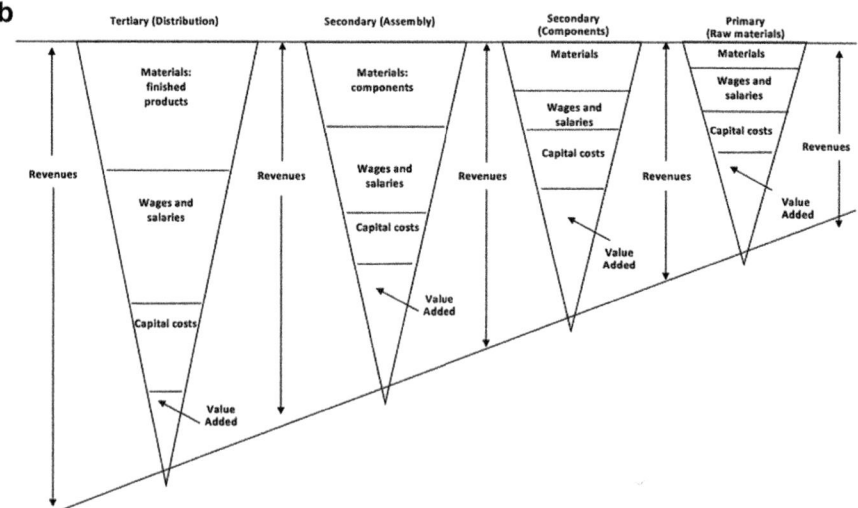

Fig. 5.3 (**a**) Kay's added value perspective. (Adapted from Kay (1993)). (**b**) Stages in the added value chain. (Adapted from Kay (1993)). (**c**) Revenue and profitability measures

c

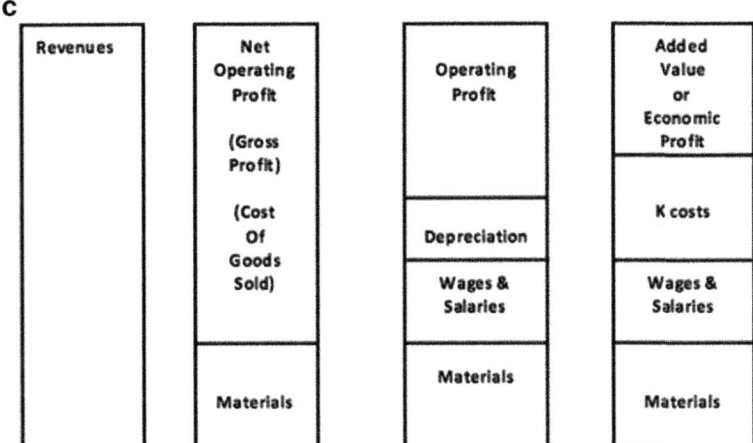

Fig. 5.3 (continued)

By creating economic profit, an organization is creating and appropriating some added value for itself and its owner(s); the value of its sales is more than sufficient to cover the cost of all the resources used by the collective. This surplus represents an addition to the value of all the resources tied up in the coalition of resources which we call the firm. What determines whether these economic profits can be sustained over time depends upon the uniqueness (or exclusivity – patentability) of the processes and activities used (and their combination) to create significant barriers to entry. If the entrepreneur's initial success is easy for others to reproduce, any success in generating profit cannot be sustained. To the extent that a particular entrepreneur creates a firm which generates profits over a longish period, then the enterprise itself must have something special.

Calculating EVA

Added value in this context includes consumption of capital assets and also provides for a "reasonable" return on invested capital; it is the economists' notion of *economic profit* (or residual income). Calculated this way added value is *less than* operating profit (NOPAT), i,e., the difference between the dollar value of the output and the dollar value of materials, labour inputs and capital costs. It has the important benefit of being easily quantified and capable of being used for comparative purposes. Stern Stewart Inc (www.sternstewart.com) introduced EVA (economic value added), a concept very similar to Kay's added value but deducted taxation from the result. *EVA is net operating profit after tax (NOPAT) less a capital charge for the invested capital employed in the business.*

The Cost of Capital: Quantitative

The EVA model uses a weighted average cost of capital in determining the "real" dollar value of its cost of capital. Given the long period of low interest rates, it is suggested that the long-term forecasts of interest rates by Central Banks and commercial lenders should be considered when making project viability assessments. The argument suggesting this to be too difficult is countered by an argument saying *any attempt* is going to be difficult. An interesting question is one that asks how much more profitable are projects that were appraised prior to the current period of low borrowing rates! A company's assets are financed by either debt or equity. The weighted average cost of capital (WACC) is the average of the costs of these sources of financing, each of which is weighted by its respective use in the given situation. By taking a weighted average, we can see how much interest the company has to pay for every dollar it finances. Arithmetically the WACC is:

$$\text{Weighted average cost of capital} = \%\text{Equity} \times \text{Interest costs} + \%\text{Debt} \\ \times \text{Interest costs} = \text{WACC}\% (\$\text{WACC})$$

An example:

Total capital employed (000)	$25,000	
Equity	$15,000	@15.00%
Debt	$10,000	@5.00%

$$\%\text{Equity} \times \text{Interest costs} + \%\text{Debt} \times \text{Interest costs}$$

$$15/25 \times 15.00\% + 10/25 \times 5.00\%$$

$$60\% \times 15.00\% + 40\% \times 5.00\% = 11\%$$

The Annualized Average Weighted Cost of Capital = $11\% \times \$25,000 = \$2750.00$

Cost of Equity

The cost of equity funds is influenced by the relative value of dividend payments to the current value of shares and the expected growth rate of dividends:

$$\begin{array}{l} \text{Return to investors /} \\ \text{Cost of equity} \\ \text{to the organization} \end{array} = \frac{\text{Future dividend payments}}{\text{Current value of shares}} \begin{array}{l} \text{Expected growth} \\ \text{rate of future} \\ \text{dividends} \end{array}$$

It is a balanced view of current and future financial performance.

Cost of Debt

It is usual (but not mandatory) for interest payments and redemption dates to be fixed prior to loan offers being accepted. It is not a difficult task to calculate the cost of debt in either case:

$$\text{Current value of debt} = \frac{\text{Annual interest payment}}{\text{Rate of return required by investors}}$$

However, if a redemption date has been established, this becomes:

$$\begin{array}{l} \text{Current} \\ \text{value of} \\ \text{debt} \end{array} = \left[\frac{\text{Annual interest payment}}{\begin{array}{l}\text{Rate of return required}\\\text{by investors}\end{array}} + \frac{\text{Redemption value of debt}}{\begin{array}{l}\text{Rate of return required}\\\text{by investors}\end{array}} \right]$$

Typically, these values are based upon balance sheet values. This requires some assumptions that, given, the rapid changes occurring within the business environment in which the organization operates, together with the "Connectivity" characteristics of Industrié 4.0, risk and return combinations of existing tangible and intangible assets are unlikely to be representative of future returns and the risks that accompany them. Furthermore, it is very likely the balance of "value" between tangible and tangible assets will change, both in their characteristics and in their value. This point was raised by Boyd (2016) commenting upon a review (by the Australian Financial Review), *CEO Outlook Poll,* in which developments of data analytics, standardization, platform applications, robotics, artificial intelligence, new metals and materials, and totally new forms of existing products (e.g., electric vehicles) will have impact. It was also argued that the valuation of tangible and intangible assets will likely be very different, with intangible assets being valued higher than they currently are.

Cost of Capital: Qualitative

A major qualitative factor for consideration when expanding the financial structure and size of the business is the impact this may have on decision-making and control. Equity owning shareholders enjoy voting rights and theoretically can influence major decisions of the organization. Increasingly shares are being held by institutional shareholders (pension funds, trade unions, etc.) and by private equity companies who can, and do, exert influence. During 2017 the General Electric Board of Directors was under pressure from a major shareholder who argued that the expansion of the company into industrial markets and away from its financial markets had gone too far and that some of acquisitions were too late and too expensive. The CEO at that time retired and was replaced. Almost immediately announcements were made suggesting the sale of what had become noncore businesses, a reduction in the "standard dividend," and, subsequently, announcements of large global redundancies. Recent changes in the direction of the organization have not met with "market" approval resulting in GE's removal from the "500 Club." The "new" CEO barely lasted a year; shareholder pressure for a more rapid increase in success led to his departure after some 12–13 months. It is interesting to comment that this change was brought about by a specific group of stakeholders concerned about the decrease in the dividend payments. No comments were made concerning the impact on the other stakeholders.

In a value chain network context, there are indirect, qualitative approaches to increasing network added value. Monitoring quality and substituting material inputs and processes can result in quantitative aspects of added value; for example, rigid adherence to quality standards reduces wastage and returns and possibly component or product failure when the product is in use; for construction equipment operating considerable distances from service points, the downtime penalties can be huge. Replacing conventional metal materials with carbon composite materials (Boeing 787) is an example of a much lighter and fuel-efficient aircraft resulting in fuel economies and longer sector range distances – with impact on increased revenues and cost reduction. Clearly the notion of added value has extensive ramifications for the product-service and process decisions throughout the value chain, and we require additional measures in order that we might evaluate alternative business models, their strategies and structures, and their operational delivery alternatives. To this end we explore additional uses of the added value measure. Before doing so, as an example, we consider the argument for preferring to measure share of market added value to market share as a performance metric.

Stern Stewart Inc (www.sternstewart.com) introduced EVA (economic value added), a concept very similar to Kay's (1993) added value but deducted taxation from the result. *EVA is net operating profit after tax (NOPAT) less than a capital charge for the invested capital employed in the business; EVA is arguably a short-term measure of performance as it is based upon short-term performance.* The inclusion of tax in the "equation" may reflect operational decisions concerning manufacturing location and for this reason should be included when making competitive comparisons:

$$Economic\ added\ value = Revenues - \left(Wages + Salaries + Materials\right) \\ - Tax - Capital\ cost$$

The model can be modified for use by networked organizations; see Fig. 5.4. Figure 5.4(1, 2, 3, and 4) reproduces the accepted EVA model with variations for value chain positioning and its applications

Figure 5.4 identifies the conventional EVA calculation. Figure 5.4 introduces a modification to the model. The value chain networks' aggregate performance is realized by the efficient, coordinated use of resource capabilities in procurement, operational efficiencies, and the use of organization's capital or by leveraging the

Basic EVA Calculation 5.4.1

Net Revenues (Sales *less* discounts allowed)
Less
Operating expenses
Equals: Operating profit before interest and tax
Less Tax
Less Capital charges (weighted average cost of annualized capital employed)
Equals Economic Value Added (EVA), Value Contribution

Network Perspective ($ Incremental Value Advantage) 5.4.2

∑Net Benefit $(Increased Revenues)
Less: ∑Incremental increases in operating expenses
Equals: ∑Operating profit before interest and tax
Less: Tax
Less: ∑Incremental investment capital charges (weighted average cost of annualized capital employed)
Equals ∑Economic Value Added (∑EVA) Value Contribution and "∑Productivity"

Customer Perspective ($ Incremental Value Advantage) 5.4.3

Net Benefits $(Revenues *less* Acquisition Costs)
Less: Incremental increases in operating expenses
Equals: Operating profit before interest and tax
Less Tax (NOPAT)
Less Incremental capital charges (weighted average cost of annualized incremental capital employed)
Equals Economic Value Added (EVA), Value Contribution *and* "Productivity"

Employee Perspective ($ Incremental Value Advantage) 5.4.4

Net Benefit $(Increased Departmental Revenues)
Less: Incremental increases in allocated operating expenses
Equals: Operating profit before interest and tax
Less: Tax
Less: Departmental incremental investment capital charges (weighted average cost of annualized capital employed)
Equals Economic Value Added (EVA) Value Contribution and "Productivity"

Fig. 5.4 (1, 2, 3, and 4) Developing network-based EVA/economic profitability models

capital capabilities of network partners. The value advantage created is calculated by deducting from its aggregate revenues, the aggregate acquisition costs, operating costs, tax, and its cost of capital used in the production of its product-service; assuming suppliers' materials sources (raw materials, components, etc.), tax, and capital used at this stage of production have been calculated, similarly the aggregate network supply response can be calculated and be used to measure productivity as well as to be used for comparing alternative value chain structures. The ultimate consumer' value advantage is shown as Fig. 5.4. Their value advantage reflects their marketing abilities (understanding of customers, market positioning, and value proposition). It can be calculated from an estimate of benefits received from its upstream supplier, procurement efficiency, and the comparative impact the product-service differentiation has on prices their customers are prepared to pay: fit4purpose, product quality, serviceability, exclusivity, etc. Figure 5.4 identifies the role of EVA as an incentive payment to managers. Managers and employees can be persuaded to adopt a long-term perspective of the firm when offered an incentive remuneration program aligned with their efforts in achieving economic profit. It appears to be more effective if "banked" and paid incrementally on an annual basis.

Market Share Versus Share of Market Added Value

Share of market added value is preferred over market share for a few reasons. While the original PIMS study suggested that high market share and high profitability have strong positive correlation, it does not follow that a strategy to increase market share by reducing prices and operating costs, together with an increase in promotional activities (and therefore costs), will maintain that momentum. More than likely it will be met with competitive responses that result in a distribution of the available market volume rather than an expansion of the market. Share of market added value considers the shift of added value (value migration) that is a current and observable phenomenon in many markets.

Kay (1993) also considers cash flow as an efficiency measure; however the impact of strategy, structure, and operational changes on cash flow is not always immediate. Operational changes to manufacturing (outsourcing labor processes; Pacific Brands, Kraft, Lego) and to distribution (use of third-party services) may show short-term changes. Changes in asset management policies create important sources of cash flow; for example, changes in manufacturing strategy such as Airbus/EADs divested six aircraft assembly sites and began the creation of a network of partners to allow Airbus to focus the cash released on resources for developing value advantage from core activities. But increasingly we are observing strategic alliances (organizations moving investment in the value chain network into new networks; Lego launched a "Classroom of the Future" project with a US university to teach children about science and technology; launched "LegoFactory.com" a "Lego Digital Designer" that offers an opportunity to design and order a unique Lego model, and a joint venture with the MIT Media Lab that introduces robotic Lego). Clearly there are significant time spans here, and perhaps we are considering *long-term effectiveness* (the impact of strategic and organizational transformation) rather than *short-term efficiency* measurements. The recent increase in industrial capital goods procurement decision-making, purchasing the output of an industrial component (e.g., an aero engine), redirects the user organization's cash flow decision towards business growth, away from asset growth.

Value Contribution: EVA as a Strategic and Operational Metric

The Volkswagen Group's financial target system centers on continuously and sustainably increasing the value of the company. To ensure the efficient use of resources in the automotive division and to measure the success of this, they have been using a value-based management system for several years, with return on investment (ROI) as a relative indicator and value contribution, as the key performance indicator linked to the cost of capital, as an absolute performance measure. Volkswagen explains their approach in the 2013 Annual Report.

For Volkswagen the economic return on investment serves as a consistent target in strategic and operational management. If the return on investment exceeds the market cost of capital, there is an increase in the value of the invested capital and a positive value contribution. The concept of value-based management (economic return on the annualized weighted average cost of capital) allows the success of the organizations and individual business units to be evaluated. It also enables the earnings power of products, product lines, and projects – such as new plants – to be measured.

Components of Value Contribution: Stakeholder Value Management

Value contribution is calculated based on the operating result after tax and the opportunity cost of invested capital. The operating result shows the economic performance of the organization and is initially a pre-tax figure. Volkswagen uses the various international income tax rates of the relevant companies and assumes an overall average tax rate of 30% when calculating the operating result after tax.

The cost of capital is multiplied by the average invested capital (equity and debt) to give the opportunity cost of capital also known as the weighted average cost of capital (WACC). Invested capital is calculated as total operating assets reported in the balance sheet (property, plant and equipment, intangible assets, lease assets, inventories, and receivables) less noninterest-bearing liabilities (trade payables and payments on account received). Average invested capital is derived from the balance at the beginning and the end of the reporting period. As the concept of value-based management only comprises operating activities, assets relating to investments in subsidiaries and associates and the investment of cash funds are not included when calculating invested capital. Interest charged on these assets is reported in the financial results.

Value Contribution to Customers: Network Contribution Margin Analysis

Value contribution margin analysis is a measure of operating leverage; it measures how growth in sales translates to growth in profits. The value contribution margin is computed by using a value contribution income statement, a network management accounting version of the income statement that has been reformatted to group together a business's fixed and variable costs.

A *value contribution margin* is different from gross margin in that a contribution calculation seeks to separate out variable costs (included in the contribution calculation) from fixed costs (allocated by using activity-based management costing) and is an *economic analysis* of the nature of the expense. The *value contribution margin* is computed by using a value contribution income statement, a network management accounting version of the income statement that has been reformatted to group together a business's fixed and variable costs.

Whereas gross margin is determined using accounting standards, calculating the network organizations' *value contribution margin* is a useful tool for managers to help determine whether to add "performance factors" that enhance customers' own value propositions and the network's value contribution; these *factors* may be added at other locations/partners in the value chain network; ideally, they have benefits for both customer and for the network profit contributions. Any network activity with a positive value contribution margin should be reviewed and possibly kept (see detail

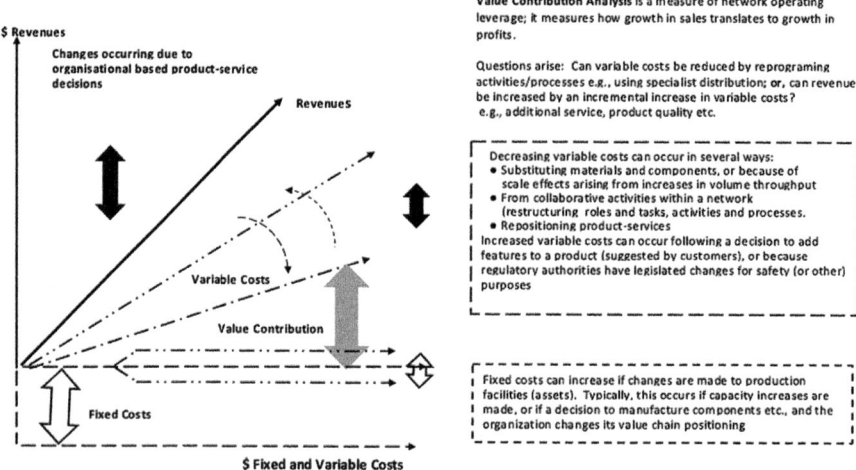

Fig. 5.5 Using value contribution analysis to evaluate the impact of added value on the vendor/customer relationship

in Fig. 5.5), even if it causes negative total profit when the contribution margin offsets part of the network partner's fixed cost. However, it should be transferred if the contribution margin is negative because the company would suffer losses from every unit it produces. The review should include consideration of the impact of the activity on the end-user customer (its value added) *and* on the overall effectiveness and/or efficiency of the network as an entity and of its individual units. It may be critical to the network's value advantage despite carrying what appears to be a disproportionate share of the variable costs. Several considerations will require resolution; there needs to be a positive impact on revenues, greater than increase in costs, and there should be a positive impact on downstream revenues (from customers' customers) if stakeholder value is to be increased optimally. The possibility that variable costs can be reduced by reprogramming or repositioning activities/processes and/or value chain activities/processes), can be achieved by reducing materials costs by substituting expensive inputs with lower cost materials, or using specialist distribution. Alternatively, it can be achieved from revenue responses that are increased by an incremental increase in existing variable costs e.g., additional service, product quality that are seen as beneficial to customers.

Examples of Value Contribution Attributes That Add Benefits to Added Value

Value contribution attributes add value by adding features to a customers' differentiation or by undertaking process activities the customer would otherwise need to undertake. As such this increases profitability and/or productivity, and this results in increased margins (for both organizations) and, therefore, quantitative added value.

Resource productivity management: e.g., collaborative procurement initiatives, such as *Covisint*, a data management system (portal) for the exchange of essential documentation between manufacturer/assemblers and their component suppliers, rather than a range of smaller independent companies, considerably reduces communication costs.

Economies of location: prior to the rapid advances in additive manufacturing technology, applications, and the reduction of operating costs, inventory was held in strategically sensitive locations. Additive manufacturing is beginning to meet demand by manufacturing components, etc. to order.

Creative RD&D: requires a broad perspective of "value engineering"; IKEA had for many years invested in, and worked closely with, its suppliers to ensure customer satisfaction. Recent developments have been focused downstream and address the serviceability issues that cause customer concern. The problem of assembly was addressed by working with small businesses who responded to customer requirements for help when "Billy" was becoming too large a problem. Customer catchments are being expanded by online facilities; this also addresses the locational availability problem in a cost-effective manner.

Collaborative RD&D: individual organizations often invest in RD&D only to find competitors are following parallel programs; this suggests that competitive advantage may be short-lived. The large automotive organizations discovered this problem many years ago and consequently work with each other on programs to develop various aspects of automotive technology, e.g., automatic transmission systems. One alternative approach is to work with component/accessory suppliers and to use standard applications; another is to monitor patent applications and awards with a view to licensing a competitive patent if it appears to have *greater value differentiation* than the organization's project and focus RD&D activities on this and discontinuing the organization's project! The critical issue concerns the size and longevity of the competitive advantage delivered; the same argument is applied to "serviceability."

Patentability: is closely allied to collaborative RD&D, i.e., investment and time; by purchasing use of a registered patent, a competitor may well be able to "catch up" and possibly overtake a competitor.

Unique or Exclusivity: products, brands, processes, and product-process combinations offer opportunities to reinforce market positioning; clearly there is a need to evaluate the options, to undertake development (involving cost and time), or to create partnerships and alliances with established organizations. This can have both tangible aspects (exclusive products to extend the existing product range) and intangible aspects (links with branded market reputation).

"Customized" performance: General Electric (with Predix) and Siemens (TeamCenter) use customer-generated data (from a GE or Siemens installed product) to suggest performance improvements for customers.

"Solutions," capabilities, and capacities: IBM, some time ago, divested its PC manufacturing to Lenovo in a move towards offering "solutions" consultancy. Clients were encouraged to bring data management and decision-related problems to IBM who would thoroughly review and identify the problem and map a pathway

for its solution. This approach has been adopted increasingly by other organizations and industries; a notable example is in agriculture where sensors and monitors are increasing product yields by monitoring animal health and soil composition and relating these with external influences such as weather.

"Experience effects" and economies of scale/MES minimum economies of scale: as operations management increases, specialist costs are being reduced and are becoming more easily managed. Additive manufacturing has increased the range of materials that can be used, and production units have increased in size. For manufacturing and distribution processes, the inhibiting factors of complexity, time, and cost are being removed by the availability of data management and analysis, thereby shrinking the advantages once offered by experience and scale.

Value Contribution Attributes That Identify Consumer Expectations and Experiences to Deliver "Service Process Quality"

"Service process management" is an important contribution made by Industrié 4.0 offering real-time interactions between suppliers and customers. The PRODUCT-service/product-SERVICE transition has facilitated M2M (machine2machine) management using digital thread and digital twin concepts. Capital goods suppliers now offer performance management as a product-SERVICE. "In-use data" is captured and analyzed (often using a digital twin) from aero engines and their surface equivalents, from rail locomotives, and increasingly from automated mining equipment and its supporting infrastructure. Factors becoming important include *economics of organizational architecture (intra- and inter-organizational resources management);* the repositioning of process activities in the value chain, for example, additive manufacturing systems, rather than inventory financing and other options, either within customers premises or nearby, reduce time (and therefore costs) and may well be operated by an intermediary (UPS have been experimenting with this "service process quality") activity.

Global strategic and operational networks: international organizations such as the automotive manufactures and industrial capital equipment companies have created networks (GE's multimodal brilliant factories, the manufacturer is adopting a range of data gathering and networking technologies to equip its operations for real-time reporting and supply-chain updates to improve process efficiency, product quality, predictive maintenance, and plant profitability).

Economics of interactions and transactions: the application of online search, ordering, and transactions facilities have been significant in both consumer and industrial markets. The data generated from these activities is providing suppliers with data that profiles customer responses and improves "service process quality"; it is also argued that it reduces total cost of ownership for end-user customers when the total cost of purchasing activities are considered.

Integration and coordination of value-adding activities: Industrié 4.0 has made significant contributions to the structure of industry and inter-organizational value chains by providing customers with real-time performance data to manage product-service output and remote serviceability; the actual time service is required, and the service activities are also required. These developments suggest that a major "service quality" advantage can be obtained from providing serviceability as and when it is necessary based upon product-service use and application rather than using fixed time schedules.

Value Contribution Attributes: Social and Economic Benefits Received

Internationally, federal and state governments struggle to provide *Infrastructure support* (road, rail, electronic communications (Internet services) as well as health and age-related welfare facilities), the attraction of public-private-partnerships (PPP), activities whereby manufacturing and reselling companies include some of these 'services' in development projects, thereby improving social amenities as well as organizational efficiencies. Applied RD&D and Industrial RD&D can offer social benefits, for example, Proctor and Gamble's packaging innovation; replacing petroleum-based packs with bio-degradable corn starch packaging, offers an opportunity to make the waste management program more cost-efficient and lowers packing costs by recycling for FMCG producers.

Applications of *applied knowledge and technology* have been beneficial to consumer markets. For example, GPS and automotive braking systems have been developed from military applications, reducing the RD&D investment required to make these and other applications commercially viable in consumer markets. The digital characteristics of Industrié 4.0 are being applied to "city management"; traffic management (achieved by integrating technology into transportation infrastructure) can reduce traffic congestion and its associated costs. Sensors embedded in roads can count vehicles in real time and communicate with traffic signals that adjust dynamically to match traffic volume. Sensors embedded within autonomous vehicles will communicate wirelessly with each other, to smart infrastructure, and to the wider internet of things to help ensure smooth traffic flow; this will be a major benefit developing demand responsive public transport services, and city illumination, a wide range of benefits are being investigated and delivered.

Attributes That Reduce Costs of Acquiring the Product-Service (Life Cycle Costs/Value-in-Use)

Reductions in acquisition costs may be realized by considering how design and development may be directed towards technical concerns (improved quality, reduced maintenance costs), economic aspects (volume discounts through scale operations), service (free installation, staff training), and social factors (pollution reduction). Some or all of these features may have such a large impact on customers' costs that as Slywotzky and Morrison (1997) suggested, they will be prepared to pay an extra premium to obtain them or to switch suppliers. While marketing principles suggest that costs should not be used to determine prices, cost-efficiencies may be used to influence prices. Other important factors include, the major benefit offered by Industrié 4.0 such as; information: accessibility, content, accuracy, and currency, the expansion of detail, its real time, on-line availability offering M2M (machine-to-machine) management of manufacturing operations equipment performance and predictive maintenance.

Product-service support: applications advice, installation, and maintenance are significant costs of ownership and introduce the notion of "payment by performance" as well as planned maintenance; these offer characteristics that may have significance for asset management policies of high-cost industrial equipment.

Access to capital for transaction financing (financial engineering): is an extension of the previous characteristic; these companies typically regard their brands and customer relations as intangible assets; the "payment by performance" facility enables them to use capital made available to invest in the development of these and other intangible assets.

Opportunity Costs of Alternative Social and Economic Benefits Forgone (Shared Value)

Location decisions by organizations can and do have impact on social and economic benefits that can be delivered. The location of manufacturing and distribution facilities onshore can improve employment and skills proficiencies in areas of high unemployment. The downside to this has been seen with a number of projects. GE repatriated its white goods business to the USA from an Asian location. It also took the opportunity to update its manufacturing processes with the introduction of some robotic activities; it found that the skills required to operate the facility were not available locally. Clearly the decision needs careful consideration as the benefits offered by Industrié 4.0 may not be fully available. Shared value is a concept discussed by Porter and Kramer (2011).

In summary, given the increasing influence of Industrié 4.0, supplier relationships are moving towards real-time interactions. Supplier/customer relationships are changing rapidly, and vendor benefits offered to customers require close analysis if cost-effective and cost-efficient procurement management is sought. Value is, therefore, based upon "net benefits"; the balance between the quantitative and qualitative value delivered less the cost of acquiring the benefits, the social costs, and the price paid by the customer. Costs involve some important concerns. This clearly is an area of interest to financial and accounting management.

Determining the Current Cost of Capital: Adding Risk

The cost of capital is the weighted average of the required rates of return on equity and debt. The cost of equity is determined using the capital asset pricing model (CAPM). This model uses the yield on long-term risk-free bonds, increased by the risk premium attaching to investments in the equity market. The risk premium comprises a general market risk and a specific business risk. (See below for an explanation of the CAPM). Volkswagen uses a general risk premium of 6.5% reflecting the general risk of a capital investment in the equity market and is oriented on the Morgan Stanley Capital International (MSCI) World Index.

The specific business risk – price fluctuations in Volkswagen preferred shares – has been modeled in comparison to the MSCI World Index when calculating a β(beta factor). The MSCI World Index is a global capital market benchmark for investors. The analysis period for the beta factor calculation spans 5 years with annual beta figures published daily and an average subsequently being calculated. A beta factor of 1.22 (1.28) was determined by Volkswagen, for 2016. Hence for Volkswagen the risk premium is $1.28 \times 6.5 = 8.32\%$.

The cost of debt is based on the average yield for long-term debt. As borrowing costs are tax deductible, the cost of debt is adjusted to account for the tax rate of 30%. A weighting based on a fixed ratio for the fair values of equity and debt gives an effective cost of capital for the automotive division of 6.2(6.8) % for 2016.

In summary, there are two recognized measurements of corporate efficiency. The tradional accounting, depreciation-based metric, and the use of economic profit (and its derivative, EVA). While we would argue the case for EVA, we see a slow but positive growth in its application. Evidence suggests economic profit is becoming accepted as a more realistic measure of the profitability (and productivity) of organizational activities; we do accept it is slow progress.

A more detailed approach to the DuPont model can now be shown. Figure 5.6 has been modified to enable the use of economic profit. Figure 5.6 adds specific EVA-based concerns. The central area of the diagram is similar to Fig. 5.1, *developing operating profit management options* and *asset management and productivity*; how this may be achieved is the additional detail. The diagram suggests the evaluation of partnerships, alliances, and possibly M&A activities. The decisions will be influenced in the extent to which the value contribution (delivered by the value proposition) can increase the value advantage of the organization/network. The major automotive networks are acquiring and developing strategic alliances with software-related businesses, car hire organizations, and sharing economy-based businesses (ride-sharing networks). Recently Tesco (a major UK food and FMCG retailer) acquired Booker, a food wholesaler to extend its market reach into the smaller, independent-based operators.

The diagram refers to capital structure decisions that are discussed in the following section of this chapter.

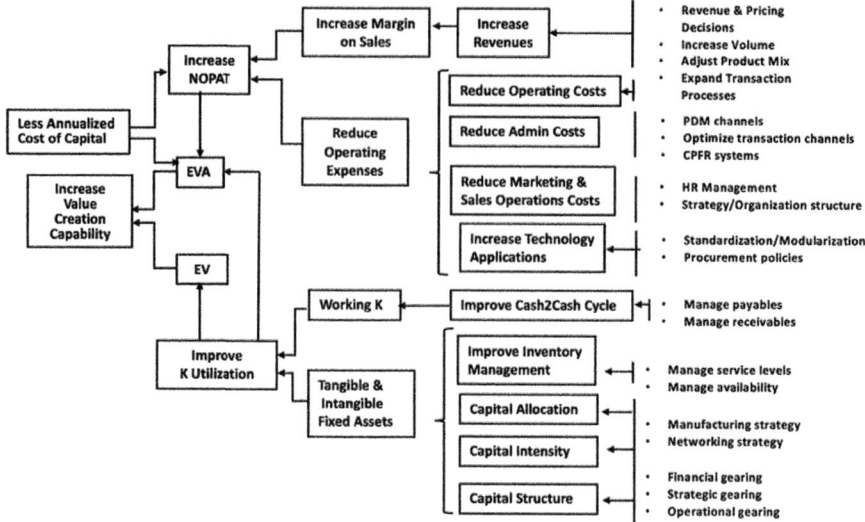

Fig. 5.6 Reflecting the impact of using economic profit in the DuPont analysis

Financial Structure Decisions: Operational and Financial Gearing and Financial Engineering – Profit Pools

Operational Gearing: Structure and Performance

The transition between formulating strategic response decisions and implementation is not clearly defined. This is not surprising because developments initiated by Industrié 4.0 are occurring very rapidly, and therefore the process of strategic decision-making becomes difficult; it is resulting in CapEx (capital intensive) business models becoming Opex (low intensity) business capital models. Some issues remain to be resolved. One of these decisions is the cost structure of the business. Before we explore this topic, we should revisit and apply the notion of risk. At this juncture we will consider two areas of risk that are common to financial structure decisions:

Business risk: any innovatory product-service or process will be surrounded by unknowns that the best of research design and implementation is incapable of identifying. Despite evidence suggesting it will be profitable and be "fit4purpose," it would not be viewed (from the risk perspective) as being as safe as an established product or production process. Questions for which answers should be found concern: what will be the overall impact on the business' capabilities? Will it change relationships with network partners? And more importantly with customers? What will be the response of competitors?

Financial risk: while this is a concern from an investment perspective, it must be considered by strategic and operational managers. Management can select from a range of alternative market opportunities and production methods, and these will vary from the manufacturing, distribution, and serviceability support considered necessary; any or all these activities may be managed internally or subcontracted out. Clearly the financial implications of the alternatives have input to the operations and marketing structures.

Current practice tends to suggest that unless the risk probability can be identified with some acceptable level of accuracy, and a decision be made concerning the financial impact on the capabilities of the organization, some level of subcontracting/outsourcing will occur. The application of platform technology, together with the adoption of network producibility (a strategic and operational operations infrastructure; i.e., the collaborative 'seamless', integrated and coordinated, processes of: research-design-manufacturing–operations-service-consumption-recycling and remanufacture), has contributed towards identifying and reducing structural risk). This adds dimensions of control and time management to the overall decision-making process.

Attitudes towards risk are reflected in the organizations' decisions concerning *operational gearing when considering the operations infrastructure;* operational gearing is "the extent to which an organization commits itself to high levels of fixed operating costs (such as dedicated locations, plant and equipment, RD&D, and management and labor expertise) as compared with variable costs (materials, proprietary components, and semiskilled labor). An organization with a view that level of risk is high is likely to adopt a relatively low capital intensity model. In Fig. 5.7a (low fixed costs/high variable costs – low operational gearing) the operational gearing is typical for manufacturers of assembled products for which specialist components are dominant inputs. Automobile manufacturing is one example of domestic appliances, and fast fashion manufactures are others. For each product type, extensive external sourcing exists. There are three possible reasons for this situation. One is the view that the organization sees itself as occupying the key role in the value chain of assembling a range of vehicles, *coordinating* the activities of a

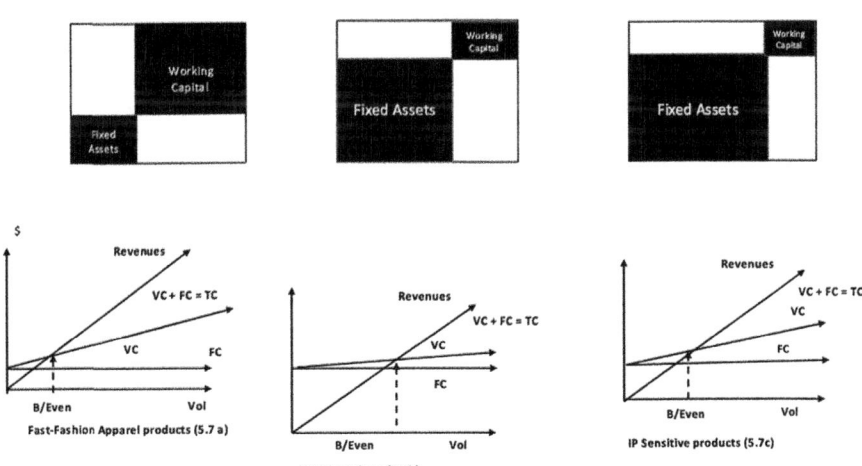

Fig. 5.7 Basic operational gearing options

network of suppliers of specialist components and accessories rather than undertaking the *development* of motor vehicle electronics and braking systems. The second reason follows; this strategy provides the opportunity of *differentiating* parts of the vehicle range with selected assortments of specialist components that are available at acceptable costs, and it is the marketing that creates the differentiation. Thirdly, the organization can diversify its risk by sharing the overall risk burden with its suppliers.

Figure 5.7b demonstrates an alternative approach (high fixed costs/low variable costs – high operational gearing). The reasoning behind selecting this model suggests a corporate view that business risk is low and therefore the organization can reduce its reliance on suppliers and internalize the supply activities, therefore earning higher margins as the supply functions are "in-house." Food processing and distribution is a typical example.

The break-even and cease production volumes are also a reflection of an organizations risk. The low operational gearing model (figure 5.7 (a)) permits production at lower levels of output than does a high operational gearing model. Often there are sound reasons for maintaining control of component manufacturing and incurring higher fixed costs. For example, technology driven organizations adopt this approach in order to protect the intellectual property of a developing component (or process) even though patents exist or may have been applied for. The innovative "intellectual property" organization is shown as Figure 5.7 (c), here the fixed assets (intangibles) are primarily patents (the pharmaceutical companies are typical examples but increasingly software companies (automotive electronics and data management and analysis) are included in this group.

The degree of operational gearing is a useful financial ratio that measures the sensitivity of a company's operating income (EBIT, earnings before interest and tax) to its sales. This financial metric shows how a change in the company's sales will affect its operating income. The degree of operational gearing is a method used to derive a quantified measure of a company's operating risk. The risk arises due to the structure of fixed and variable costs. Fixed costs do not allow the company to adjust its operating costs to affect its sales. Therefore, operating risk rises with an increase in the fixed-to-variable costs proportion. Generally, low operational gearing indicates that the organization's variable costs are larger than its fixed costs (VC > FC). It implies that a significant increase in the organization's sales will not lead to a substantial increase in its operating income. At the same time, the company does not need to cover large fixed costs. High operational gearing indicates that the organization's fixed costs exceed the variable costs (FV > VC). It suggests that the organization can boost its operating income by increasing its sales. In addition, the company must be able to maintain the relatively high sales to cover all fixed costs. Operational gearing can be calculated using the formula below:

$$\text{Degree of operational gearing} = \frac{\text{Contribution margin}}{\text{Operating profit}}$$

If an approach to profitability is taken that is based upon the use of capital (i.e., the annualized cost of capital, an EVA approach) the formula required is:

$$\text{Degree of operational gearing} = \frac{\text{Contribution margin}}{\text{EVA}}$$

The "convergence of Industrié 4.0, Value Chain Network 2.0, and stakeholder value-led management" is beginning to change these traditional structures. Digitization of products and processes (additive manufacturing, digital threads, and twinning) has reduced (even eliminated) the amount of inventory within the supply chain response, thus changing the traditional views held concerning operational gearing and risk.

Financial Gearing: Structure and Performance

Financing a business involves senior management in determining the financial structure of both the organization and to identify benchmarks for partnership arrangements. These decisions have implications for the level of *financial risk* proportional to the financial gearing of the firm, the measure of the comparative levels of equity and debt that finance the revenue generating activities of the organization. Financial gearing is measured by the ratio of total debt to total assets. Debt funding is a loan at an agreed interest rate by institutions and individuals whose involvement with the organization is simply to receive agreed interest payments, at agreed times, and the repatriation of the loan at an agreed date; these creditors have no influence in the organizations strategic direction. Equity funds are raised by selling ordinary shares in the organization. Ordinary shareholders receive dividends, the amount being determined by the organization's directors. These, together with appreciation in the share price, are the immediate returns to the shareholders.

Gearing implies borrowing to finance the business, using debt as well as equity funding. A major consideration is the fact that debt funding involves the organization in an ongoing commitment to meet regular interest payments regardless of the level of success of the business. Interest payments are considered as costs of creating revenues and as such are tax deductible. Clearly if the return generated by assets funded by debt exceeds the cost of borrowing, the profit benefits the shareholders. However, if the returns are low, a situation could occur where there is insufficiency to meet the interest payments and the organization has problems.

There is a view of gearing that suggests there is an optimal range of capital structure (equity/debt combinations) over which the cost of capital (the interest rate required by the investors) will vary only by very small amounts, and the market value of the organization, calculated by multiplying shares issued by their current market price, may be increased by increasing the financial gearing of the company. See Fig. 5.8. Clearly responses to market opportunities requiring large amounts of funding may have serious implications for the organization. A project involving the

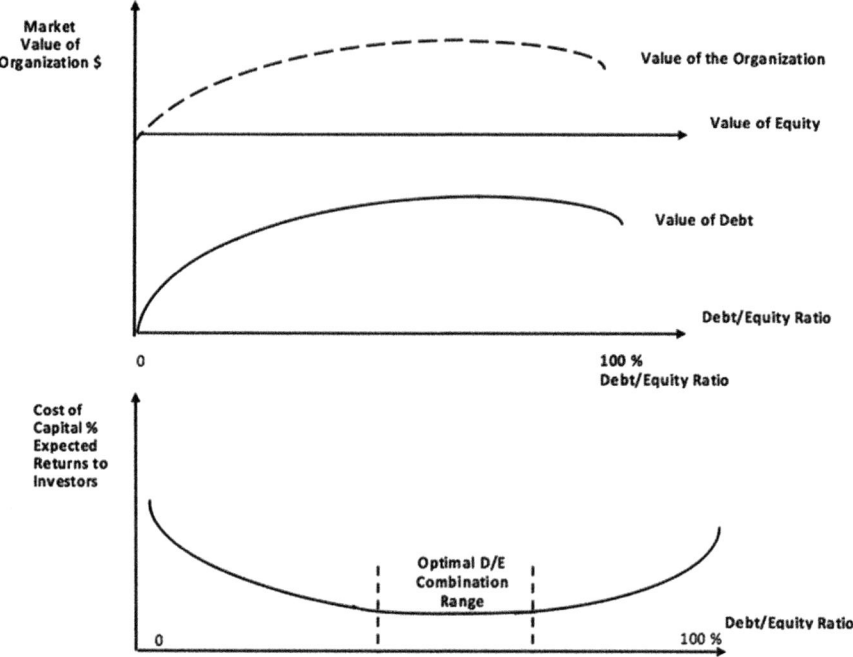

Fig. 5.8 Combination of debt and equity that minimizes the cost of capital

organization in new product-services or perhaps new demand technology life cycles/strategic business areas will have larger levels of both business risk and financial risk.

It follows that gearing implies borrowing to finance the business rather than raising and using only equity capital; the level of financial risk is directly proportional to the financial gearing of the organization. However, there are qualitative factors to be considered. Shareholders can, and do, influence organizational decisions.

The original argument suggests shareholders have increasing expectations of return as the level of borrowing increases; this also obtains for the entire group of stakeholders. The influence of the increased levels of debt (i.e., as the debt/equity ratio increases) is to lower the cost of capital up to a point (increased volumes increase the productivity of tangible and intangible assets, the tax effect), but beyond these investors consider levels of risk to be increasingly unacceptable. While there are clear benefits to the organization from the deduction of tax due to debt interest, this can only obtain only if there are taxable profits. A point can be reached where the increase in risk and potential failure make the tax benefits increasingly less attractive. Furthermore, as an increasing level of debt is added to the capital structure, so the organization may find its lenders impose conditions upon courses of action it wishes to pursue.

Financial Engineering: Focusing Asset Management

Financial engineering has a number of interpretations. Financial engineers design, create, and implement new financial instruments, models, and processes to solve problems, taking advantage of new financial opportunities. Financial engineering can also refer to the strategies companies use to maximize profits or other important performance metrics. We suggested financial engineering to be:

> Working with customers and network partners to focus their capital investment on core-revenue generating activities often by repurposing their existing tangible assets, activities and processes. This may involve structural changes internally or with network partners.

There are numerous ways in which this performance response capability can be successful. For example, joint RD&D programs that create unique/exclusive solutions to production problems within a value chain (Amcor's LiquiForm technology uses the contents of the bottle – be it soft drink, juice, shampoo, or BBQ sauce – to "blow" the bottle shape, simplifying the manufacturing and filling process). By combining the bottle forming and filling into one step, on a single line at a single site, the game-changing technology should reduce energy, handling, and transport costs for companies such as The Coca-Cola Co, PepsiCo, Schweppes, Unilever, and Heinz (Mitchell (2014)), transforming the PRODUCT-service into a product-SERVICE (Rolls-Royce Engines and General Electric). Caterpillar's value proposition offers customized product and service packages to meet specific customer usage applications.

Networked organizations can reconfigure retired production units to become competitive when introducing new product concepts; for example, General Motors has resurrected a "retired" manufacturing facility to produce an electric vehicle (the Bolt); other benefits were also realized and applied:

> GM can spread costs and revenue over a fleet of about 40 vehicles and four separate brands. It buys parts by the trainload and sourced parts and engineering solutions from across the company. The Bolt's gear-selector comes from Buick. The rear-view mirror, which is essentially a camera most of the time, is courtesy of Cadillac. GM didn't need to go on a building spree either. It's had an assembly plant outside Detroit since 1983, and it's been building Chevrolet Sonics there for five years. Batteries, meanwhile, are just another part that can be outsourced. GM was able to source its power-packs from LG Chem in Korea (see Fig. 5.9).

Product-services that are in low volume categories but which "fit" the product-service brand may be manufactured completely by network partners for whom economies of scale are engineered to occur and much lower production volumes.

> Magna International is contemplating a new assembly plant after securing a deal to build luxury sedans for BMW AG, which will help fill capacity at its flagship factory in Graz, Austria. Production of BMW's 5-Series will begin next year and adds to an agreement to manufacture vehicles for Jaguar Land Rover Ltd. and Daimler AG, Magna said in an e-mailed statement. Magna, which said it expects to produce 200,000 vehicles per year by 2018, is considering adding production capacity in China, the U.S. or Europe within the next two years. (Bloomberg, 15 September 2016)

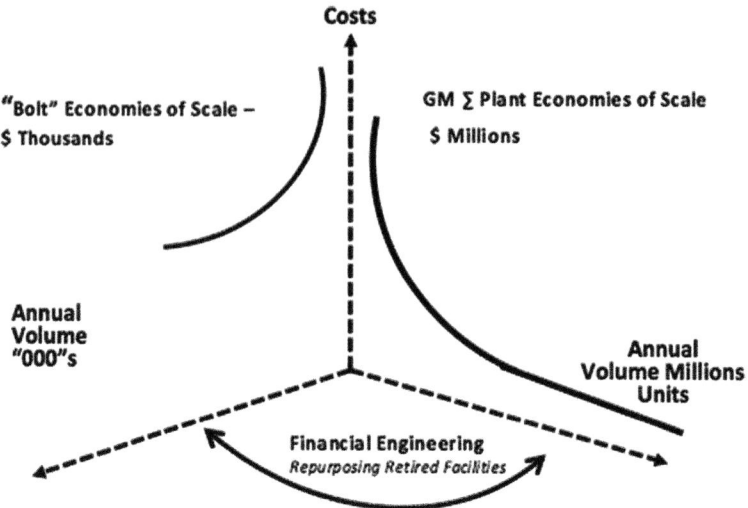

Fig. 5.9 Financial engineering and the benefits of scale: GM's Bolt

These automotive industry examples suggest financial management is contributing to strategic issues by matching asset resource capabilities to current opportunities with long-term prospects.

Profit Pools

Changing patterns of *value production* suggest that the structure of the added value contributions from value chain production processes results in a shift (or "migration") of value in the industry value chain. The significance this has for organizations that work in partnership networks is that due to these changing patterns of value migration, some firms may find it necessary for themselves to reposition within the value chain, leave the network, or perhaps develop more relevant processes and capabilities.

Gadiesh and Gilbert (1998) offered a model (the profit pool) with which to analyze the impact of value migration based upon the notion that "Successful companies understand that profit share is more important than market share." *A profit pool is defined as the total profits earned in an industry at all points along the industry's value chain.* The pool may be "deeper" in some segments of the value chain (and therefore offers new (perhaps larger) returns) than in other network positioning. See Fig. 5.10. Variations may also be due to customer, product, and distribution channel differences (e.g., online transactions), or perhaps there may be geographical reasons. Often the pattern of profit concentration differs markedly from revenue concentration.

Gadiesh and Gilbert (1998) recommended *the model be used to identify profit trends and to create an awareness of the implications of future structural shifts.*

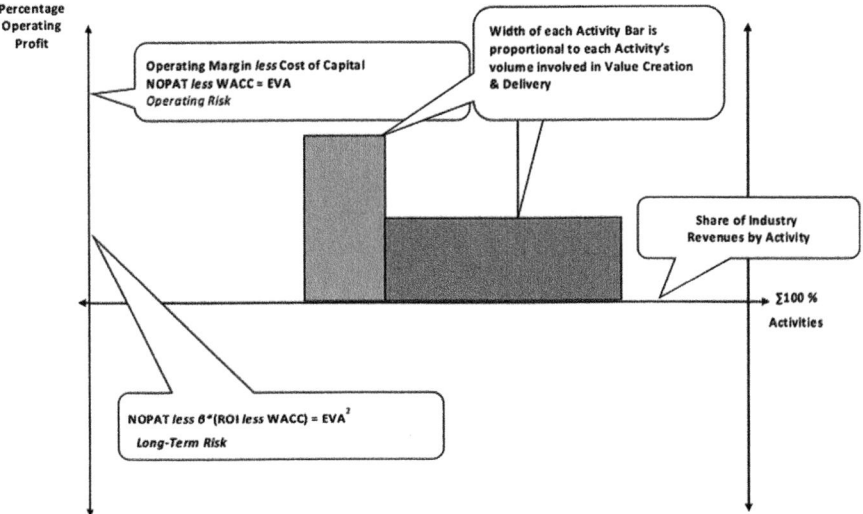

Fig. 5.10 Profit pool mapping: Risk

The profit pool approach does explain why a number of the large automotive component manufacturers are questioning their long-term viability (just as component manufacturers) and are researching the feasibility of involvement elsewhere in an alternative value chain network production system. The same argument can be used when considering the activities of the automobile assemblers who are redefining "mobility" to include the economics of sharing. Profit pools can be used to identify structural options (e.g., online activities) and to question them in terms of optimal stakeholder value delivery. Additionally, we can identify trends, establish scenarios, and identify where relationship strategies and structures need to be changed and identify the financial implications of each of the response capability options (e.g., *industry dynamics (and investment requirements) and value builders and their strategic cost drivers).*

Figure 5.11a illustrates a view of the healthcare industry in the 1990s. Prospective entrants attempting an immediate entry into the industry would have been attracted by the large activity share and the profitability of the pharmaceutical activity. However, a closer examination would have identified the locations of core profitability, activity, in the manufacturing and distribution of the so-called blockbuster drugs (statins and antidepressants), requiring considerable investment in tangible assets (manufacturing facilities) and intangible assets (brand support advertising); given the volumes produced by major competitors, any new entrants would have faced considerable difficulty in making an impact.

Another disincentive would have been the anticipated development pathways of the future activities within the healthcare industry. Based upon ongoing views of the *industry drivers*, there is evidence to suggest the industry activity structure will change, offering opportunities in activities that are responding to digitization and

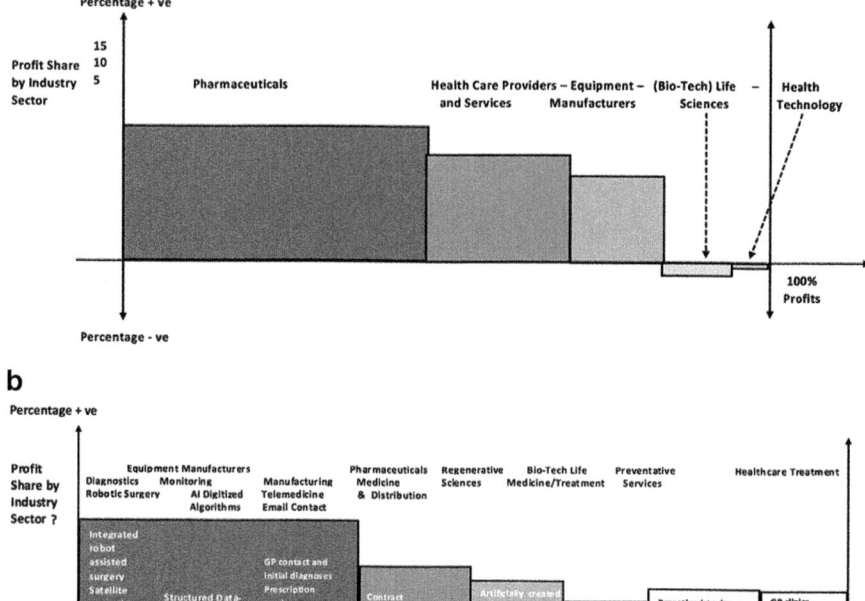

Fig. 5.11 (**a**) Profit pool mapping healthcare as an industry – 1990s. (Adapted from Gadiesh and Gilbert (1998)). (**b**) Profit pool mapping; an example – Healthcare (2025?)

connectivity aspects of Industrié 4.0: diagnostic and monitoring equipment, applications of artificial intelligence (digitization, algorithms), robotic surgery, and enhanced telemedicine developments. Figure 5.11b suggests how these may develop in the next few years. Clearly the activities will differ (to what precise levels is being determined), and medical technology applications and relationship developments will have significant impact.

Managing Cash Flow: "Pickers" and "Polishers" and "Pathfinders" – The Role of Cash Flow Management in Developing the Capability Response Business Model

Cash Flow as the Firm's Primary Objective

Even within the confines of traditional accountancy, it is clear that the notion of "profit" is quite an artificial one, being derived from the application of various "rules" and having potentially different meanings in different contexts. In this text it is generally proposed that to the extent "value" is measured in purely monetary terms, then, as has often been quoted, "profit is a matter of opinion, cash is a matter of fact" (Ellis (1999)). Simple cash measures have however often failed to consider the fact that cash is generated in different manners over different timeframes. This has particular implications when considering what a firm's key success resource capabilities are and how these should be managed, a topic explored further below. For current purposes it is proposed that in quantitative terms value in a firm is best measured in terms of anticipated free cash flow ("AFCF").

Operating cash flow management is the traditional measure of cash flow starting with the firm's earnings from which direct and indirect costs associated with performing its activities are deducted and are based upon short-term alliances and partnerships that establish operational partnerships to:

- Access/reinforce dominant operational drivers
- Reduce costs
- Add capacity
- Access skills
- Access capabilities
- Access process expertise
- Increase market/product/service application

Strategic cash flow management introduces long-term alliances and partnerships that offer:

- Capability to access/reinforce dominant strategic drivers
- Access to "enablers"
- Related markets
- Related industries
- Access to complementary markets

Strategic integration/transformational cash flow management encompasses the cost of fixed assets, long-term working capital requirements to develop inter-organizational response capabilities that achieve long-term strategic value advantage and growth in an alternative market and entry and exit costs associated with performing the processes of developing:

- Long-term-performance value builders
- Unique/exclusive PRODUCT-service/product-SERVICE solutions
- Unique/exclusive product/process technologies in concentrated and complex industries (e.g., automotive, aerospace, pharmaceuticals)
- Emerging/developing markets
- Trend markets ("fashion" and "fad" markets for which it is difficult to forecast)

Note: One important qualification needs to be added to any formulation of a firm's anticipated free cash flow and that is the influence of taxation, which is not a constant and which varies from jurisdiction to jurisdiction not only in quantum but also structurally in how it is levied.

See Fig. 5.12 for some examples of cash flow sources and uses.

Case examples are appended to this chapter for discussion.

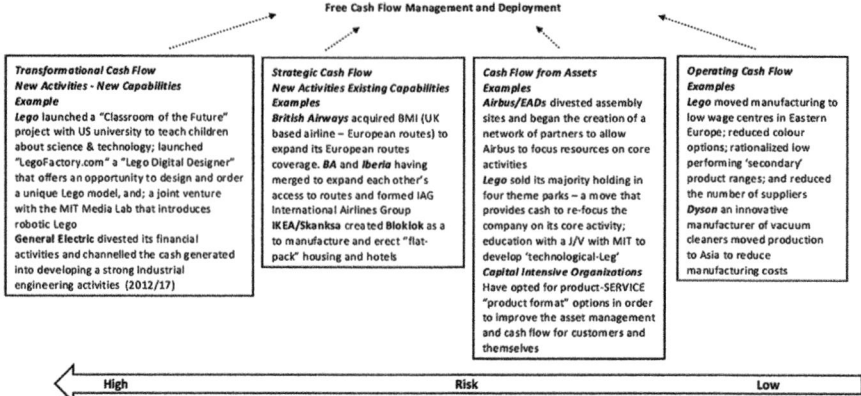

Fig. 5.12 Managing cash flow: Partnerships, alliances, mergers, and acquisitions

Risk and Return Management: Determining Realistic Risk Objectives

Response Capability Fit and Performance: Qualitative Aspects of Risk

A measurement of the free cash flow profiles appears to be a favored approach to evaluating potential and actual performance of partnerships. The proposed model (earlier in this chapter) identified four levels of cash flow: operational cash flow, cash flow from assets, strategic cash flow, and the resulting free cash flow. The model permits an analysis of optional response alliance structures at the important decision levels. The topics suggested by Bamford and Ernst (2004) are readily incorporated. They suggest that any revenues that accrue to one or other of the network partners are included. These are "cash flow from earnings of joint venture"; "transfer pricing, royalties, fees, and other cash flows"; and "value to parent not recognized in joint venture." Costs that are to be included are "parent management costs" and "operating costs."

Clearly while each is important, it is likely that *strategic fit* dominates the capability response model. It gives *effective direction* by identifying opportunities for growth from strategic alliances that are unlikely to be available to nonaligned organizations. Given that the purpose of an alliance is to identify opportunities for growth that may not otherwise be accessible and also given that these are sought on the basis of "high" return/"minimal" risk, then it follows that strategic fit will have priority. *Financial fit* is important here because not only is enhanced and improved financial performance a basic business motive; it is essential that the business model that emerges is financially viable. Concerns here are the overall impact of decisions on not just revenues but on investment and costs. Any alliance model that results must have *operational fit*. Simply put the structures that emerge must ensure efficient implementation of the business model. Essentially, we are concerned to be able to achieve key operational objectives cost-efficiently.

Measurement of *financial fit* will indicate the viability of the response model. The important metric here is free cash flow. *Operational fit* metrics are clearly outputs and service based. Operational performance decisions can be evaluated and monitored against the cash flow model and time. The model offers a means of exploring the impact of delivering customer added value expectations. Dominant issues here would concern the use of working capital (as indicated by an aggregate cash2cash cycle), but clearly some thought should be given to the longer-term concerns of relationships that indicate signs of permanence.

However, while there are clearly benefits available from a business model that incorporates partners who bring expertise and economies to the alliance, there is also a degree of risk:

Market growth risk will influence stakeholders' growth expectations, and the network will identify opportunities each will require close evaluation.

Structural risk factors constitute a few variables. These are best seen as operating at four levels:

Business Response Model Risk: is a measure of the overall risk to its viability that an organization attracts (or avoids) by owning, or acquiring (or delegating the ownership and management), of a core process. It reflects either the avoidance of or increase in internal transaction costs and the impact on process integration within the firm's business model generally. For example, a firm may have what seems a relatively unimportant manufacturing process such as a paint shop that can be outsourced with low capability risk, but in doing so this may disrupt the entire manufacturing flow and incur substantial hidden internal transaction costs.

Strategic Risk: reflects the importance of the process and the degree of direct control over it needed to impact the firms overall strategic positioning in the market. For example, outsourcing a process that is driven by unique intellectual property that is key to the firm's market positioning and competitive advantage would entail a high strategic risk.

Alliance Risk: reflects the introduction of virtual integration. It is more than simply the viability and dependability of particular partners, which is inherent in the notion of capability risk but reflects the strength of the alliance (or holonic network) as a whole. Alliance risk may be influenced by a number of factors. For example, an established alliance undertaking a new or expanded venture will increase the risk previously identified. A new alliance will be "riskier" than one built from known partners. Equally an alliance built around a new business model and one that has members from a diverse group of industries (that have yet to prove they can work together) also attracts high levels of risk.

Capability Risk: is the most obvious and important measure of risk and is the actual ability to operationally perform the process in question – can you actually make the product or deliver the service? High capability risk should be a significant discount factor to any anticipated free cash benefits.

Structural risk then is the aggregation of these risk factors. The comparative weighting of the risk factors will obviously be driven by the particular circumstances and will be to some degree subjective. In this sense the purpose and usefulness of a relevant list of "risk factors" are as much to act as a checklist of critical factors as anything else.

Response Capability Fit and Performance: Quantitative Aspects of Risk

The "returns spread" (return on assets managed *less* the current cost of capital) is both a performance measure *and* an indication of risk for current activities (investment in tangible assets such as manufacturing facilities and intangible assets such as brand positioning):

Returns Spread = ROE/EROI less Annualized WACC (Used for Current Activities & Capabilities)
ROI: Return on Equity
EROI: Economic Return on Investment
Annualized WACC (Weighted Average Cost of Capital)

However, for assessing *future activities*, several methods of quantitative risk exist; the most favored, discounted cash flow (DCF) methods use the cost of borrowing by taking interest charged and the borrowing time into consideration in the appraisal. From the *market opportunity analysis*, reliable *revenue forecasts* would be developed. Using these, value delivery alternatives would explore the implications for alternative *strategic cash flow, asset-based cash flow decisions*, and *operating cash flow*. DCF methods are based upon the present value of these future cash flows using a discount rated based upon an estimated (typically desired) rate of return. The present value is arrived at by discounting the forecast net cash flows for the planned life span of the project discounted at a rate of interest that reflects the cost of a loan (cost of capital) of equivalent risk on the capital market. It follows that an investment is *value creating* if its NPV is positive (the aggregate present value of the net income streams is greater than the initial investment cost of the project). It is *value destroying* if the aggregate present value of the income streams is negative. However, to be sure about the amount of value to be created (or destroyed), the impact of risk should be assessed. The *capital asset pricing model (CAPM)* can be used for this purpose.

The Capital Asset Pricing Model (CAPM)

The CAPM assumes that investors are risk-averse, and the greater the risk of variable return on an investment, the greater will be the actual return expected. This suggests there is a trade-off between risk and expected returns which must be reflected from the investment alternatives. The trade-off between risk and return is

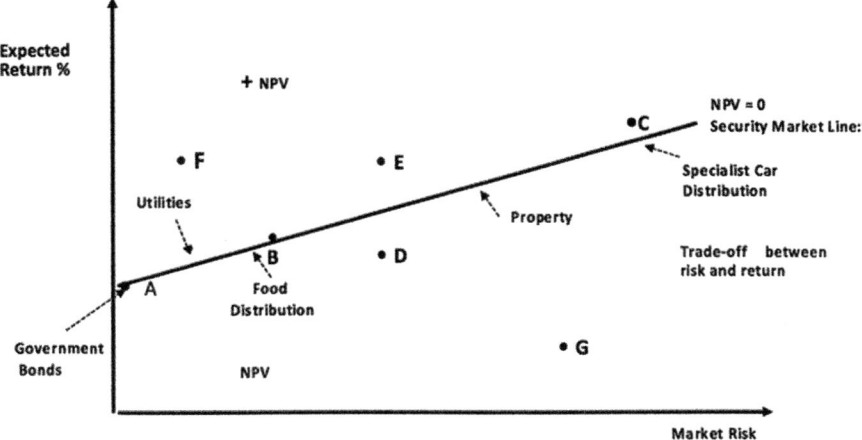

Fig. 5.13 Risk/return relationship across industry sectors

represented graphically as Fig. 5.13. The "trade-off line" indicates the trade-off between expected return and market risk combinations. Therefore, activities A and B and C fall on the trade-off line, and the expected return compensates for their risk as the NPV equals zero. Activities F and E offer good opportunity to create stakeholder value, while G and D, particularly G with high risk and low return, should be avoided. The capital asset pricing model was developed to evaluate securities in the investment market. The diagram suggests how different industry activities might be perceived from an investment perspective.

The underlying principle of the CAPM is that the expected return on an investment comprises two components: a risk-free return and a risk premium:

$$\text{Expected economic return} = \text{Risk free return} = \text{Risk premium}$$
$$EER = Rf = \beta(Emr - Rf)$$

where ER = expected return, Rf = the current short-term risk-free interest rate (e.g., government bonds), β = a coefficient measuring the "market risk" of the individual investment, and Emr = expected "market return" for the next period. The β-factor is determined by calculating the relationship between the returns on a specific asset (product/project) and those of the market. For our purposes we can consider it to reflect the relationship between the product-service-market and a market or an SBA and an industry sector. The beta value for "minimal risk" as a whole is equal to one ($\beta = 1$). Investments with beta values greater than one are sensitive to market changes (e.g., value migration). If the economy expands, the product/project will perform well; but if the economy declines, the product/project will reflect the lower performance. The impact of *alternative response capabilities* may be evaluated by assessing their relevance, their cost of ownership, and the time period before they can be implemented.

The CAPM is of interest because it provides a forward-looking, "project-based" variable interest rate evaluation model, in which the required rate of return and, therefore, the rate used for discounting can be related to risk as suggested by the beta value. The weighted average cost of capital (WACC) is, as we have seen, a cost of capital that reflects capital structure and investor-stakeholders' perceptions of risk based upon this current structure. However, this too could be adjusted by applying a beta factor if the organization was considered to moving into a higher-risk market/market segment as a means of risk assessment. Clearly the current capability resource base would require to be factored into the decision.

Examples of Cash Flow-Led Business Models: Pickers, Polishers, and Pathfinders

What is more important from a capability response perspective is *how cash flow is managed*; this has two considerations: how can cash flow be increased and how can it be made more productive. Managing *operating cash flow* requires clear policies on inventory management, particularly field levels of service parts inventory, service product obsolescence and obsoleteness, as well as supplier and customer payment cycles. *Producibility* decisions also play an important role; PRODUCT-service and product-SERVICE design decisions also influence the inventory management decision, using standard industry products (e.g., automotive industry's use of proprietary braking systems, in-car "infotainment" systems, air conditioning, etc.) that offer low procurement prices for assembly and service part purchasing; furthermore, they are usually readily available and therefore do not require excessive levels of inventory. *Cash flow from assets* raises questions of ownership versus partner leverage. Asset management is becoming increasingly important from a financial productivity perspective. The recent change in leadership at General Electric has focused attention on returns to tangible assets (recent entries by GE into a number of industrial markets) and the profitability and productivity of these acquisitions, resulting in the new CEO announcing intended divestment proposals. Network activities offer an opportunity to audit the response capability requirements and to locate assets in relevant and response-sensitive positioning in the value chain.

Strategic integration/transformational cash flow decisions differ from those of operating cash flow decisions; they typically involve an outflow of cash and as such require close scrutiny (it may well be that GE's merger and acquisition offer some lessons). Typically, strategic cash flow concerns *related product-service-markets where alliance partners contribute related capabilities that, combined, offer extended network capabilities;* transformational cash flow develops synergistic capabilities *where partners build capabilities that in isolation are unrelated but combined offer unique/exclusive value advantage.* Clearly timing is important; investment into energy markets requires questions concerning the geopolitical role oil, as well as its production cost alternatives; fluctuations in market prices make considerable differences to the return on investment. User applications are also important; the projected growth of electric vehicles and the cost of energy and the investment cost in recharging facilities are also important considerations. In aerospace construction materials and techniques are important (the increasing orders for the Boeing 787 reflect the impact of carbon compound materials on fuel use efficiency and increased range; by contrast the adoption of the Airbus A380 has slowed). Examples of cash flow-led response capability-led business models are explored next.

"Pickers," "Polishers," and "Pathfinders"

Examples can be found of business models designed around cash flow management. We suggest three current cash flow-based business models: "pickers," "polishers," and "pathfinders."

The "picker" model can be applied to General Electric during the Immelt era. Pickers create substantial CVA∗ (competitive value advantage) by identifying opportunities in new product-service developing markets. They offer opportunity to use managerial expertise developed in adjacent markets; they expand into new strategic business areas by creating the next-generation product-service in the demand technology life cycle, for example, customer-performance management (digital thread and twin concepts). They monitor and leverage owners of relevant *industry dynamics* areas and observed value migration opportunities. They also identify potential *value builders* and create specialist networks for future competitive value advantage (CVA) growth. They create dominant positioning based upon establishing "high-entry/catch-up" costs (patents and investment barriers) and by capturing volume by M&A activities and product-service and process innovation. See Fig. 5.14.

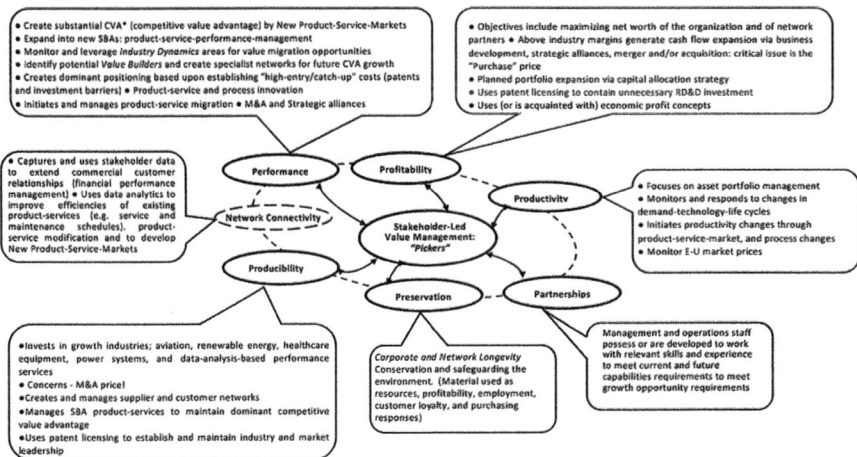

Fig. 5.14 The *pickers* business response capability model: GE oilfield services

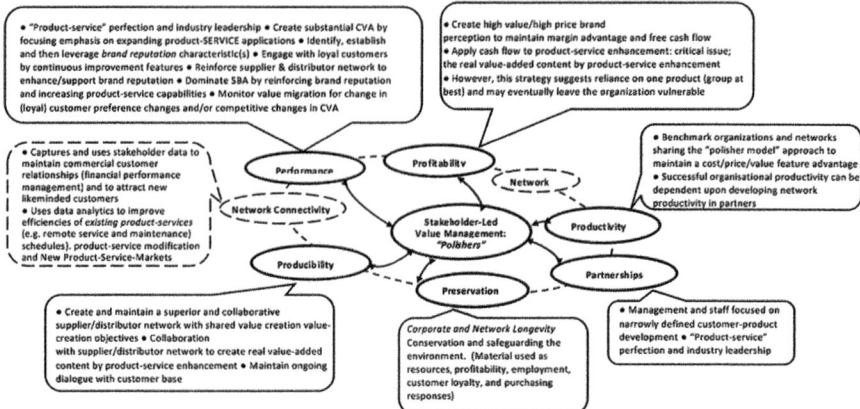

Fig. 5.15 The *polishers* business model: *Apple*

"Polishers," for example, Apple, build strategy around product-service perfection and industry leadership. They create substantial CVA by focusing emphasis on existing product-service development by leveraging their existing *brand reputation* characteristic(s). They engage with loyal customers by continuous improvement features and by reinforcing supplier/distributor networks to enhance/support brand reputation; this dominates their strategic business areas (SBAs) by reinforcing brand reputation and increasing product-service capabilities. "Polishers" monitor value migration for changes in (loyal) customer preference changes and/or competitive changes in CVA and respond with stylish innovation (aesthetics) that appeals to the "millennial" target market, accessible using "service provider plans" (see Fig. 5.15).

Amazon is a "pathfinder" with a value proposition based upon an opportunity identified from an analysis of broad-based data that can be pursued using the benefit of the organization's experience with well-tried platform-based online third-party and customer collaboration. Online transactions and distribution systems provides low-cost market entry. Market entry into adjacent markets (e.g., food retailing) is helped by its market strength and published record of efficiencies; the use of data analytics to capture and use stakeholder data to maintain commercial customer relationships is an important aspect of the success. See Fig. 5.16.

A review of each of the operations of these organizations suggests strong financial disciplines particularly in the management of network cash flow generation and application.

Convergence...

The convergence of Industrié 4.0, Value Chain 2.0, and stakeholder value-led management has significant implications for financial decisions, and these are inevitably linked to other response capabilities. Figure 5.17 considers the increasing impact of

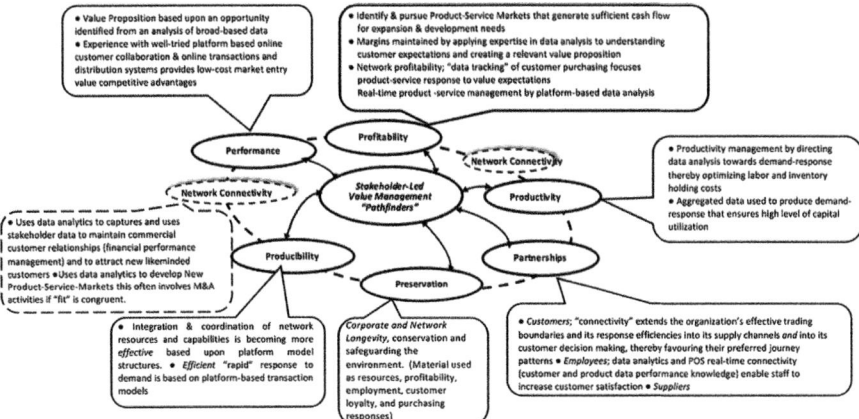

Fig. 5.16 The *pathfinders* business model: *Amazon*

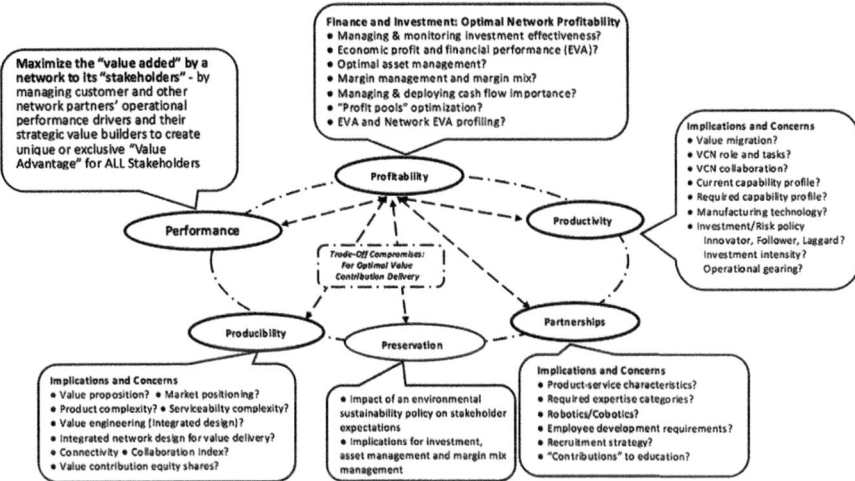

Fig. 5.17 Capability trade-off considerations and increases/decreases in value contribution outcomes

financial decisions on stakeholder value-led management and the responses of the other response capabilities. Stakeholder value-led management requires optimal performance by all stakeholder entities matched with equitable rewards. Accordingly, the response capabilities should represent an interactive approach.

Profitability: Finance and investment: Optimal network profitability is achieved by managing & monitoring investment effectiveness. Using, economic profit and EVA as measures of financial performance, optimal asset management leveraging network assets to achieve optimal performance e.g., network partners focus on key assets to produce an aggregate economic profit from either tangible or intangible assets (Rolls Royce Engines manufacture engines, Airbus the airframes, and airlines the branded travel experience), measuring and deploying cash flow into important projects, identifying "profit pools" characteristics, network EVA profiling Impact on the remaining Response Capability Centers is:

Performance: performance decisions have implications for strategic capital effectiveness performance and also for operating working capital efficiency on intra- and inter-organizational levels. Strategic decisions are based upon long-term viability of network partners, providing scope for maintaining ongoing, successful activities but giving the network membership financial flexibility to respond to opportunities. Ongoing success requires a product-service portfolio value proposition that meets customer expectations (fit4purpose, price/service/value equivalence, time2market, time4service response, etc.). Long-term success is based upon identifying the "changes and challenges" that opportunities in the business environment offer. With this in mind, financial managers should be conscious of the requirement to manage the investment activity; financial and operating leverage targets should therefore reflect the possibility of a flexible/agile response to opportunity.

Productivity: a standard and transparent measure of productivity using economic value added (EVA) will be useful in the assessment of increasing alternative production methods. Activities that have impact on profitability include strategic and operational procurement and collaborative activities that result from shared RD&D and "connected" operations in the network.

Partnerships: Network relationships that strengthen network partners' financial sustainability and intra- and inter-organizational partners possessing creative management skills and operations staff that possess (or can obtain quickly) the relevant skills and experience to meet current and future tasks. Focused relationships and network partnerships will improve customer focus, collaboration, and transparency – and consequently network profitability and cash flow.

Producibility: By investing in the development of a network "infrastructure," profitability will improve as collaboration among value chain partners will improve RD&D productivity and IP and patent development and management. Network productivity can be increased by increased collaboration through membership in global value chains. Using EVA as a profitability performance metric will offer "realistic" measurement and facilitate competitive comparison.

Protection: Sustainability policies impact the business organization, the network stakeholders, and the environment. Participation in the circular economy has implications for profitability and design for "serviceability-based-product-services," and the use of finite resource inputs is requiring an organization-wide review of future response capabilities, their application in the value chain, and careful decisions concerning investment and funding. Thus far examples reporting adoption of the circular economy/value chain suggest they have a positive impact on resource reuses on productivity.

Network Connectivity: Real-time network connectivity and collaboration are reducing downtime, production changeover time, operating cycle time, and therefore "profitability" and improve cash2cash performance. Data management and analysis capabilities will include improved management of hitherto variable influences and add to profitability improvements.

References

Bamford, J., & Ernst, D. (2004, August). Managing an alliance portfolio. *McKinsey Quarterly, 3*, 28–39.

Bloomberg. (2016. Retrieved from: https://business.financialpost.com/transportation/magna-mulls-new-car-plants-after-securing-bmw-5-series-deal-at-flagship-austrian-plant.

Boyd, T. (2016). CEO outlook poll. *The Australian Financial Review*. Retrieved from: https://www.afr.com/chanticleer/leading-ceos-upbeat-with-a-focus-on-china-20161209-gt8.

Ellis, J. (1999). *Doing business in the knowledge based economy*. Amsterdam: Pearson Education.

Freeman, R. E. (1984). *Strategic management: A stakeholder approach*. Boston: Harper Collins.

Gadiesh, O., & Gilbert, J. L. (1998, May-June). Profit pools: A fresh look at strategy. *Harvard Business Review, 76*, 139–147.

Kay, J. (1993). *Foundation of corporate success*. Oxford: Oxford University Press.

Mitchell, S. (2014). Amcor says new plastic packaging technology cuts costs. *The Australian Financial Review*. Retrieved from: https://www.afr.com/companies/amcor-says-new-plastic-packaging-technology-cuts-costs-20140730-j6y07.

Neely, A., Adams, C., & Kennerley, M. (2002). *The performance prism: The scorecard for measuring and managing business success*. London: Financial Times/Prentice- Hall.

Porter, M., & Kramer, M. (2011, Jan-Feb). Creating shared value. *Harvard Business Review, 89*, 62–77.

Quirk, B. (2011, July 4). How smart companies avoid getting burned by wild dollar swings. *Fortune, 164*, 32.

Slywotzky, A. J., & Morrison, D. J. (1997). *The profit zone*. New York: Wiley.

Volkswagen Group. (2013). Volkswagen Group Annual Report. Retrieved from: https://annualreport2013.volkswagenag.com/servicepages/welcome.html.

Walters, D., & Rainbird, M. (2007). *Strategic operations management*. Basingstoke: Palgrave.

Chapter 6
Productivity

Abstract The concept of productivity needs to be reviewed by arguing that the product-service "product" that current productivity is measured against is no longer the norm; for many economies the industry mix is dominated by "intangible service products" and "tangible manufactured products" that are produced by networks of organizations that are both interorganizational and international in their nature. It has been suggested that governments look to develop a more realistic method for calculating productivity, to identify a methodology reflecting the twenty-first century, rather than the approaches taken since the mid-twentieth century, if not earlier. A Boston Consulting Group (BCG) document *Productivity Now* offers an interesting approach to calculating the productivity of the organization, but this needs to be taken further to include productivity profiles of networked organizations. One important consideration concerns the treatment of intangible assets, another concerns the international and increasing complexity of manufacturing. The BCG approach uses economic profit generated and calculated using the aggregate EVA (economic value added) contributed by the network component organizations.

Keywords Productivity · Multifactor productivity · Productivity pools · Platforms

Introduction: Productivity – Time for an Alternative Approach?

The concept of productivity needs to be reviewed by arguing that the product-service "product" that current productivity is measured against is no longer the norm; for many economies the industry mix is dominated by "intangible service products" and "tangible manufactured products" that are produced by networks of organizations that are both interorganizational and international in their nature. Bean (2016) suggested that governments look to develop a more realistic method for calculating productivity, to identify a methodology reflecting the twenty-first century, rather than the approaches taken since the mid-twentieth century, if not earlier.

© Springer Nature Switzerland AG 2020
D. Walters, D. Helman, *Strategic Capability Response Analysis*,
https://doi.org/10.1007/978-3-030-22944-3_6

A Boston Consulting Group (Rose et al. 2016) document *Productivity Now* offers an interesting approach to calculating the productivity of the organization, but this needs to be taken further to include productivity profiles of networked organizations. One important consideration concerns the treatment of intangible assets; another concerns the international and increasing complexity of manufacturing. The BCG approach uses economic profit generated and calculated using the aggregate EVA (economic value added) contributed by the network component organizations. BCG recommended an alternative, new approach; the use of digital technology and big data and analytics as being essential to enhance the full set of productivity levers. They (BCG) identified Industrié 4.0. as the most underutilized (and least appreciated) lever and specifically cited four technologies that *are revolutionizing* production: augmented reality, advanced robotics, additive manufacturing, and the Industrial Internet of Things (advanced connectivity). BCG have developed an approach that focuses on increasing the *value contribution* from procurement-based activities: part numbers and prices, supplier management and contracting, and orders and plant execution.

Clearly, productivity measurement has some serious problems. In many respects, these problems are accounted for by the methodology used to calculate productivity and possibly because of the variable nature of the locations used to make comparisons. *The Economist* report "The World Economy" (2014) commented, "The digital revolution has yet to fulfil its promise of higher productivity and better jobs … if there is a technological revolution in progress, rich economies could be forgiven for wishing it would go away." The article referred to the lack of productivity growth from investment in technology. "The rich world is still trying to shake off the effects of the 2008 financial crisis. And now the digital economy, far from pushing up wages across the board in response to higher productivity, is keeping them flat for the mass of workers while extravagantly rewarding the most talented ones".

Solow, a Nobel economist, in a book review for the New York Times, of *The Myth of the Post-Industrial Economy*, argues Cohen and Zysman (1987) were lamenting the shift of the American workforce into the service sector and exploring the reasons why American manufacturing seemed to be losing out to competition from abroad. The authors suggested, America was failing to take full advantage of the magnificent new technologies of the computing age, such as increasingly sophisticated automation and much-improved robots. Solow commented, "like everyone else, they are somewhat embarrassed by the fact that what everyone feels to have been a technological revolution… has been accompanied everywhere…by a slowdown in productivity growth." The Australian Bureau of Statistics suggests a steady increase in labor productivity over the years 1949–1999 through 2009–2015, where multifactor productivity (labor, capital, etc.) has, after some increase during 2000–2009, levelled off and capital productivity has declined since 1994–1999. Returns from other sources suggest this is common to most western economies. More recently, in a review for the UK Government of its economic statistics (with an emphasis on productivity reporting), Bean (2016) commented that the statistics were beginning to show their age having been designed some 50 years ago, when the structure of UK industry and the economy was dominated by goods not services.

Furthermore, the author argues, "...the digital revolution and fast technological advancements of recent years, have changed the way many businesses operate (Amazon, Skype), given rise to new ways of exchanging and providing services (Airbnb, TaskRabbit), have muddied the waters between work and leisure, and made it far harder to measure accurately economic output. Many businesses also operate across national boundaries and depend on intangible assets, which adds to the complication of accurate measurement." The report suggests that if the digital economy was fully captured by official statistics, it could add between one-third and two-thirds of a percent to the growth rate of the UK economy.

To this could be added the comments of other academic and industry experts reminding us that the structure of business engagement has changed dramatically over this period with the growth of international networked specialists resulting in product manufacturing crossing many international borders before reaching its destination, the end-user consumer. This raises a question concerning the validity of national productivity reporting. We suggest productivity is becoming the domain of the networks and individual component organizations and that perhaps a more effective means of measuring performance would be to calculate the value added at each node of the production process, its value contribution, such as used by Volkswagen.

A recent article in *The Economist* (2017) comments on recent OECD research showing a productivity performance divergence among its membership. The divergence varies between sectors: in manufacturing, for example, top-tier firms (labelled "frontier firms") saw their labor productivity increasing by 2.8% a year, against 0.6% a year for the rest. The gap was even bigger in services: 3.6% compared with 0.4%. The "frontier firms" appear to have certain things in common; being ahead in technological application and making much more intensive use of patents; They are frequently part of multinational groups and constantly benchmark themselves against other frontier companies across the globe. International technological transfer spreads more rapidly across countries rather than intra-country transfer. It is argued that digital technology is creating a phenomenon of "winner-take-most" markets due to a combination of low marginal costs (encouraging first movers to expand quickly) and network effects. The OECD also noted that the information technology industry is producing a class of super-frontier firms: the productivity of the top 2% of IT companies has risen relative to that of other elite firms. Other studies show that this is not because the top tier is investing more in technology *but because they are investing more intelligently* to enable their workers to do new things and to reinvent their business models. A *second* explanation is that frontier firms (the 5%) have identified their own unique "distinctive value competitive advantage." Some have learned how to develop equally unique management techniques. Examples are given: 3G Capital, a Brazilian private equity group, which takes over mature businesses and reduces costs that no one else can; Amazon mixes digital expertise with just-in-time logistics. Others have devised rare material inputs; BMW, the carmaker, is using a special carbon fiber, stronger and lighter than steel, for its i3 and i8 electric cars. A *third* reason is suggested; technological diffusion has stalled, cutting-edge ideas are not spreading through the economy in the way that they used to, leaving "productivity-improving ideas" with the frontier firms. This suggests diffusion may be harder in a knowledge-intensive economy because frontier firms can hire the most talented workers and cultivate relations with the best universities and consultancies. But it is also made worse by bad policy. The OECD notes that divergence in productivity is particularly marked in sectors which have been sheltered from competition and globalization, most notably services.

In a later contribution, "Schumpeter" in *The Economist* ("Homespun economics" 2018) commented on the debate created by Robert Allen who argued for a "high-wage hypothesis." Allen's argument that the key to Britain's industrialization was due to the expansion of commerce and trade that preceded it. The expansion increased wages for British labor at a time when elsewhere in Europe pay rates were static. As Industrial Revolution 1 "took off" in Britain, coal was cheap relative to labor wage rates, thus making a strong argument for investment in "coal-fired technology" which enhanced labor output (an exercise in economies of scale); not until some years of mechanization and innovation in Britain had demonstrated the efficiencies of the new equipment was it worth adopting the "new technology" elsewhere in Europe. This argument reverses the traditional argument that of a disruption based upon low wage rates and the application of technology to expand the efficiency of labor. *The Economist* comment suggests that to treat real wage growth as a function of technological growth, rather than an influence on it, may not be helpful. "Schumpeter" suggests that innovation based upon the creativity of engineers, of which there is insufficient, leads to a "network partnerships collaboration" issue and points to the inadequacies and the rigidity of the education system that the adoption of STEM curriculum is beginning to correct.

How Important Is It?

The collaborative study by Boston Consulting Group and the National Association of Manufacturers undertook to investigate the impact on competitiveness of the recent downward trend in productivity in US manufacturing. The report concluded that what underlies the downward trend would appear to be reliance upon traditional, familiar *productivity levers*, rather than utilizing cross functional levers. The traditional levers they identified are working capital reduction, lean manufacturing R&D efficiency, supply chain optimization, procurement, and pricing and incentives, thereby confirming the views of Bean (2016) and others. BCG found respondents were less familiar with design-to-value (producibility), organizational efficiency, complexity reduction, and Industrié 4.0. BCG recommended an alternative, new approach: the use of digital technology and big data and analytics as being essential to enhance the full set of productivity levers. They (BCG) identified Industrié 4.0. as the most underutilized (and least appreciated) lever and specifically cited four technologies that *are revolutionizing* production: augmented reality, advanced robotics, additive manufacturing, and the *Industrial Internet of Things* (advanced connectivity).

There is, perhaps, another explanation, a focus on accounting profitability rather than economic profitability. Measuring "profitability" by deducting an annualized cost of capital reflects (economic profitability) offers an understanding of decision-making realism and, it can be argued, can also be "seen" as also having an impact on the "nation's economic welfare," the ultimate *stakeholder.*

Clearly, by utilizing the benefits now available from big data and analytics and digital technology and applying these to the BCG proposal (above) suggest greater knowledge concerning productivity performance (and how it may be improved) is readily available. Clearly the approach requires structuring and in-company leadership for it to be successful; however, we suggest the issue should be approached from an inter-company perspective (the value chain network with which the "company" belongs) and suggest a model that explains the approach taken by successful organizations later in this chapter.

Managing Resources and Productivity: A Multifactor Approach

Productivity features large in the equation that balances customer expectations and a managed resources response; Fig. 6.1 identifies this link. Given that the essential role of the value proposition is to communicate a response to a target customer group indicates that the value chain network understands the customers' expectations. The multifactor productivity model, then, seeks the most likely combination

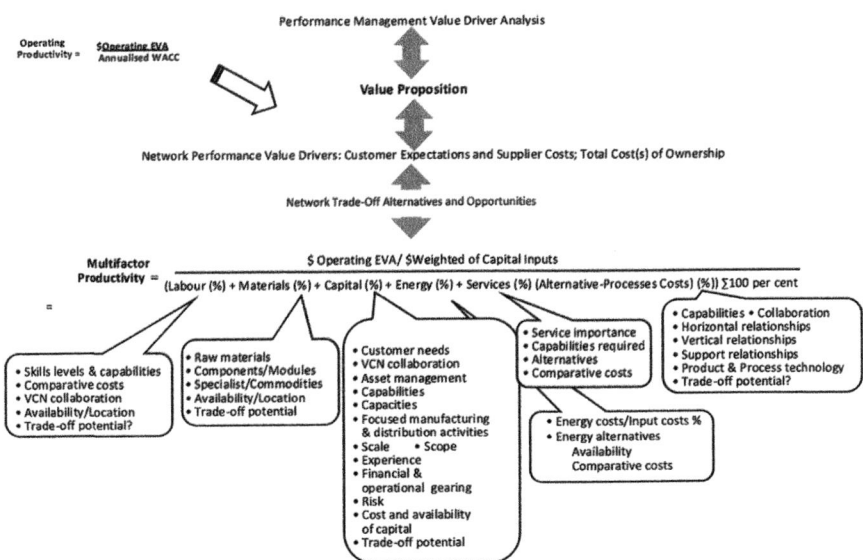

Fig. 6.1 Managing multifactor productivity using EVA

of resources that will successfully meet the expectations of both customers and suppliers: an optimal stakeholder solution. The well-known reverse/frugal innovation examples of General Electric's white goods and Panasonic's brown goods successes in Asian markets are an example of the impact of how multifactor productivity analysis can assist decision-making. Levi (Denizen Jeans in China and India), Coca-Cola (Pulpy), and Procter & Gamble (a lower-priced diaper brand in Brazil) are other examples.

Figure 6.1 identifies possible solutions for the decreasing decline in productivity; all the variables are not included in the analysis, and the available data may be doing "industry" a disservice! In the General Electric example not only was labor cost an inhibiting factor but so too was the assumed "fit4purpose," a product range with many features that were unnecessary in the target markets eyes, as well as organizational issues, concerning manufacturing and operations together with marketing and sales activities. See Immelt and Govindarajan (2009). Panasonic's problem was emphasized by the high value of the yen. For both organizations, a redesigned (simplified specification) and locally manufactured product was the solution, labor costs in both exporting countries, and organizational structure and marketing infrastructures being overly expensive. Both reassessed the importance of location.

Figure 6.1 includes the need for a value proposition that identifies a clear understanding of the local implications of customer needs and expectations of their resolution. Multifactor productivity analysis and planning (MFP&A) presents several what-if? scenarios that can be reviewed on an impact basis over factors influencing the range of input factors:

Labor cost and skills capabilities are a common reason for reviewing productivity. Some options are available beyond the usual outsourcing alternatives. Increasingly value chain tasks and roles begin to overlap, change, or become redundant. Constant monitoring of value migration and its causes are essential. Some years ago, the automotive manufacturers realized that certain components and subassemblies were seen by end-users as essentials and sought other features when considering individual makers' value propositions.

Materials; selection of materials and components is important. Factors such as endurance, speed, and payload can be greatly improved based on the choice of basic construction materials. Airbus and Boeing both use carbon composites for constructing airframe components; Airbus opted for 20% on their A380 aircraft, while Boeing opted for 80%; the associated manufacturing and assembly problems delayed the Dreamliner's commercial introduction.

Capital needs; the previous chapter identified and discussed a number of aspects and considerations; some have been added; one important characteristic that is common to each of these MFP&A "total cost of ownership" factors concerns the extent to which *collaboration* is considered to be shared and seen as an essential feature for the success of the network.

Energy; the cost and availability of energy (and the costs involved in changing alternative energy input types); sustainability and desirability issues are becoming political as climate change views drift endlessly. Clearly energy is a major input concern at all levels, organizations, networks, industries, and federal governments.

Services; service is becoming a major "value competitive" factor. The increasing technological complexity of the response requirements of industries such as aerospace has led to the growth of specialist small-medium enterprises who supply product services from fuselage and wing surface finishes to high-accuracy performance fuel management systems for aircraft. The automotive industry has a number of organizations moving towards being automotive software companies; those that have not moved that far have acquired or have strategic alliances with relevant expertise. The automotive industry has members who consider the "sharing" of vehicles to be a developing segment and have financial ownership or shared interests in vehicle rental companies.

Alternative value delivery and product process formats; many of the new business models appearing currently are based upon new "product-service-market' formats, such as; the generic format of existing drugs, biosimilars, walk-in clinics currently operating within retail stores, networked consortia of neighborhood-based GPs (or family physicians) operating out of a variety of retail formats.

Complexity of multiple manufacturing locations; the Apple case study has been cited on numerous occasions to demonstrate not only the interorganizational network structure of the manufacturing of its products but also its international network structure. *Apple* doesn't make the iPhone itself. It neither manufactures the components nor assembles them into a finished product. The components come from a variety of suppliers, and the assembly is done by Foxconn, a Taiwanese firm, at its plant in Shenzhen, China. *Samsung* turns out to be a particularly important supplier. It provides some of the phone's most important components: the flash memory that holds the phone's apps, music, and operating software; the working memory or DRAM; and the application processor that makes the whole thing work. This puts *Samsung* in the somewhat unusual position of supplying a significant proportion of one of its main rival's products, since *Samsung* also makes smartphones and tablet computers of its own. *Apple* is one of Samsung's largest customers, and *Samsung* is one of *Apple's* biggest suppliers. This is part of *Samsung's* business model; acting as a supplier of components for others gives it the scale to produce its own products more cheaply. For its part, *Apple* is happy to let other firms handle component production and assembly, because that leaves it free to concentrate on its strengths: designing elegant, easy-to-use combinations of hardware, software, and services, "Slicing an Apple: How much of an iPhone is made by Samsung?" (2011).

Accounting systems; developments in accounting theory and technological applications are moving management accounting into a predictive capability away from its current reactive capability; together they can benefit from the real-time data collection, analysis, and decision-making facility of Industrié 4.0. The value chain network is well placed to benefit from a move away from reactive accounting data towards a proactive approach; its established collaborative and interactive transparent structure is already partway there; real-time performance data from end-user product applications, together with attribute costing techniques, is facilitating the construction of algorithms capable of evaluating the optimal network productivity response to ongoing product operational performance, maintenance, modifications, and product development.

A problem with multifactor productivity measurement is that it is often a "finished" product service that may be subject to "serial" added value; many consumer electronic products cross numerous international borders and as such accumulate added value inputs from each location, thereby making productivity measurement based upon dollar values of inputs difficult, may be impossible, to achieve. OECD statistics can help, but, in many instances, it is difficult to identify with any reliable accuracy the changes in output that are combined inputs. For example, during manufacturing, mobile telephones cross a number of international boundaries – it is difficult to calculate the specific added value at each stage of production.

Economic Value Added: As an Alternative Quantitative Measure of Productivity

EVA is a metric based upon the notion of economic profit and has been developed commercially by Stern Stewart & Co. We see this as offering not only the facility to develop a more useful measure of productivity but also a possible contribution to a more useful competitive measure of GDP (gross domestic product). For a full discussion on added value and EVA, see Chap. 5.

It is suggested that the notion of (and use of) economic profit as a measure of productivity adds reality, consistency, and continuity. Rose et al. (2016) suggested a corporate productivity metric based upon economic value added (EVA) data that is readily available from publicly available financial and available data. They define productivity growth as the percentage change in the ratio of two factors:

> Company value added, defined as the difference between revenue and the cost of direct and indirect materials and components.

and

> The sum of labor costs plus the annualized cost of capital employed (maybe this should reflect intangible assets managed and include brand building costs, etc.?)

But if maybe this is not enough, should the impact of organizations', networks', and industries' responses to the changing business environment be incorporated? By using economic profit (measured as economic value added) as a basic metric, the modified operating statement would comprise:

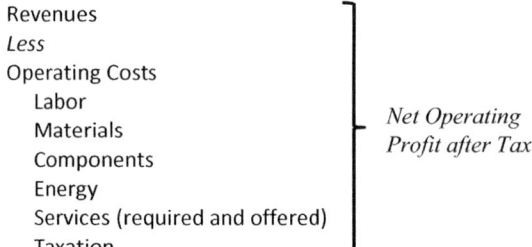

Less: The annualized cost of capital (WACC) used in asset management to produce the revenues. As well as the traditional fixed asset items of real estate, buildings, dedicated operations equipment etc.; intangible assets, such as, investment maintenance, brand development support, sales generation, supplier collaboration and development (RD&D activities), distributor development, "serviceability" costs (inventory holding 'services' such as vendor managed inventory), and network "connectivity" development costs should be included.

Equals: Economic Value Added (EVA)

It can be argued that Bean's (2016) suggestion that governments should be looking to develop a more realistic method for calculating productivity identifies a significant problem for industry, as well as government, and that economic profit generated over a reporting period is a more logical basis from which to start.

A suggested approach, taking a lead from Volkswagen, is to use EVA performance as a measure of productivity. Given that EVA is calculated from ongoing activities (operating statement after tax and capital charges/NOPAT), it provides an initial report on the operating efficiency of *current operations* – and therefore, year-on-year comparisons can be made. Given the amount of data being made available by the "Convergence" (the Industrié 4.0, Value Chain 2.0, and stakeholder value-led management), it is now possible to forecast EVA performance on a project/product/process basis. However, we suggest that EVA alone does not provide for full comparisons to be made and that some further measure of the efficiency (and the effectiveness) of the asset management within organizations and networked value chains should be incorporated in productivity analysis. This is suggested by Fig. 6.2 aligning some relevant management performance ratios; the ratios identify aspects of current, and potentially long-term, asset management. Clearly it is important to look beyond the traditional metrics and consider the impact of asset intensity and utilization when analyzing productivity importance.

This is particularly important for organizations that are value chain network components; for example, without collaboration and connectivity, it is quite feasible

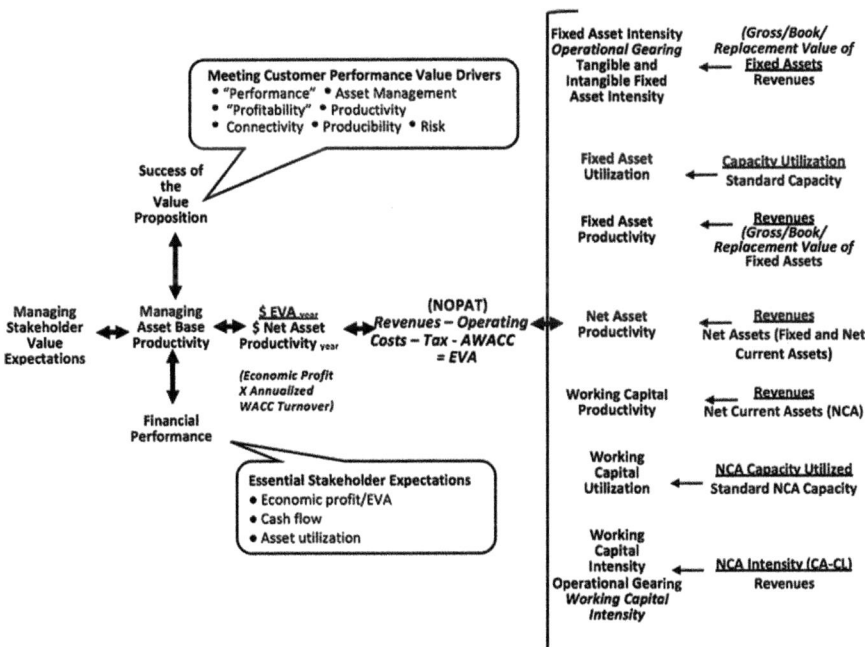

Fig. 6.2 Financial capabilities and capacities influence productivity performance

for there to be unnecessary investment in both fixed and working capital, process activities may become expensive in terms of excess capital and Time2Market. We suggest the inclusion of the relevant impact of both asset intensity and asset utilization as quantitative measures. Furthermore, we suggest it is important to recognize risk; clearly levels of risk differ across industries, and this is reflected in the β (beta) factors assigned by the "market." The β factor used should relate to the industry within which the opportunity resides; however, this may require some thought as shifts in value migration, together with changes in industry dynamics, may well be occurring. This information is in the public domain; we have yet to see it combined to make realistic performance comparisons. Our approach can be described by the following relationships:

Operational Productivity

$$= f \left(\begin{array}{l} EVA_{cop};Fixed\,Asset\,Productivity_{cop};Fixed\,Asset\,Utilization_{cop.}\,Fixed\,Asset\,Intensity._{cop}; \\ Working\,Capital\,Productivity_{cop};Working\,Capital\,Utilization_{cop};Working\,Capital\,Intensity_{cop)} \end{array} \right) * \beta_{cop})$$

Strategic Productivity

$$= f \left(\left(\begin{array}{l} EVA_{t+fop};Fixed\,Asset\,Productivity_{t+fop};Fixed\,Asset\,Utilization_{t+fop}; \\ Fixed\,Asset\,Intensity._{t+fop};Working\,Capital\,Productivity_{t+fop}; \\ Working\,Capital\,Utilization_{t+fop};Working\,Capital\,Intensity_{t+fop} \end{array} \right) * \beta_{cop} \right)$$

where

- t+cop = current operations period
- t+fop = future operating period
- β = beta value - stock market assessment of industry risk

Positive productivity occurs when operational productivity>net operating profit after tax (NOPAT) and the annualized WACC have been deducted.

Clearly a negative value indicates zero productivity. Management is interested in "relationships," for example, productivity as a measure of efficiency, year-on-year changes in productivity trends, competitive comparisons (within and across industries), and, increasingly, comparisons across business network alternatives. Productivity efficiency can be expressed as an index:

$$\frac{\$EVA}{\$\,Operating\,Costs + Capital\,Used + Taxation}$$

(Local rates of taxation vary, and the impact on the choice of "location" can influence manufacturing location decisions). It follows that production efficiency can be easily calculated by expressing $ EVA/$ operating costs + taxation paid on revenues less $ operating costs.

For evaluating long-term opportunities and planning purposes organizations can, and do, use a projection of future EVA returns using NPV (net present value) analysis. Forecasting future cashflows is a complex activity, for example, the developments in

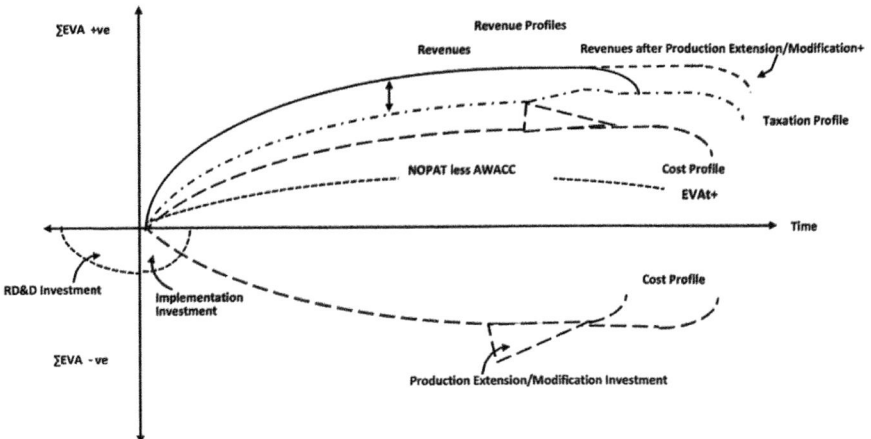

Fig. 6.3 Using Eva projections to evaluate response options: Evaluating responses to long-term opportunities

Level	Metric	Components	Process	Value Contribution
	EVA & Asset Management Based Productivity	NOPAT *less* WACC	Provides a return on capital used	Positive EVA identifies Value creation; a negative EVA indicates value is being destroyed
Individual Organization	EVA & Asset Management Based Productivity	Identify revenues less discounts. Identify individual component costs; • Labour costs • Materials costs • Components • Energy costs • Services • Taxes payable • The annualized cost of capital used to produce revenues	• Use MFP for analysis • Rank order on impact basis • Cross impact and trade-off analysis • Compare offshore labor rates • Explore robotic options • Investigate alternative materials & components • Alternative sources and/ or locations	• Identify intra-cost trade-off potential • Identify optimal cost structure that maximizes ∑added value contribution to the network • Uses "use of capital" and value created, or value destroyed as primary measure of the management of resources used performance
Network	∑ EVA & Asset Management Based Productivity	• Establish view on market positioning • Identify each partners VCN positioning • Identify ∑Individual partners EVA performance. • Identify alternative structures, roles, tasks against the expectations of the Value Proposition • Identify optimal VCN structure	•Establish view on market positioning •Identify each partners VCN positioning ∑Identify ∑Individual partners "productivity performance". •Identify alternative structures, roles, tasks against the expectations of the Value Proposition.	• Identify current value adding profiles (and their costs) of network partners. • Calculate their EVA performances • Compare ∑network added value output (quantitative and qualitative) with customer expectations and the value proposition. • Select optimal VCN – compare with competitive VCN structures
Industry	∑∑EVA Asset Management Based Productivity of all competing VCNs in an industry	• Calculate/estimate total Industry EVA • Calculate Network ∑EVA/Total ∑EVA Industry %	•Identifies value added to "international products"	• Identifies comparative performances across industries

Fig. 6.4 Quantitative productivity has an impact on individual organizations, networks, and national productivity (detailed)

industry dynamics and potential developments in value builders; equally difficulties can occur when attempting to assess discounting rates and the potential life span of projects (particularly for new technologies), as such the model should be used as a *guide for decision-making*; further discussion with specialists will provide reliable and usable estimates. See Fig. 6.3.

Such an approach identifies the sources and amounts of added value at the levels at which they occur and would identify *productivity efficiencies*. Figure 6.4 uses the EVA concept to develop productivity metrics applicable at the level of the individual organization, network structures, industries, and national GDP. At one time this would have proven to be a nightmare to plan, manage, and analyze the scant data captured. The data capture and analysis facilities of Industrié 4.0 now make this possible. *Predix* (General Electric) and *Teamcenter* (Siemens) offer real-time data collection, analytics, and algorithm designs that make the real-time "rolling value proposition" a possibility. Both companies offer product-service performance analysis (data threads) and interpretation leading to a digital twin of the product service that produces algorithms to enhance performance and to plan maintenance cycles.

Productivity has a wider impact than currently perceived; Fig. 6.4 is suggesting an alternative perception of what comprises "productivity." It is proposing a matrix approach within which individual organizations have dedicated response capabilities and assets that together with relevant operating costs are used to calculate *an economic operating statement*. Given the increasing availability of "data," a matrix approach offers an opportunity to identify and explore what could be described as "optimal productivity and the application of resources," not only at the level of the individual organization but the network within which it predominantly works.

Figure 6.5 extends the analysis. Given the current range of arguments concerning the causes, measurement, and therefore the usefulness of productivity as a useful metric it suggests a pathway through to a quasi-measure of productivity at a range of levels, from the individual organization, networked organizations, industry groups, and towards a quantitative added value approach to measuring GDP. It is argued here that together Industrié 4.0, Value Chain 2.0, and the increasing focus on stakeholder value management are able to facilitate the data collection, analysis, and interpretation required to provide a realistic measure of individual and aggregate business activity performance.

Figure 6.6 considers productivity as added value contribution to national productivity (GDP). Value contribution can be measured as both dollar value and percentage and as a one-off value as well as year-on-year trend values.

Level	Metric	$ Value Capital Utilised	$ Value of Added Value	Productivity Coefficient
	EVA & Asset Management Based Productivity			
Individual Organization	EVA & Asset Management Based Productivity	• $\sum\sum$ Cost of Capital Used by Value Adding Activities • $\sum\sum\$$ Capital Used by Organization to Create Value	• $\sum\sum$ Value Added by Value Creating Activities • $\sum\sum\$$ Organisation Value Added	• $\sum$ $ Value Added by value creating activities % • $\sum\sum$ $ Operating Costs + Capital + Taxation (Use 82/20 analysis) • $\sum\sum$ $Value added by Organization % • $\sum\sum$ $ $ Operating Costs + Capital + Taxation
Network	$\sum\sum$ EVA & Asset Management Based Productivity	• $\sum\sum\$$ Aggregate Capital Cost of Capital used by Network members to Create Total Value Added	• $\sum\sum\$$ Value Added by Individual Network Partnership • $\sum\sum$ $ Value Added by Value Creating Activities of the Network Partnership	• $\sum\sum\$$ Value Added by Individual network partnership% • $\sum\sum$ $ Operating Costs + Capital + Taxation of the Individual Network Member to Create value • $\sum\sum\$$ Value Added by Value Creating Activities of the network partnership % • $\sum\sum$ $ Aggregate Operating Costs + Capital + Taxation of the Network Members to Create Total Value Added
Industry	$\sum\sum\sum$ EVA Asset Management Based Productivity of all Competing VCNs in an Industry	• $\sum\sum\sum\$$ Aggregate Capital Cost of Capital used for cCeating value Added by an Industry	• $\sum\sum\sum$ $ Value Added by an Industry Grouping	• $\sum\sum\sum$ $Value Added by an Industry Grouping % • $\sum\sum\sum$ $ Aggregate Operating Costs + Capital + Taxation of the Industry used for Creating Value Added by an Industry

Fig. 6.5 Quantitative productivity: Using EVA to produce individual organizations, networks, and (GDP) national productivity – "productivity coefficients"

Level	Metric	Components	Process	Value Contribution
	EVA & Asset Management Based Productivity	NOPAT *less* WACC	Provides a return on capital used	Positive EVA identifies value creation; a negative EVA indicates value is being destroyed
Industry	∑∑EVA Asset Management Based Productivity of all competing VCNs in an Industry	•Calculate/estimate total Industry EVA •Calculate Network ∑EVA/Total ∑EVA Industry %	• Identifies value added to "international products"	• Identifies comparative performances across industries
National GDP	Network and Industry EVA Asset Management Based Productivity Contributions to GDP	•GDP ≈ ∑EVA of ∑Industries (Current) •GDP ≈ ∑EVA of ∑Industries Y/Y	• Calculates an economic-led approach to GDP	• Replaces traditional GDP method of calculation with one based upon the "use of capital" and value created, or value destroyed

Fig. 6.6 Quantitative productivity has an impact on individual organizations, networks, and national productivity

Identifying Sources of Productivity: Productivity Pools and Platforms

Productivity Profiles

The Gadiesh and Gilbert (1998) argument (previous chapter) suggested that "profit pools" differ within segments and by value chain and that profit pool profiles shift as structural changes occur within both segments and/or value chains. Plotting the revenues, profits, cost, and productivity profiles of alternative value chain formats enables the options to be identified in the knowledge of critical financial *and* structural criteria. It follows that the approach could be useful to identify productivity profiles enabling *value engineering options* (strategic and structural effectiveness) and *value delivery options* (customized PRODUCT-service and/or product-SERVICE customer journeys) to be identified and to question them in terms of locating optimal stakeholder value delivery locations in value chain networks; these options may result combining value chain activities that offer either improvements in the product service delivered or reducing the costs of the current value delivery. Additionally, we may also identify trends, establish scenarios, and identify where value chain relationship strategies and structures need to be changed and identify the financial implications of each of the options.

Gadiesh and Gilbert (1998) considered a situation within which an organization operates in just one location, thus making the task of identifying network partners and, specifically identifying their volume contributions and costs, much easier to identify. However, Industrié 4.0 has led to a "connected," interorganizational approach to strategy that will differ from that of a single company and relate to working with networking partners, recognizing the needs and contributions of all its stakeholders. And, there is another assumption; given the dynamic nature of value generation in many industries, the most likely "common denominator" will be the aggregate added value generated by value chain. Figure 6.7 illustrates this approach.

Given the increased activity of networked structures, it follows that productivity profiles should be projected across both the network and the industry to ensure that competitive options are identified and compared, thereby ensuring that "bests practice" is adopted. See Fig. 6.8.

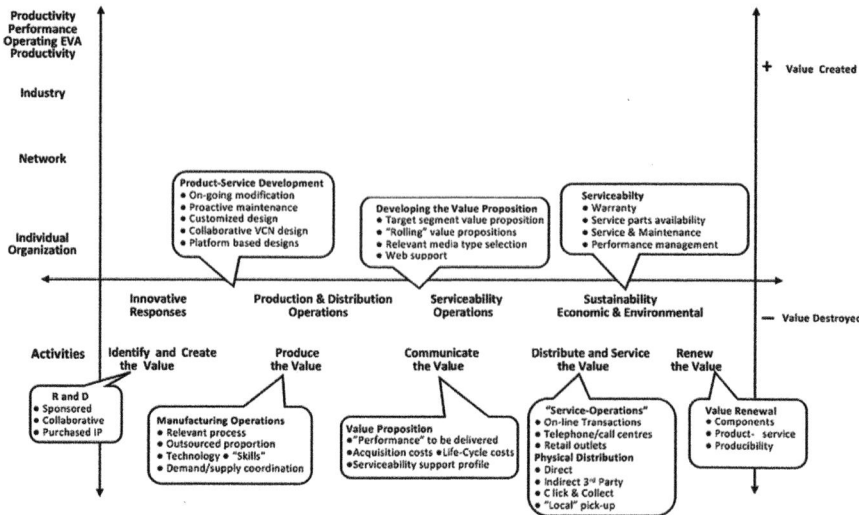

Fig. 6.7 Create an EVA productivity model based on the value chain network activity sequence

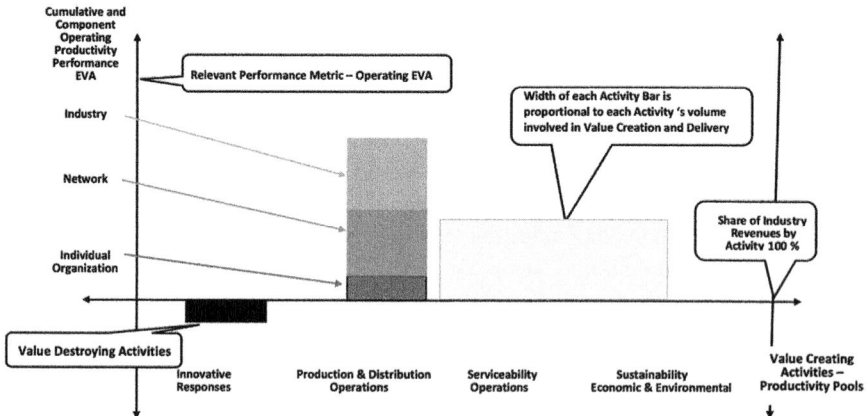

Fig. 6.8 Productivity pool mapping: Network and industry perspectives

Connectivity and Productivity Pools

The *productivity pool* approach considers the importance of strategic and operational productivity performance management. In the networked organization, the strategic implications are becoming more significant; the increasing expansion of manufacturing and serviceability activities of networked organizations in the context of collaboration, integration, and coordination suggests a similar approach be taken with *productivity pools* as that taken with profit platforms to ensure optimal stakeholder satisfaction. Initially it is more important to identify productivity performance by the value-generating activity rather than by a specific organization.

Figure 6.9a suggests the current and recent past profile of productivity pools. The two primary contributors to the situation indicated in Fig. 6.9a have been the lack of accurate, detailed, real-time information and the collaborative behavior this encourages.

By contrast Fig. 6.9b suggests the emerging profile. This is based upon the notion that successful companies understand that share of aggregate added value is more important than market share and the requirement for the network to create a strong productivity-based response model; this is being made possible by the integrated (and coordinated) impact of Industrié 4.0, Value Chain Network 2.0, and a wider recognition of stakeholder interests that have also expanded the range of market/industry opportunities becoming increasingly available. Figure 6.9b illustrates these developments.

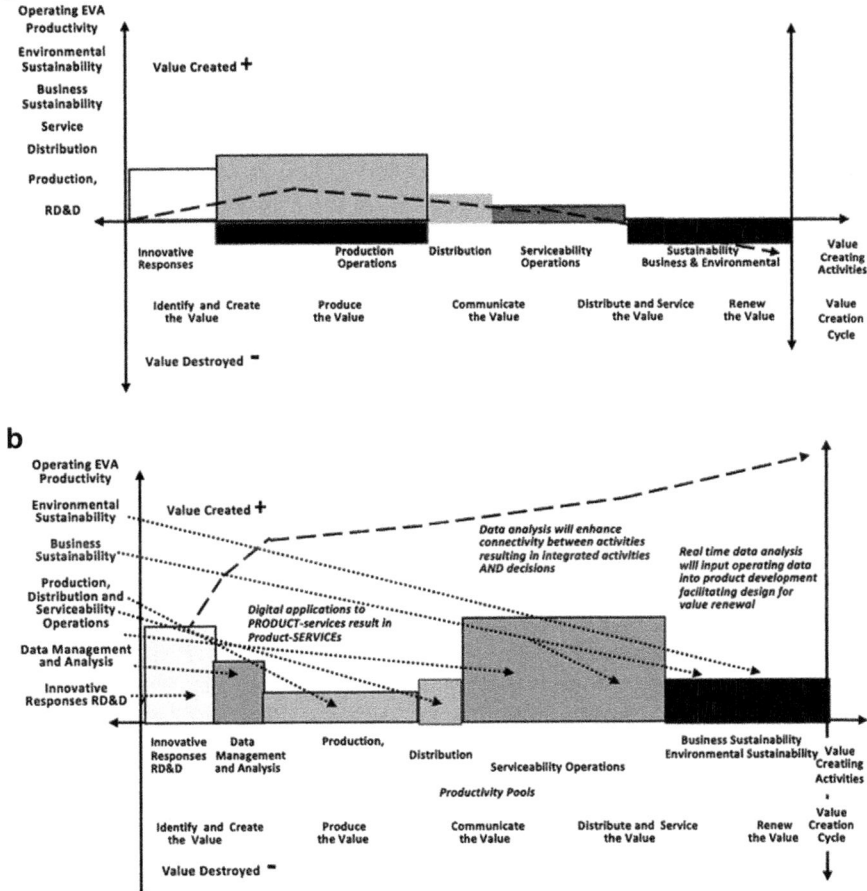

Fig. 6.9 (a) Added value and asset management and productivity performance pre-Industrié 4.0 (b) Anticipated added value and asset management and productivity performance post-Industrié 4.0

A *profit pool* is defined as *the* aggregate added value generated by an industry at all points along the industry's value chain. A *productivity pool* is defined by the effectiveness of both strategic productivity asset management *(value engineering)* and efficient operating productivity asset management *(value delivery)*.

Examples of strategic asset management are investment in market and value chain network positioning by shifting the emphasis of value delivery towards serviceability (and subsequently to customer performance improvement) by focusing investment on a product-SERVICE strategy (rather than the traditional PRODUCT-service strategy) and the acquisition of software companies (and employees) by consumer durables manufacturers in order to improve data availability and analytics, thereby adding "informatics" to the product-service range, and to obtain performance data to improve product maintenance and development capabilities. Examples of efficient operating asset management can be seen in inventory deployment activities such as vendor-managed inventory programs and the customized surgical packaging offers of pharmaceutical consumables. Both strategic and operational asset productivity management overlap as additive manufacturing is used to produce "service parts" in field locations. This development adds emphasis to the importance of "time" in supplier/customer relationship management where the penalties of downtime and delivery are of importance.

The *productivity pool* approach considers the importance of strategic and operational productivity performance management. In the networked organization, the strategic implications are becoming more significant; the increasing expansion of manufacturing and serviceability activities of networked organizations in the context of collaboration, integration, and coordination suggests a similar approach be taken the same with *productivity platforms* as that taken with profit platforms to ensure optimal stakeholder satisfaction. Initially it is more important to identify productivity performance by the value-generating activity rather than by a specific organization.

Productivity pools can be used to identify *value engineering options* (strategic and structural effectiveness) and *value delivery options* (customized PRODUCT-service and/or product-SERVICE customer journeys) and to question them in terms of optimal stakeholder value delivery. Additionally, we can identify trends, establish scenarios, and identify where relationship strategies and structures need to be changed and identify the financial implications of each of the options. The Gadiesh and Gilbert (1998) argument suggested that profit platforms differ within segments and by value chain and that profit platforms shift as structural changes occur within both segments and value chains. Plotting the revenues, profits, cost, and productivity profiles of alternative value chain formats enables the options to be identified in the knowledge of critical financial *and* structural criteria.

Connectivity and digital-led changing patterns of value production suggest the structure of the added value contributions from value production processes can result in a shift (or "migration") of value generation in the industry value chain. Demand life cycles are consistently enhanced by value migration with well-defined demand technology cycles offering alternative or improved solutions to customers' needs or reducing the costs of existing methods. The significance this has for organizations

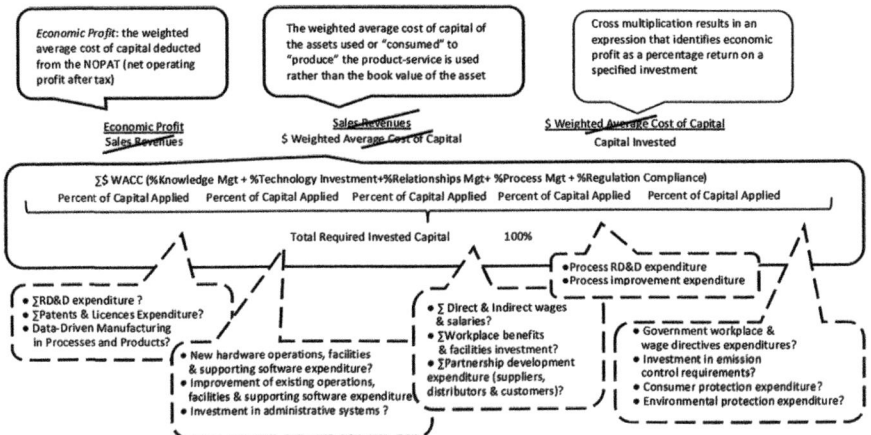

Fig. 6.10 Productivity – the "optimal utilization" of *relevant network* assets, capabilities, capacities, processes, and relationships positioned in "accessible" locations within the network *to* achieve stakeholder performance expectations; ownership or location of resources is no longer relevant

that work in partnership networks is that due to these changing patterns of value migration, some firms may find it necessary to reposition within the value chain, leave the network, or perhaps develop more relevant processes and capabilities.

Figure 6.10 develops a relationship between productivity analysis and asset management, suggesting that productivity analysis is a combination of aggregate network economic profit ($\sum$Value Chain Network Value Added) and network asset management ($\sum$Value Chain Network weighted average cost of capital of asset management). The model suggested by Fig. 6.10 currently is a "think about this" approach; the arithmetic of the financial relationships identify the financial implications of alternative value engineering and delivery options, as suggested by the fixed capital and working capital intensity/operational gearing measures.

Platforms and Productivity

Platform Typology

A platform is a business model that creates value by facilitating exchanges between two or more interdependent groups, component suppliers and original equipment manufacturer (OEM) assemblers, and consumers and producers. In order to make these exchanges happen, platforms harness and create large, scalable networks of users and resources that can be accessed on demand. Platforms create communities and markets with network effects that allow users to interact and transact. Platforms don't own the means of production; they create the means of connection.

It follows that a "platform business" isn't necessarily about creating the underlying technology (this is platform technology as applied to products and services), rather it is about understanding and creating the whole business and how it will create value for and build a network. In the twenty-first century, the supply chain is no longer the central aggregator of business value. What a company owns matters less than the resources it can connect and connect to. In the old model, scale was a result of investing in and growing a business's internal resources. But in a networked world, scale comes from cultivating an external network built on top of your business. A platform ultimately enables this value creation by facilitating transactions. While a linear business creates value by manufacturing products or services, platforms create value by building connections and "manufacturing" transactions.

In February 2000, GM, Ford, and Daimler Chrysler announced a plan to combine resources to create an integrated supplier exchange connecting a range of businesses from the automakers to independent retailers. The notion was that "jobs" could be "bid for," supply chains better managed, and RD&D cycle could include input from downstream suppliers. Later in 2000 Nissan and Renault joined the venture. The venture was labelled COVISINT ("Co" suggests cooperation and communication; "vis," visibility and visions; and "int," the Internet and international). Covisint was designed primarily for electronic procurement of parts, supply chain management, and product collaboration between the manufacturers and their entire supply chain. This type of platform is a business model that creates value by facilitating exchanges between two or more interdependent groups, usually consumers and producers. In order to make these exchanges happen, platforms harness and create large, scalable networks of users and resources that can be accessed on demand. Platforms create communities and markets with *network effects* that allow users to interact and transact.

Network effects are the incremental benefit gained by an existing user for each new user that joins the ecosystem; these are direct and indirect network effects. For platform businesses, network effects are key since there are two types of users: consumers and producers. As more consumers join the platform, the more useful and valuable it is to producers. You can see network effects at work when a platform attains a critical mass of users, at which point the cost of joining the platform is outweighed by the value of joining derived from the power of network effects.

Product-Process Platform Technology

Volkswagen, which has partnerships with nearly 30 other manufacturers in its group, manages to save costs without reducing the visual appeal of any of their vehicles. In fact, they have one of the most diverse-looking vehicle range of any major auto group. Their secret lies in their MQB program an example of product-process platform technology. These modular vehicle bases allow for a huge number of different possible combinations while sharing a common family of parts. The only fixed measurement is the distance between the front axle and the firewall.

Everything else, from the wheelbase to the front and rear overhangs, can be altered dramatically. For the Volkswagen brand alone, the MQB is used in the Polo, Golf, Scirocco, Jetta, Tiguan, Touran, Passat, and several others.

"Connectivity" is extending the busines2business capital goods market and the automotive markets, transforming them into product SERVICE into the service economy-of-services, subscriber-based revenue models and positive ownership experiences. The connected product provides a wide array of data to help OEMs make better decisions, improve safety, and increase productivity and, therefore, profitability. OEMs in both industrial and consumer-related markets are looking for a competitive value edge while delivering on customer expectations of value-added services. Benefits emerging are based upon focused serviceability. For example, Hyundai has utilized a Covisint IoT Platform; they have developed a mobile application that allows owners to monitor vehicle health, communicate conveniently with service representatives, and control their vehicle remotely, from just about anywhere. The Covisint IoT Platform seamlessly integrates the numerous systems associated with the vehicle and manages the identities of all drivers, devices, and cars across the business ecosystem.

Specialist Networked Platforms

Kleindorfer et al. (2009) identified Li and Fung Ltd. as an example of a global network that has developed an expertise that understands and embraces the complexities of the net-centric strategy model to harness global resources and expertise. They suggested Li and Fung's success is based upon an ability to identify and connect with the right competencies (capabilities) around the world. The argument suggests there are new capabilities required for success: the ability to identify, integrate, and coordinate capabilities for innovation and learning the expertise and resources of stand-alone organizations into an effective network. There are examples other than Li and Fung in which individual capabilities are being integrated and coordinated; the authors point towards the pharmaceutical and apparel industries. They argue that "The key is to enable, not to control, to allow emerging profit structures to come forth rather than to dictate." The Li and Fung business is driven by seven principles, and these have become the organizations core capabilities and reflect the attention to overall network productivity (Li and Fung and customers):

- Being customer-centric and market demand driven
- Focusing on one's core capabilities and outsource non-core activities, in order to develop a positioning in the supply chain
- Developing a close, risk- and profit-sharing relationship with business partners
- Design, implement, evaluate and continuously improve the work flow, physical flow, information flow and cash flow in the supply chain
- Adopting information technology to optimize the operation of the supply chain
- Shortening production lead time and delivery cycles; and
- Lowering costs in sourcing, warehousing and transportation

(Chang Ka Mun, Li and Fung Research Centre)

These characteristics identify the Li and Fung core response capabilities and offer a productivity-based means to acquire added value for both Li and Fung and their clients. The added value manifests itself for the clients in a number of ways. Clearly the economies of scale across a range of activities and processes are shared with clients; these may be in investment or in additional product-service features. They may also add access to aspects product-service development that may be costly in the context of Time2Market. Li and Fung's approach is illustrated as Fig. 6.11.

Improving Productivity EVA Analysis and Planning

The Rose et al. (2016) contribution focuses on the importance of understanding the impact of productivity in both the short-term and long-term. Figure 6.8 extends their proposals by adding additional "leavers." Value chain visibility, connectivity across the value chain, is delivering increased asset utilization rates. Recently an IHS Markit presentation on the impact of Industrié 4.0 on business models of productivity downtime estimates some 40 percent in specialist manufacturing networks and costs of $20,000/30000 per minute in automotive manufacturing. It is statistics such as these that create awareness of the requirement to consider the importance of productivity and the need to review the primary performance relationships identified by DuPont many years ago:

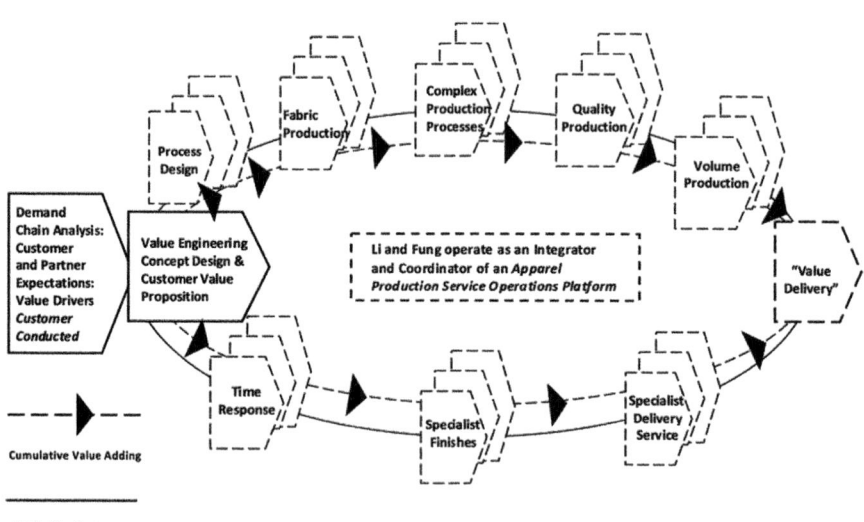

Fig. 6.11 Process productivity: Li and Fung operate a production process that utilizes some 12,000 partners across 30+ international borders

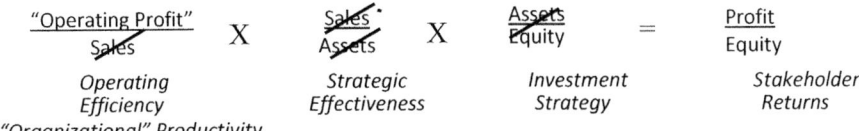

DuPont were very aware that successful outcomes across the organization required attention to integrated decision-making if optimal performance outcome was to be achieved. Figure 6.9 extends the DuPont and BCG contributions by linking the customer (delivering the value proposition) and the value chain network partners (productivity management driver responses), with an optimal positive performance outcome (economic profit-based EVA-led value engineering and delivery responses).

The point being made by BCG based diagram is to identify the *contemporary sources of productivity* available to organizations/networks that can contribute to total stakeholder performance satisfaction. It is not suggested all are essential, rather they be evaluated to identify individual impact. Unlike the traditional sources of productivity that had a more direct impact, the contemporary sources offer indirect and collaborative opportunities to enhance performance. For example, *product-service complexity reduction* and *producibility (design-to-value)*, together with *organizational efficiency* inputs, may possibly result in closer understanding of customer requirements necessitating a shift in the format of the value proposition. Both Rolls-Royce Engines and General Electric reformulated their value propositions to be service products; their customers purchase a product SERVICE ("uptime") rather than the engines, PRODUCT service, and hardware products. Michelin and Pirelli incorporate sensors to predict tire wear for preventive maintenance notification and energy management to customers. These developments are changing the business model structures of user customers; they are moving away from the "Capex" model (capital intensive–high fixed cost model–high operational gearing) towards an "Opex" model (low capital intensity–variable cost model–low operational gearing).

Figure 6.12 suggests a model for productivity planning activities that are becoming interorganizational; manufacturers are offering customers performance management capabilities using the real-time data to improve existing product service offers a "rolling value proposition" as well as for PRODUCT-service/product-SERVICE development and transition.

Performance Measurement Considerations: Deriving Relevant Response Capabilities

If productivity measurement is to be of use to management, it has to be pragmatic and relate to their decision-making requirements. Figure 6.13 responds to this expectation. Based upon the application of the "convergence," specifically the data collection and analysis features of digitization, connectivity, data transparency, and

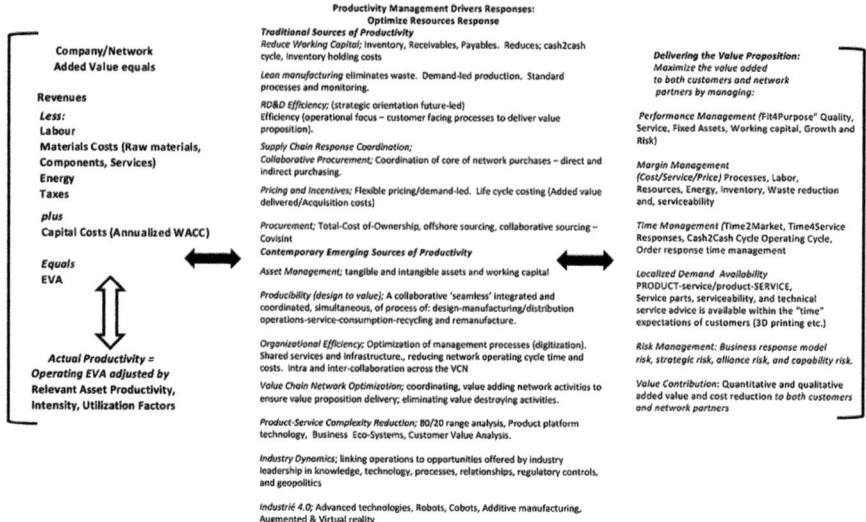

Fig. 6.12 Productivity is led by meeting value proposition delivery expectations. (Adapted from: Rose et al. (2016))

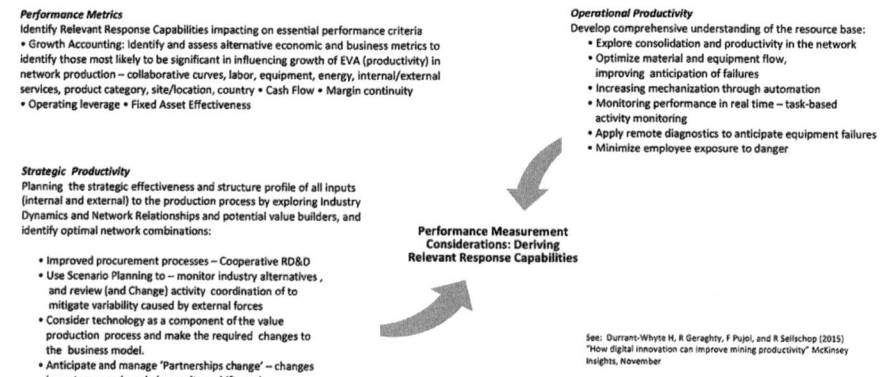

Fig. 6.13 Pragmatic strategic and operational performance measurement: Deriving effective and efficient response capabilities

collaborative value chain networks (collaborative curves that extend the BCG experience curves), it is possible to see the potential of network-wide productivity measurement; Fig. 6.13 approaches this by setting productivity metrics that can be applied on an intra- and interorganizational basis. Productivity is both operational and strategic. Operational productivity develops comprehensive understanding of the resource base at the organizational level and across the network, thereby providing the basis for improving operational efficiencies by modifying the network structure and member roles and tasks. Strategic productivity identifies relevant response capabilities impacting on essential performance criteria that require to be developed. Figure 6.10 provides details.

Gross National Happiness and Productivity

Gross Value Added and Quantitative Productivity

Gross value added is a productivity metric that measures the contribution to an economy, industry, network, and an individual organization. Gross value added provides a dollar value for goods and services that have been produced, less the cost of all inputs, and that are directly attributable to that production. At the company level, this metric could be calculated to represent the gross value added by a product or service the company currently produces or provides. Gross value added reveals how much money the product or service contributed towards meeting the company's fixed costs potentially creating a bottom-line profit. Once the consumption of fixed capital and the effects of depreciation are subtracted, the company identifies the net value the operation has created. On a national level, gross value added is the output of the country less the intermediate consumption, which is the difference between gross output and net output. Gross value added is important because it is used in the calculation of gross domestic product (GDP), which is a key indicator of the state of a nation's total economy *(Investopedia Terms)*. However, this measure includes depreciation (an accounting estimate of capital consumption, typically agreed with the tax authorities) and, as identified in the previous chapter, is not acceptable as an "economic" measure. It is argued that while the taxation authorities are satisfied with conventional accounting practice, it is not in the best interests of stakeholders. The EVA model discussed earlier in Chaps. 5 and 6 provide a pragmatic approach to identifying how effective organizations and the networks they create are in producing value within the context of capital made available; what is the productivity of the organization and of the organization's network structure; and how does their productivity contribute to a nations GDP. Using EVA at a number of organizational levels makes a significant, and measurable, contribution to the organization's stakeholders.

Gross National Happiness (GNH): A Qualitative Indicator of Influences on Productivity/GDP Performance

Gross national happiness (GNH) attempts to measure the sum total of the environmental impacts, the spiritual and cultural growth of citizens, mental and physical health, and the strength of the corporate and political systems. Not surprisingly, any measure of GNH is going to be dominated with estimates such as the depletion of environmental resources, for example, most of which are impossible to determine with accuracy. Other aspects, such as spiritual or cultural growth, can only be measured subjectively (see Fig. 6.14). As economists work with the core values of GNH, new measures should allow investors to more accurately view the long-term effects not only of economic growth but also the depletion of natural resources and the overall well-being of citizens. Green economics shares the same ideology and many of the same values as GNH proponents. The term was first attributed to *Jigme Singye Wangchuck*, the King of Bhutan in the early 1970s. Bhutan sought a measure of growth that reflected the nation's deep commitment to maintaining cultural, spiritual, and environmental sustainability standards. Updated measures of GNH try to incorporate variables that can be objectively measured in the hopes of applying them to the advanced economies of the Western world.

As the business environment continues to undergo disruption, it is essential to identify and measure the impact of GNH topics on partnerships, not only their efficiencies as a workforce but as members of societies. Figure 6.14 considers a range of topics, each of which has an influence on the level of productive output of partnerships within the organization, the network, the industry, and the nation. GNH is likely to have an increasing impact on quantitative output productivity given the changes Industrié 4.0 is introducing to workplace practice and attitudes. Of particular importance are economic security and a materials-based standard of living (wage increases are at a low level internationally), health (access to health services is becoming increasingly difficult due to low/poor investment by governments), education (the relevance of secondary and tertiary education curricula is seen as poor), balance of time use (organized labor in Germany is seeking a 28-hour working week), cultural diversity (while being seen as broadening understanding of international cultures, immigration has been seen as a reason for low/zero wage growth in a number of countries), and the environment and sustainability (growth of waste

GNH Components	Component Characteristics	Indicator
Economic Security and a Materials based Standard of Living	Household income per capita (50%) Assets (25%) Housing facilities (25%) Total = 90 per cent	Real income year-on-year. Robotics and continuity of income Mobile phone, PC, refrigerator, washing machine, television Toilet, electricity, construction quality, persons per house, etc.
Health	Number of healthy days	Mental health, disability restrictions, self-reported health status
Education	Amount of education, relevance and literacy	"Knowledge Awareness/ Index"; local/national, international, critical issues "Values"; social mores
Balance of Time Use	Work/family/leisure /"health-activity" balance	Allocation of time (percentages); work, family, leisure, sleep etc. Health awareness
Voice and Governance	Participation, "knowledge and pursuit of "rights", "performance index," "service index"	"Participation"; voting, involvement in governance "Rights"; freedom of speech, voting, political activities, equal value of work, freedom from discrimination "Performance"; "equality" – income, education, opportunity, environmental protection, prevention of corruption,
Cultural Diversity	"Cultural" participation, knowledge, attitudes relationships	Attitudes to migration, involvement in socio-cultural events Specific local activities Importance, recognition and acceptance of change
Personal Security/Law and Order	Responsibilities , government and individual	Expectations and attitudes towards "Codes of Behaviour"
Environment and Sustainability	Responsibilities, concern, and compliance	Knowledge and acceptance of "Codes of Behaviour"
Social Connectivity Community Vitality	"Community awareness", involvement and responsibilities	Family; involvement, care and concern, harmony Community; sense of belonging, trust Donations of time and money; proportion of household income donated Total days of volunteering completed per year
Plus ...		
Work and the Workplace	Individual average "tenure" of employment Individual: number of organizations worked in Employment longevity Individual "job satisfaction"	Workforce turnover

Fig. 6.14 Qualitative measurement of productivity; Gross national happiness

recycling and remanufacturing is becoming mandatory rather than being embraced overall). Commentators suggest difficulties when attempting to measure GNH; however, it is suggested by these authors there may be some future interesting comparative studies to be made concerning the levels of an EVA-based measure of productivity and GDP, Industrié 4.0 adoption, and economies with relatively high GNH measures.

Convergence

Industrié 4.0 has introduced several technological and relationship management options into the process activities of the value chain. A convergence of technologies has expanded the production and productivity options available from flow line mass production and batch manufacturing (mass customization) to a batch size of one. Data management/analytics (collection, analysis, sharing, and collaborative decision-making) is increasing "organizational effectiveness and efficiencies"; component suppliers locate closer to their assembler customers, abandoning the one-time advantage of low costs and increasing quality reliability and time4response. Connectivity has brought M2M (machine2machine) facilities, M2P (machine2partnerships-employees), and M2NC (machine2network collaboration) enabling information transparency. Figure 6.15 considers the impact of implications and concerns of productivity on stakeholder value-led management and the responses of the other response capabilities. Stakeholder value-led management requires optimal performance by all stakeholder entities matched with equitable rewards. Accordingly, the productivity response capabilities should represent an interactive approach:

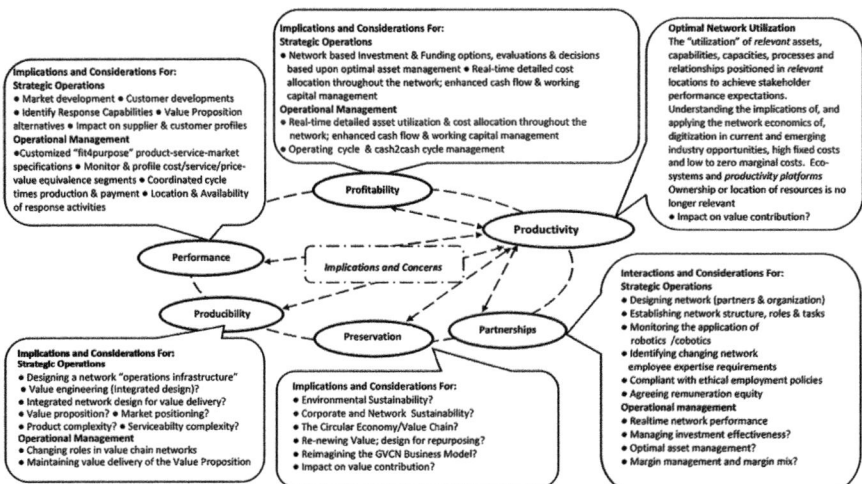

Fig. 6.15 Profitability: Implications and considerations for response capability interactivity

Productivity: The optimal utilization of network assets. The "utilization" of *relevant* assets, capabilities, capacities, processes, and relationships positioned in *relevant* locations *to* achieve stakeholder performance expectations. Management should understand the implications of applying the network economics of digitization and connectivity in current and emerging industry opportunities, rationalize high fixed costs within a network context, and structure CapEx and OpEx *business response capabilities. Ecosystems* and *productivity-based platforms* are becoming significant features of Value Chain Network 2.0 reinforcing that ownership or location of resources is no longer relevant.

Performance: Performance decisions have implications for strategic effectiveness performance (value engineering) and also for operating efficiency (value delivery) on intraorganizational and at interorganizational levels. Strategic decisions are based upon collaborating with network partners, providing scope for maintaining ongoing, successful activities but giving the network membership financial flexibility to respond to opportunities efficiently. Ongoing success requires a product-service portfolio value proposition that meets customer expectations (fit4purose, price/service/value equivalence, time2market, time4service response, etc.) to which the network is committed to deliver. Long-term success is based upon identifying the "changes and challenges" that opportunities in the business environment offer. With this in mind, financial managers should be conscious of the requirement to manage the investment activity; financial and operating leverage targets should therefore reflect the possibility of a flexible/agile response to opportunity – productively.

Partnerships: Network relationships that strengthen network partners (suppliers distributors, serviceability suppliers, and end-user customers), effectively (asset management) and efficiently (value delivery). Other partner characteristics include financial sustainability and intra- and interorganizational partners jointly possessing response capabilities such as creative management skills and operations staff that possess (or can obtain quickly) the relevant skills and experience to meet current and future tasks. Focused relationships and network partnerships will improve end-user customer focus, collaboration and transparency, and consequently network profitability and cashflow.

Producibility: By investing in the development of a network "infrastructure," productivity will improve as collaboration among value chain partners will improve RD&D productivity and IP and patent development and management. Network productivity can be increased, by increased collaboration through membership in global value chains. Using EVA as a profitability performance metric will offer "realistic" measurement and facilitate competitive comparison.

Protection: Sustainability policies impact the business organization, the network stakeholders and the environment. Participation in the circular economy has implications for productivity; design for "serviceability-based-product services" and for the reuse of finite resource inputs requires intra- and interorganizational-wide review of future response capabilities, their application in the value chain, and careful decisions concerning the potential economic return on investment and funding. Thus far, examples reporting adoption of the circular economy/value chain suggest they have a positive impact on resource reuses on productivity.

Network Connectivity: Real-time network connectivity and collaboration are reducing downtime, production changeover time, operating cycle time, and therefore "productivity" and improved operating and cash2cash cycle performance. Data management and analysis capabilities (digital thread and digital twinning) will include improved productivity management throughout the value chain.

References

Bean, C. (2016). *Independent review of UK economic statistics*. March. Retrieved from https://assets.publishing.service.gov.uk/government/uploads/system/uploads/attachment_data/file/507081/2904936_Bean_Review_Web_Accessible.pdf

Business (2017). The curious case of missing global productivity growth. *The Economist*. 11 January.

Cohen, S, & Zysman, J. (1987). Manufacturing matters: The myth of the post-industrial Economy. New York: Basic Books.

Gadiesh, O., & Gilbert, J.L. (1998). Profit pools: A fresh look at strategy. Harvard Business Review. May–June.

Homespun economics (2018, 4, August) *The Economist*.

Immelt, J., & Govindarajan, V. (2009). How GE is disrupting itself. *Harvard Business Review*. October.

Kleindorfer, P., Wind, G., & Gunther, R. (2009). *The network challenge: Strategy, profit, and risk in an interlinked world*. Upper Saddle River, NJ: Wharton School Publishing.

Rose, J., van Duijnhoven, H. Kopp, C. & Van Wyck, J. (2016). Productivity now: A call to action for US manufacturers. BCG-Perspectives. December 14.

Slicing an Apple: How much of an iPhone is made by Samsung? (2011, 10, August). *The Economist*.

Chapter 7
Producibility: The Networked Organizations' Strategic and Operational Infrastructure

Abstract Producibility is the process by which value is created and captured by the network and industry infrastructure. It is a total design activity that includes all relevant activities within the value chain network and creates intra- and inter-organizational partnerships to achieve stakeholder satisfaction. It becomes the network organizations' strategic and operational infrastructure and a process. It is a "total management activity," whereby the *product-service-design process* is integrated with the design of manufacturing processes and with the structuring of the subsequent operational processes of physical distribution, service support management, and product hardware recovery for remanufacturing/recommissioning while meeting customer specified expectations and stakeholder performance requirements. However, the dynamism accompanying Industrié 4.0, Value Chain Network 2.0, and the reinforcement of a stakeholder value management focus has brought about significant changes.

Keywords Producibility · Design to value (DTV) · Design for value and growth (D4VG) · Fulfilment execution system (FES)

Introduction: Producibility – A Connected Value Ecosystem Network

Producibility is the process by which value is created and captured by the network and industry infrastructure. It is a total design activity that includes all relevant activities within the value chain network and creates intra- and inter-organizational partnerships to achieve stakeholder satisfaction. Producibility becomes the network organizations' strategic and operational infrastructure, a process whereby the product-service-design process is integrated with the design of manufacturing processes and the subsequent operational processes of physical distribution, service support management, and, increasingly, remanufacturing. It is a "total management activity," whereby the *product-service-design process* is integrated with the design

© Springer Nature Switzerland AG 2020
D. Walters, D. Helman, *Strategic Capability Response Analysis*,
https://doi.org/10.1007/978-3-030-22944-3_7

169

of manufacturing processes and with the structuring of the operational processes of physical distribution, service support management, and product hardware recovery for remanufacturing/recommissioning while meeting customer specified expectations and stakeholder performance requirements. However, the dynamism accompanying Industrié 4.0, Value Chain Network 2.0, and the reinforcement of a stakeholder value management focus has brought about significant changes.

Examples abound, healthcare providers are trialing knowledge-based AI (artificial intelligence) algorithm models (Murgia 2017) to replace NHS (National Health Service) 111 call centers to improve urgency assessment and reduce costs, in the UK. Medical practitioner/patient relationships are managed by email systems. Surgical processes are making operations safer and more efficient. In emerging markets, surgical processes have imported automotive production processes and have applied them to heart and eye surgical procedures. Real-time physiological data is enabling new collaborative relationships between healthcare providers, pharmaceutical and consumer electronics companies, and patients. Digitization and the application of sensors to capital equipment are changing the role of supplier customer relationships by suppliers becoming customer asset performance management companies. The technology of augmented reality has been applied to aerospace manufacturing, distribution (warehouse order management), and flying training programs (simulation) resulting in increased operator efficiencies and reduced operating costs. Regulatory compliance management systems such as smart city infrastructure management, online tax submissions, voting, and civic services are increasing – providing additional examples.

Changing Industry Dynamics, Boundaries, and Management Approaches

The notion of *producibility (or design for value)* is a "total design management" activity that includes all relevant activities within the value chain network and creates intra- and inter-organizational partnerships to achieve end-user satisfaction Boothroyd and Dewhurst (1990) who pioneered DFMA (design for manufacturing and assembly). It is an integration of management processes – a 'seamless', integrated and coordinated   process where; the product-service design process is integrated with manufacturing, marketing and sales operations, physical distribution, customer service, refurbishing, recycling and remanufacturing and product-service retirement – to develop an integrated approach across the value engineering/delivery cycle.

Clearly not all the activities are required to be performed by any one organization (indeed often the value advantage produced is created, but given the assets, capabilities, and capacities of current and potential network members, often the value advantage is enhanced by involving industry specialists focused on components (e.g., Leica lenses in Huawei phone/camera products and Intel processors in computers)).

Together with an understanding of the end-user's expectations, it is now possible to integrate the design of the product-service, the manufacturing processes (procurement, component, module build, and final assembly), the physical distribution processes, the maintenance service activities, and reuse of selected components (based upon intellectual property, cost, or durability) concurrently. By identifying the "value-added intensity/productivity profiles" within each network member, it is possible to construct *industry value-adding chains*. This is becoming economically viable with the more recent ideas of Barkai and Manenti's (2011) *Fulfilment Execution System* (FES) and IDC Manufacturing Insights' *Global Plant Floor* model introduced in October of this year. Together these concepts propose a coordinated international multiplant operation that may be located anywhere by using the connectivity facilities of Industrié 4.0.

Almost There: Design to Value, Design for Value and Growth, and the Design-to-Value Advantage – But Not Quite

Design to Value (DTV) McKinsey

Design to value is described by the McKinsey authors (Fritzen et al. 2013) as "a portfolio-optimization approach that helps them (management) understand precisely which product features are important to consumers and whether consumers are willing to pay for those features." DTV can result in cost reductions or cost increases by designing products that better reflect customer preferences.

The authors suggest successful DTV has three common factors:

Cross-functional governance: a cross-functional group comprising R&R, marketing, finance, manufacturing and product development.

Standardized tools, processes: typically, each DTV group has differing methodology, processes, activities, and metrics.

A dedicated working team comprising: "creative thinkers," technically competent staff, with portfolio governance responsibilities and efforts (this is understood to mean access to senior management).

Design for Value and Growth (D4VG) McKinsey

Design for value and growth (D4VG) is a subsequent visit to the topic by McKinsey (Agrawal et al. 2016). DTV was essentially a model to evaluate opportunities to manufacture products at lower costs while retaining the features required for competition. D4VG extends DTV to provide "exceptional customer experiences" for value generation and growth by producing products with characteristics that create customer loyalty (and therefore repeat purchases). D4FG is defined by the McKinsey authors as:

> Investment in design that includes extra features or uses more expensive materials, based upon design involving a closer understanding of consumer motivations and innovative design enabling follow-up design/re-design improvements focused upon clever cost reduction leading to successive (serial) generations of products.

The authors cite Apple with the iPhone as an example of a design program offering a product having "incremental features and improvements" together with cost reductions (this suggests that the benefits/cost ratio to purchasers offers an increasing benefit/cost ratio).

Enabling characteristics of D4VG for design-driven growth comprise:

- Knowledge of competitive activities (and consumer responses) facilitating the capture of the product space.
- Insights about competitive products to understand alternative consumer propositions (typically through "tear-down" analysis).
- Input from customers to determine customer expectations (their price/value equivalence assessments and the impact on their performance value drivers; i.e. do they save money (and or time) in the medium/long term?).
- Identification of production capabilities and impact on cost drivers.
- Design teams that integrate knowledge into viable product-options.

The authors also identified "next-generation" customer insight tools:

- Systematic identifying and matching of customer expectations with "valued" features responses.
- Use of social media to elicit consumer word-of-mouth comments and/or recommendations.
- Product-market testing, researching purchasing in purchasing situations.
- Identification of "relative importance" of product features before, during and after purchases.
- The use of conjoint analysis (or other suitable techniques) to determine price/value equivalence and trade-off decisions.

Organizing the approach requires:

- Defined (clear) boundaries between D4VG and R&D, by this it is suggested that D4VG groups should emphasize applications that lead to profitable growth.
- D4VG groups should be accountable to senior management such as the COO.
- A remit that encourages design groups to extend their thinking beyond the 'design-to-cost' mentality and to focus upon issues concerning added-value.

Clearly D4VG was developed to fit into a changing business environment that was (and continues to be) fueled by demands for more accuracy, variety, experiences, etc.

Design-to-Value Advantage (D2VA) Boston Consulting Group

The design-to-value advantage (D2VA) response by BCG (Myerholtz et al. 2016) emphasizes an approach based upon the relationship between an NPD project's impact on design objectives and its ease of implementation. In making an argument, they refer to the capability of identifying the roles of both *high value/low volume manufacturing* and *low value/high volume manufacturing* and the potential for a combination of both approaches in the pursuit of a challenging opportunity. They suggest four attributes for successful implementation of the D2VA model:

- Question the efficacy of the value proposition in response to changing consumer preferences, regulatory compliance requirements and competitive innovations.
- Use a holistic approach. Apply 'teardown' research (products and production methods) methods of competitive products to understand how they create value for customers.
- Focus on value opportunity. Analyze companies successfully pursuing D2VA and D4VG programs. Understand fully the opportunity and how the RD&D activity creates customer satisfaction. Identify opportunities to expand on competitors added-value offers. Here the BCG authors discussed the disruption of the disposable-razor market by quality durable competitive products.
- Search for new information and insights by creating data bases using customer inputs; work with suppliers. For example, the electric vehicle development program is very dependent upon development of more powerful lighter weight batteries; the industry has a considerable level of strategic alliances and M&A activities.
- Establish collaborative ongoing implementation processes. Invest in dedicated DTV/D2VA etc. resources.

It can be concluded that to create a design-to-value approach, the successful companies will be those that apply a "futures perspective" to product-service design within cross-functional, intra- and inter-organizational collaborative activities. There is scant mention of how procurement, manufacturing and distribution operations, serviceability, and remanufacturing can add value.

However, we suggest producibility now has the ability to extend its focused responses. We explore this in the remainder of this chapter.

The Impact of Changing Industry Dynamics

The increasing impact of Industrié 4.0, Value Chain Networks 2.0, and the impact of stakeholder management provides opportunity to rethink the business. The growth of *product outputs* as "products" is becoming responsible for a reconsideration of product-market positioning as PRODUCT-service "products" become products in their own right. Industrial investment in "hardware" products that provide "means to ends" and convert CapEx business models into OpEx (capital light) business models is becoming commonplace as product-SERVICE products replace traditional product-services and offer organizations an opportunity to direct cash flow into intangible core assets which have greater impact on growth. Repositioning the PRODUCT-service as a product-SERVICE, repositioning the organization in the value chain network, the organization within the industry, and becoming part of an emerging or reimagined industry are all choices becoming available to growth-seeking companies. An understanding of the powerful influence that producibility makes available to companies is a starting point.

There are some interesting examples. Siemens excels in product design and factory automation: it already has experience in digitizing the entire life cycle of an industrial product, from design to fabrication, the "connected-producibility" approach extends the activity into management of the entire value chain network where Siemens is active with connectivity links through digital thread and digital twinning, both of which are integral to a connected value chain network; the company offers performance management services to customers who have purchased (product-SERVICEs) from Siemens. Bosch Australia, another example, has repositioned itself in entirely new networks following the departure of automotive manufacturing in Australia, where Bosch supplied braking and stability control systems to all three carmakers, Toyota, Ford, and Holden; the company now produces about 120 million power diodes each year used in vehicle alternators; 20 per cent of the total global market and production rates were on the rise. Bosch has entered other markets; demand is rising from a range of companies in medical devices, transportation systems, consumer goods, and the food industry as they remodel their own plants with robotics and automation. A team of Bosch in-house engineers design and build specialist robotics and visual inspection equipment, anticipating the expansion of the "smart factory." Bosch Australia is also making advances in agricultural technology and the design and construction of specialist equipment in that industry. Such changes require an understanding of the connected, collaborative value chain networks in these new applications from RD&D through to end-users and the network stakeholders (Evans 2018). Boundaries between suppliers, producers, distributors, and customers in the value chain are changing and will continue to undergo change; automotive manufacturers are acquiring *knowledge-b*ased software companies to provide "infotainment" centers within the vehicles. They are undertaking similar strategies with vehicle sharing and mobility service organizations (PwC, 2015), impacting on the supplier/distributor/customer *relationships* of Toyota/Uber, VW/Gett, and GM/Lyft.

The "rate of change" is having serious impact on response capabilities. General Electric has experienced massive change in its targeted markets in the past 18 months, resulting in two changes of CEO and a massive rethink on its portfolio of industries. In announcing the second incumbent in the period (Culp's appointment on 1 October 2018), *GE cited the way he transformed Danaher from "an industrial manufacturer into a leading science and technology company.* Under his watch, GE said, Danaher "executed a disciplined capital allocation approach, including a series of strategic acquisitions and dispositions, a focus on investing for high-impact organic growth and margin expansion, and delivering strong free cash flow to drive long-term shareholder value" (General Electric news announcement October 2018). It is interesting to note the implications of these comments: they are suggesting that the product-SERVICE approach whereby software solutions rather than hardware solutions (plant and equipment) is becoming an important response capability, led by RD&D and process applications, to offer a faster return to stakeholder interests. This example emphasizes the need for "organizations" to maintain an ongoing review of *response capabilities* and being prepared to make major changes – quickly!

Producibility: An Operating Infrastructure Network

Industrié 4.0 makes possible the connection of industries, business networks, businesses, and business processes offering real-time data availability across the spectrum of business processes and activities. The increased trust among network *Value Chain Network 2.0* partners has increased the levels of collaboration and transparency that facilitate data collection and analysis at relevant locations in the network that in turn increases the managerial efficiency of the network and its component organizations. An emphasis on *stakeholder value management* provides a level of governance that ensures optimal and equitable financial relationships throughout the network.

Consider the future possibility of managing a product-service through the value delivery cycle of a series of activities from product-service conception through to its retirement. For example, consider a "white good" – a refrigerator. Manufacturing capabilities are available such that the refrigerator can now be built to meet individual specifications at acceptable levels of cost. The end-user can use Internet facilities to "discuss" the product characteristics with designers to ensure a 100 per cent "fit4purose." The design specification is comprehensive, detailing size, capacity, required features, materials, and production process specification; these data will prescribe the manufacturing route the product will follow. Here the principles *of high value/low volume* versus *low value/high volume* will be applied; customized products (likely to be ordered for inclusion in a new house or possibly a restaurant) will be manufactured as high value/low volume products, and those with less differentiation will be produced using low value/high volume facilities. When ready for installation, the refrigerator will be collected by a specialist distribution/installation organization that may well manage warranty and maintenance activities, and these are very likely to include an ongoing arrangement to update the specification of the product; alternatively, specialist insurance organizations will offer the owner a comprehensive policy that, in some cases, may include a replacement product, should the owner's product be beyond repair. The "connectivity" facility of Industrié 4.0 offers the ability to replenish the refrigerator inventory (contents) on a "use-by" basis: online service is a committed feature of the retail product-service mix, and indications are that this will be extended; Waitrose (part of the John Lewis Partnership, UK) is trialing an extension of the delivery service with an unpacking-putting-away feature ("Waitrose unpacks groceries while you're out" 2018). As the refrigerator becomes obsolescent (and then obsolete), it will be removed and dismantled for component recycling. It is very likely that the B2B customers will purchase an intangible product output rather than the tangible product (a product-SERVICE – rather than a PRODUCT service), i.e., purchase the cool/cold output of the refrigerator rather than the refrigerator, creating a serviceability activity.

The automobile and aerospace industries are moving towards this concept. The automotive industry has moved towards mass customization, offering selective customization facilities with rapid delivery time, some offer a completely customized product! The industry is aware of changes towards vehicle ownership as indicated by the partnerships entered with mobility services by major manufacturers (Toyota/Uber, VW/Gett, and GM/Lyft) cited above. Aerospace offers a product-SERVICE activity; airlines no longer purchase hardware items, such as engines, rather they purchase the capability of the supplier to provide reliable engine performance (power output, maintenance, and ongoing performance monitoring and management). The concept of the digital twin was introduced over two decades ago. Today's technology now makes it possible to feed operational data back into the complete value chain, creating the opportunity to continuously optimize both product design and production activities in a closed-loop decision-making process that touches all three components – the product digital twin, the production digital twin, and the performance digital twin.

Producibility therefore has the ability to become the networked organizations' strategic and operational infrastructure, a process whereby the product-service-design process is integrated with the design of manufacturing processes and the subsequent operational processes of physical distribution, service support management, and, increasingly, remanufacturing.

Producibility: A Seamless Activity

To be effective as an integrated system, the primary requirement is for the capability of the system to respond to the expectation requirements of all of its stakeholders (see Fig. 7.1).

Four activities form the basis of producibility: first, *identifying market opportunities and network structures* that are most able to respond to them or will be able to do so with additional capabilities. Currently we are witnessing interesting related events in the automotive industry. While Ford Motor Company competitors have embraced the idea of "mobility plus," Ford have been slow to respond to changing market expectations for sharing rather than owning vehicles, for more exciting vehicles (SUV options), and an array of software-based options. Collingridge (2018) suggests the company has lagged behind its competitors in European markets quoting comments by Morgan Stanley; "Ford has persisted with a tired line-up of cars, while only belatedly latching on to the huge popularity of

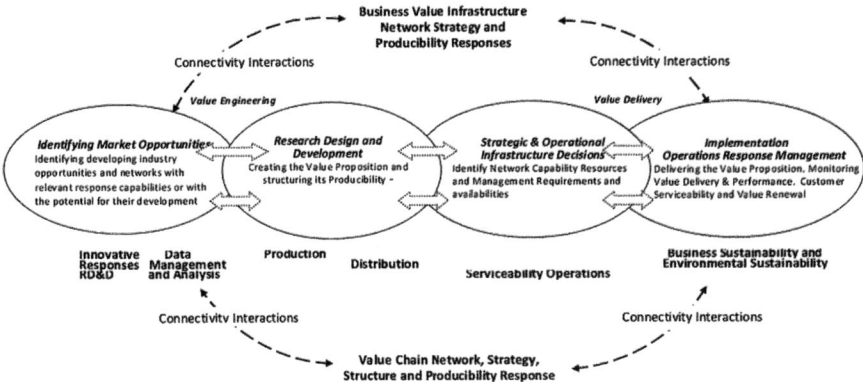

Fig. 7.1 Producibility: A sequence of seamless value creating activities

SUVs." Its market share in 2017 was 11.3 percent; in 2008 it was 14.5 percent, and 2001 17.2 percent. "Commentators suggest its manufacturing activities in Germany and Spain is underutilized, operating well below capacity. Morgan Stanley opines; "The segments they are exposed to are in decline. The Ford brand has questionable to minimal value in most segments in Europe. They have too many dealer[s] … … too many nameplates … …too much capacity … …too many compliance costs." The Economist ("Rocky road ahead" 2018) adds the problems beginning to occur with the impact of tariffs. Supply chain partners are forecasting significant problems creating price increases throughout the value chain. The Economist ("Rocky road ahead" 2018) reports a Peterson Institute forecasts that tariff increase costs could inflate the price of a typical small car by between $1408 and $2057. Clearly by *identifying market opportunities and network structures*, both large and small organizations are able to (and should) consider product-service, production processes and locations, and serviceability concerns holistically, a primary feature of the producibility model.

Second, research design and development: creating and communicating the value proposition. It has been argued that creating the value proposition is one of the most important activities an organization can do; in which case structuring its infrastructure is alongside it. A value proposition must identify for the target customer how *customer expectations* are to be met. It is an explicit statement of customer value attributes (both value criteria and how acquisition costs are minimized for the customer). It includes both product or service characteristics *and* service support features, and it becomes the basis of a set of agreed performance processes and metrics for network partners. Haier's market value proposition has gradually broadened since Zhang Ruimin became CEO in the mid-1980s. The company first took the role of a category leader, maintaining top market share because of its reputation for quality in China. It then became a customizer (adapting its products to customer demands) and a solutions provider (helping consumers manage issues like water quality and home design). Haier now sells not just home appliances but related services, adapted to consumer demand in China and, increasingly, other markets. Haier excels by developing differentiating capabilities.

Third, strategic and operational infrastructure decisions: identify network resource capabilities and management requirements and availabilities. Industrié 4.0 represents a major shift in operations management capabilities that has been driven by four major coinciding events: an extraordinary rise in data volumes, cost-efficient computing power, "connectivity," and the emergence of analytics and business-intelligence management capabilities. Add to these capabilities, *collaborative relationships,* whereby value chain networks combine the distributed expertise of organizations into collaborative *product-service-market-value propositions.* These organizations commit transparent collaborative developments (RD&D, joint ventured process applications, serviceability program, and value remanufacturing). The result is a highly "value competitive network" structure. Structured cyber-physical work flows ensure optimal capital investment, working capital management, margin management, and free cash flow at both strategic and operational levels of activity. *Product life cycle management (PLM)* is an information management system that can integrate data, processes, business systems, and, ultimately, people in an extended enterprise. PLM software allows you to manage this information throughout the entire life cycle of a product efficiently and cost-effectively, from ideation, design, and manufacture through service and disposal, thereby creating a "connected" strategic and operational producibility infrastructure. PLM can be viewed as both an information strategy and as an enterprise strategy. As an information strategy, it builds a coherent data structure by consolidating systems. As an enterprise strategy, it lets global organizations work as a single team to design, produce, support, and retire products while capturing best practices and lessons learned along the way. It empowers your business to make unified, information-driven decisions at every stage in the product life cycle. The automotive industry is an example of the application of product life cycle management and is a digital approach that connects people and processes, across functional silos, with a digital thread for innovation. "Teamcenter" (a Siemens product) and "Predix" (General Electric) provide the means to solve the challenges required to develop and maintain successful products throughout the cycle from product development, operational service, and, eventually, remanufacturing or retirement by building a digital twin of the hardware product, providing the supplier and the end-user with the facility of managing strategic product-service development and operational performance.

Fourth, implementation operations response management: delivering the value proposition monitoring, value delivery and performance, customer serviceability, and value renewal. In an operations management context, collecting, collating, and managing data through business intelligence cloud-based analytics systems allow a manufacturer to quantify exactly how, where, when, and why its products are being used. This facility creates insights that drive improvements across every business function, from R&D and design, through production and distribution, to service and maintenance. Customers become involved in the direction of the business, rather than the vendor's marketing and sales team. Competitive value advantage can be enhanced by the application of telematics that analyzes customer usage and performance data. Servitization, as a component of the operational offer, provides companies with the capability to forge stronger, deeper customer relationships, based on long-term, high value propositions, and a significant competitive differentiator. Customers are placed at the center of a business' growth strategy, with data generating meaningful insights to create a superior customer experience and more responsive, agile internal processes. Caterpillar is an example of how services have played an important role in the business for more than five decades, predominately asset finance-based. The big change in more recent years has been a move towards capability, rather than machinery. Understanding what capabilities our customers' need and would value would be essential without this insight; it can be difficult to identify the best areas to focus attention on when implementing operations response management and the servitization content of the value proposition.

Figure 7.1 provides evidence of the logic and practice underlying producibility. Figures 7.2, 7.3, 7.4, 7.5, 7.6, 7.7, and 7.8 explores the processes by which the producibility infrastructure is developed around the value creation, production, and delivery sequence to identify and create the value, produce the value, communicate the value, deliver and service the value, and renew the value while maintaining business and environmental sustainability.

Producibility: A Response Capability Networked "Producibility Infrastructure"

Figure 7.2 details the process of deriving the producibility infrastructure. Parallel paths of research are required. From published experiences it appears that an understanding of the macro and micro business environments is essential. General Electric experienced difficulties when attempting to market a "standard" healthcare product range in Asian markets but very soon found that it was overpriced and over

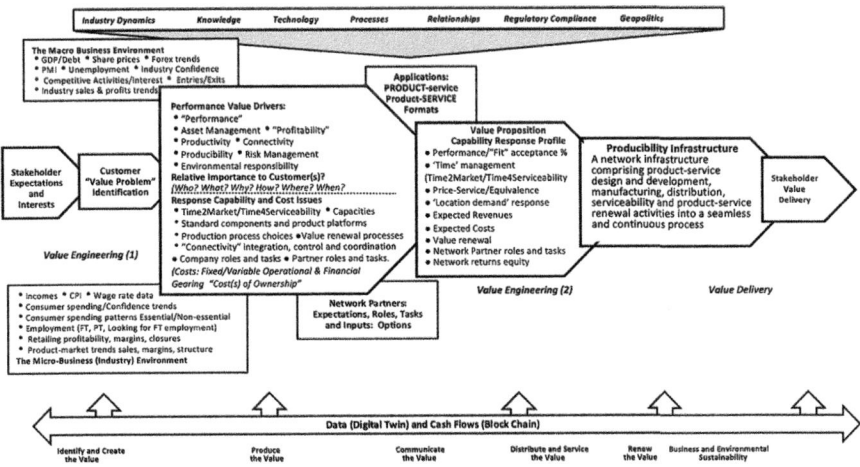

Fig. 7.2 Producibility: A response capability networked "producibility infrastructure"

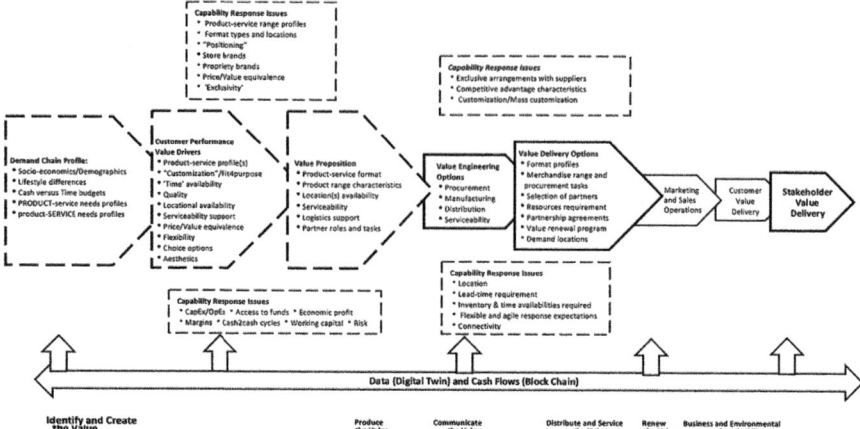

Fig. 7.3 Producibility: Using demand chain analysis to identify and develop producibility infrastructure response to meet stakeholders' expectations

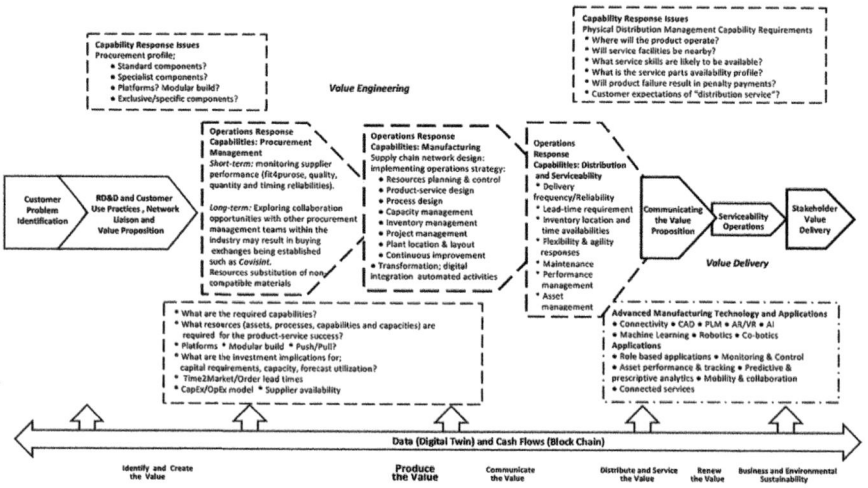

Fig. 7.4 Producibility: Producing the value

"featured." Research into levels of funding, "affordability," and medical practitioners use practices identified there to be an opportunity for healthcare equipment but with products priced at lower levels and with fewer features. GE liaised with local medics and manufactured locally. The problems were resolved; local stakeholder expectations were identified and incorporated in product-service design; local manufacturing resulted in acceptable margins; a value proposition reflecting end-user (medical practitioners) expectations proved to be relevant and acceptable.

Relating this example to Fig. 7.2 identifies activities that, if they had been considered, would have prevented the initial outcome. Reviewing the "indicators"

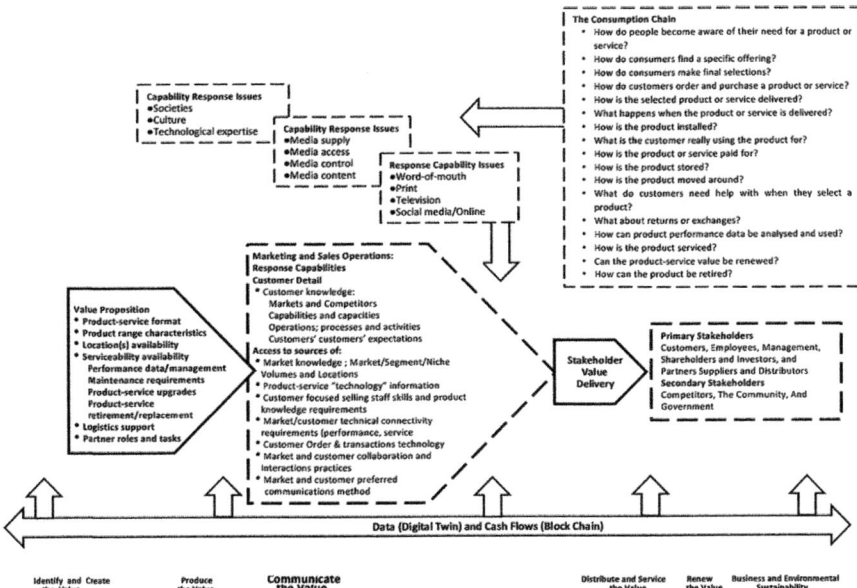

Fig. 7.5 Producibility: Communicating the value

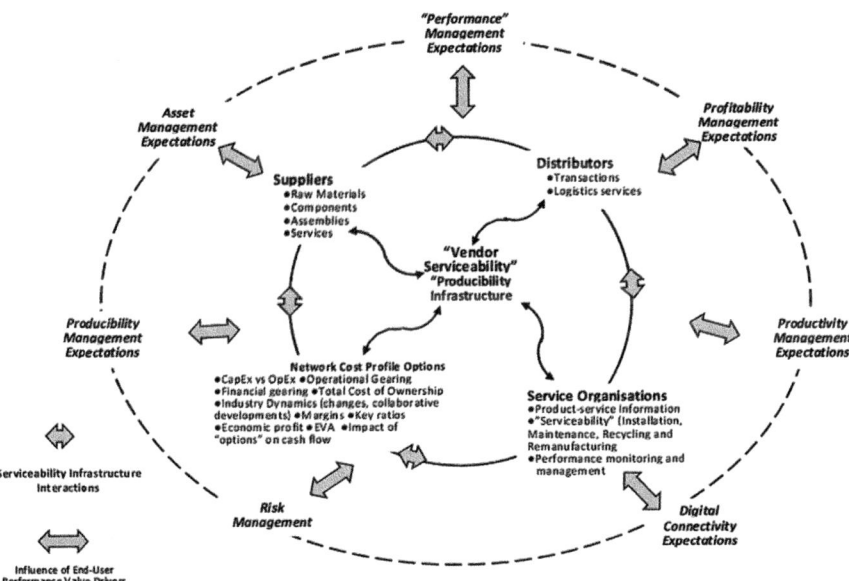

Fig. 7.6 Producibility: Serviceability – managing the total product-service/service component mix

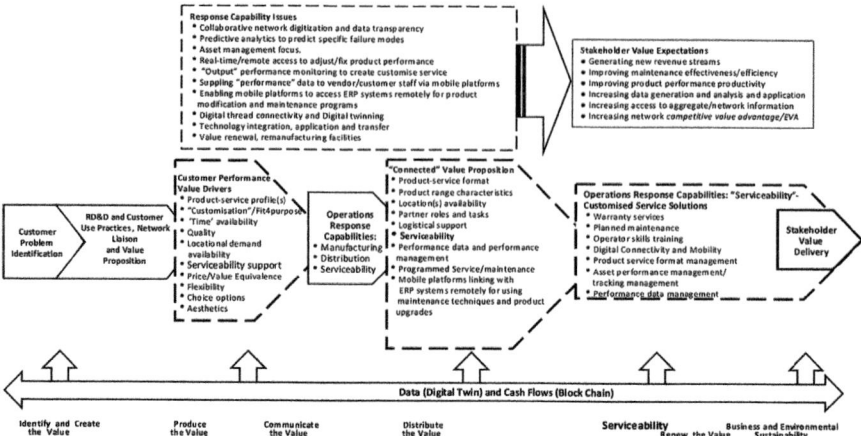

Fig. 7.7 Producibility: Serviceability

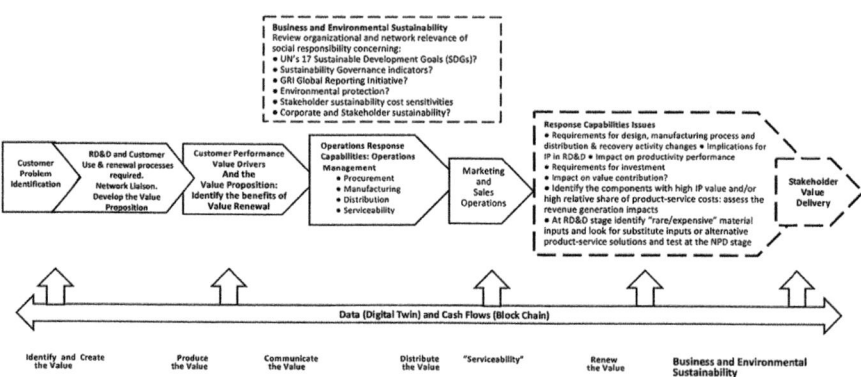

Fig. 7.8 Producibility: Managing sustainability

identified in the macro and micro business environments would have indicated a number of planning response considerations such as; availability of funds to purchase equipment, end-user use of the product-service (costs and frequencies etc.) that are critical issues to be resolved prior to deciding upon a full analysis of the opportunity. Research of the appropriate macro environment topics would have indicated growth and NPD introduction viability. A comparative review of *industry dynamics* across product-markets can indicate the application of developments in any of the important topics and thus indicate the level of product-service sophistication being applied, expected, and affordable. A study of network activity and comparative developments would identify workable structural options. PRODUCT-service and product-SERVICE applications are also an indication of product-market sophistication and of an acceptable product-service format.

Given the output of the suggested research, the identification of *performance value drivers,* the *value proposition response,* and the structure of *producibility infrastructure* can proceed with confidence.

Producibility: Using Demand Chain Analysis to Identify and Develop Producibility Infrastructure Response to Meet Stakeholders' Expectations

Tesco's "Fresh & Easy" project was meant to be a controlled experiment, a low-risk way to discover whether there was room in the US grocery market for a new style of store. At launch, in 2007, Tesco spoke about initial start-up costs of £250 m a year and breakeven at the trading level within 2 years; what has happened is that a cool £1.3bn has been spent on infrastructure, and there has never been a sniff of profit. Add up the losses and the capital sunk, and some £1.6bn has been thrown at the venture. Despite the boasts about extensive market research – executives were sent to live with American families to study their habits and preferences – Tesco was surprised to discover Americans prefer their fruit and vegetables without the cling film. The starkly designed stores also received a thumbs-down, and Tesco was obliged to fit doors to fridges and generally soften the look. Those adjustments were said to help, but Fresh & Easy never got close to achieving the sales densities of veteran west-coast retailers such as Trader Joe's and Whole Food Markets. As for the supposedly low-risk nature of the experiment, that claim always looked hollow. As losses mounted, Terry Leahy, who launched Fresh & Easy, and Phil Clarke, his successor as chief executive, argued the fresh-food-at-low-prices model would work only if done on a big enough scale. Tesco had to build a massive support network, including two distribution centers and a huge kitchen, before knowing whether the stores would ever earn a dime.

The plans came after 2 years of intensive research that involved, as mentioned above, Tesco sending senior executives from the UK to live with Californian families, to assess the way they shopped and ate and to build secret test stores. But their research proved faulty. Almost every aspect of the shops, from their interior decoration to the pack sizes and self-serve tills, have been changed. Fresh & Easy also opened as the subprime mortgage crisis and subsequent economic downturn took hold, hitting US consumer confidence and spending power. The chain also faced opposition from US trade unions (Pratley 2012). Clearly there was a lack of planning for so many problems to be encountered.

Figure 7.3 suggests a detailed approach to the research and analysis of the demand chain. As the Tesco example suggests, identification of *customer performance value drivers,* the *value proposition, value engineering activities,* and *value delivery options* are critical. Researching the *demand chain profile* appears as an important activity in all industrial sectors and product-service format offers (i.e., PRODUCT-service, product-SERVICE applications). There are important issues to be considered. Clearly the importance of time2market and time4service response in determining competitive value advantage. There are other important factors; producibility options may require consideration of options that are incompatible with the current business model, such as location of demand points that are currently not covered, new manufacturing practices, increases in inventory service cover because of distance, and therefore time4service response expectations, unexpected customer behavior responses, etc. Demand chain analysis can identify the impact of customer-led activities on operating and financial capabilities.

Producibility: Producing the Value

Producibility is an extension of the *design for manufacture and assembly* (DFMA) and *total cost of ownership* (TCO) concepts. DFMA has been mentioned in the American business literature since the 1940s. Boothroyd and Dewhurst (1990) and Ulrich and Eppinger (2008) have developed the DFMA concept and have demonstrated the benefits of product simplification and application of new costing methodologies (activity and attributes-based costing), in defining the approach to engineering management during the late 1980s. Each is a proven business method. DFMA and TCO are low-risk applications; both need to be "institutional" for global companies' operations.

DFMA has been practiced with notable success for some time, integrating the product-service design process with manufacturing operations to create *strategic effectiveness* and coordinating both the design and manufacturing processes to achieve production *efficiencies* (i.e., the manufacturing model, quality specifications, volume, and delivery targets at target costs). John Deere and Harley Davidson used DFMA over the years to achieve enviable results, including cost reductions of 50 per cent, shortened product development cycles in the neighborhood of 45 per cent, and part count decreases of nearly half. Whirlpool management deemed DFMA as central to the firm's strategy to be the "number one cost leader" in all of its product categories at each of its price points. Years into the program, DFMA is now part of the company's stage-gate process for product development and is used equally to redesign existing products as well as to optimize new product designs. Using DFMA (aka product life cycle management) software engineers are able to evaluate individual products part by part, in addition to documenting the assembly process step by step. The software helps generate three key Pareto charts (cost, part count, and assembly time) which establishes a baseline that allows the team to measure its success, not to mention, and identify the parts and processes where there is the greatest opportunity for efficiency improvements. Whirlpool's focus on the importance of performance – efficiency (part count), cost, and time – suggests these as important drivers for organizational success. It can be argued DFMA was the first move towards an integrated marketing/operations business model with the objective of increasing customer satisfaction significantly but, at the same time, containing manufacturing costs. DFMA has been augmented by the "circular economy/value chain" concept, whereby by core product, components are identified as remanufacturing candidates and production processes established to ensure cost-efficient recycling and subsequent remanufacturing. It can be concluded that Design for Manufacturing and Assembly (DFMA) techniques, which are used to minimize product cost through design and process improvements, are essentially an optimization process. DFM creates a product with minimal parts but is capable of the required output efficiency. DFA minimizes the number of assembly operations. Boothroyd (1994) gave an example of a DFMA project by Ingersoll Rand: the project development time was reduced from 2 years to 1 year; the design team reduced the number of parts in a compressor radiator and cooler assembly from 80 parts to 29, decreased the number of fasteners from 38 to 20, reduced the number of assembly operations from 159 to 40, and reduced the assembly time from 18.5 minutes to 6.5 minutes.

Industry dynamics have provided a combination of capabilities with Industrié 4.0, Value Chain Network 2.0, and the reinforcement of a stakeholder value management focus that enable an extension to the DFMA model that is focused on manufacturing/production efficiencies. The connectivity and collaboration now existing in networked organizations together with the pioneering principles introduced by Boothroyd and others can now be applied to the entire value chain. The *New South Wales Government Advanced Manufacturing Industry Development Strategy* (Australia) recently commented upon the fact that: "Advanced manufacturing refers to highly specialized products and processes in areas such as medical technology, biopharmaceuticals, mining, agribusiness, aerospace and defense, where expertise is the source of competitive value advantage. It identifies the process by which knowledge-intensive value is added in both the pre- and post-production phase in research and development, concept design, planning, engineering and after-sales service." And, "As much as 41 per cent of global trade is now in intermediate goods, for example, components and research. As a result, there is a growing market for advanced manufacturers that not only create finished products but add value at every stage of and within the global value chain" (Advanced Manufacturing: Government of New South Wales).

Haier's focus upon four differentiation capability characteristics has relevance for developing a producibility infrastructure (Leinwand et al. (2016)). It will be recalled that these are *customer-responsive innovation,* initially seen as a *customizer,* focusing on local customer needs but moving into the role of a *solutions provider* (helping consumers manage domestic issues such as water quality and home design). *Operational excellence* is pursued by creating a workplace culture of internal competition through a zero-defect continuous improvement. *Management of local distribution networks* has resulted from overall market size that has created a decentralized value chain. *On-demand production and delivery* is based upon a "pull-oriented distribution system" and a "zero-based" inventory management policy. This has been achieved by "being a consistently coherent and capable company: staying true to its core identity as a company dedicated to solving problems for consumers, while continually reinventing itself with imagination and verve" (Leinwand et al. 2016). Haier demonstrates a requirement for success that is based upon a policy of identifying a framework of core capabilities that underly its strategic responses. These have clearly been successful by modifying them to meet changing customer needs and competitive challenges.

Digitization, in the shape of *Industrié 4.0,* has resulted in an extensive application of "connectivity" across the manufacturing and servitization delivery of product hardware and serviceability software. The *value chain* has matured, and *Value Chain Network 2.0* has enhanced collaboration, coordination, and communication among value chain partners. *Stakeholder value-led management* is becoming important to enlightened CEOs who are publicly mindful that they are answerable to more than just shareholders; their role is one of being answerable to a much broader group as pressure from customers, suppliers, employees, government and other regulators, environmental groups, as well as shareholders make them answerable to "everyone" – the outcome is they must deliver a balanced performance.

Figure 7.4 reflects these developments and, based upon industry observations, suggests the development of a network producibility infrastructure model.

Producibility: Communicating the Value

Digitization has transformed the relationships between suppliers, OEMs, and end-user customers; the value proposition identifies details of ongoing serviceability benefits available post-transaction such as product-service performance data and management, product-service upgrades that occur digitally, and advice on cost-effective replacement. Traditionally the value proposition identified benefits targeted customers could obtain from suppliers; the most sophisticated organizations included tasks and roles of suppliers that were necessary in achieving customer satisfaction. The role of the value proposition within the producibility infrastructure network includes an effort to create liaison, transparency, and collaboration between vendor suppliers, and with the OEM vendor, between vendor and customers, and vendor with customers' customers. Digitization and connectivity have the capability of pan-value chain network communications – given the decreasing "lead times" being expected (and now feasible and viable) in the demand chain "total-connective-network-communications" may well be developed in "hi-tech-specification" industries.

There is a larger community with "value expectations." It will be recalled that Chap. 1 commented upon the increasing practice of adopting a wider perspective of "community" arguing for two tiers of interested parties: primary and secondary stakeholders. Primary stakeholders are customers, employees, management, shareholders and investors, and partners suppliers and distributors. Secondary stakeholders comprise competitors, the community, and government. Primary stakeholders welcome open, close, and ongoing relationships with organizations, involving the exchange of creative ideas, and innovation, management approaches to change, and performance improvements. The interests of secondary stakeholders have specific focus: competitors monitor progress and have interests in opportunities for collaboration (typically co-opetition, licensing patents, and using any excess capacity they have to manufacture and distribute competitors product-services); the community is interested in employment opportunities and expansion of revenues (and, therefore, in local taxes); government has interests that include employment, innovation, exporting activities, and a contribution to GDP. See Fig. 7.5.

The automotive market is transforming to an economy-of-services, subscriber-based revenue model, and positive ownership experiences. The connected vehicle communicates a wide array of data to help OEMs make better decisions, improve safety, and increase profitability. OEMs are looking for a competitive edge while delivering on customer expectations of value-added services.

The digital thread is the concept of seamlessly integrating information throughout the value chain from planning to disposal, and as such has been used by capital goods and expensive consumer durables manufacturers to establish an ongoing connected communications link with the equipment hardware. GE, Siemens, and Bosch receive data from the hardware product and monitor the overall performance and the state of serviceability of the hardware product components. Data analytics communicates information to the owner of performance efficiencies, service and maintenance requirements, and upgrades available. The Covisint platform enabled Hyundai to build new connected vehicle services rapidly, allowing the OEM to communicate continually with customers throughout the entire life cycle of their vehicle ownership. Hyundai leverages Covisint's foundational technologies to create a composite view of the customer and shares information across the extended vehicle ecosystem, in a secure and scalable way.

Producibility: Serviceability

The attention of management concerning the revenue/profitability benefits accruing from a focus upon service aspects of the product service portfolio was identified some time ago (Vandermerwe and Rada (1988)). They argued that an increasing number of corporations throughout the world were adding value to their core corporate offerings through services; and the trend was becoming apparent in almost all industries; it was customer demand-driven and perceived by corporations as sharpening their competitive edges by offering fuller market packages or "bundles" of customer-focused combinations of goods, services, support, self-service, and knowledge. At that time, services were beginning to dominate.

The authors Sandra Vandermerwe and Juan Rada termed the activity "servitization" which was regarded as a powerful new feature of "total market strategy" being adopted by the progressive companies. It led towards new relationships between them and their customers.

Giving many real-life examples, the authors assessed the main motives driving corporations to servitization and point out that its cumulative effects are changing the competitive dynamics in which managers will have to operate. The special challenge for top managers is how to blend services into the overall strategies of the company. It was suggested by Neely et al. (2013) that the approach extended the notion of the product towards becoming a *product-service system* which designs, builds, and delivers "value-in-use."

The "collation" of Industrié 4.0, Value Chain Network 2.0, and the reinforcement of a stakeholder value management make for a positive and perhaps more focused approach. To distinguish between servitization and a solution-based approach that emanates from customer-focused RD&D, we suggest "serviceability" is used to describe the management of the total product service and service component mix. Figure 7.6 illustrates the concept.

Serviceability is a complex role of managing the product-service portfolio and the service component mix which commences with customer liaison and RD&D. It requires current and ongoing knowledge of the relevant customer operational activities. This is essential to ensure the value creation, production, and delivery are optimal and that not only is the customer level of satisfaction optimal but so too is the stakeholder performance outcome. For this to be achieved, the serviceability capability includes close collaborative working arrangements with suppliers, distributors, the service organization(s), as well as customers. The success of the servitization strategies are reinforced by a planned overall serviceability approach that is designed into the product-service and then the value proposition that evolves from customer liaison, to be integrated within operations activities, product-service, and value production and delivery that effect customer value delivery expectations. Figure 7.6 argues that successful organizations demonstrate flexibility (even agility) when opportunities are identified. Returning to recent (2018) events at General Electric, there is a suggestion that existing response capabilities were considered to be adequate when the acquisitions made in earlier years were undertaken and, subsequently, following the divestments of the 2017–2018 John Flannery tenure. It would appear that for serviceability to be successful, the performance value drivers of customers should be the point of departure for strategy (Slywotzky and Morrison 1997). Ongoing tracking of developments of the relevant industry dynamics is clearly essential as is the identification of synergies; integration of specific activities, for example, technology management; and relationship (collaborative) management, which result in significant change in product-service format, i.e., the expansion of the PRODUCT-service into product-SERVICE "products." The *Cambridge Service Alliance* project (https://cambridgeservicealliance.eng.cam. ac.uk/) suggests "the top five technologies," selected by a group of capital equipment manufacturing experts and a group of academics, that will influence servitization and that were considered by both groups as important (Dinges et al. 2015):

- Predictive analytics to predict specific failure modes
- Remote combinations to adjust/fix products remotely
- Consumption (on-going production) monitoring to create customer-specific service offerings
- Pushing information to employees or customers via mobile platforms
- Mobile platforms to access the enterprise resource planning (ERP) system remotely for maintenance techniques, product details, etc.

Figure 7.6 suggests these form part of an ongoing analytical activity in order that response capabilities remain "in-tune" with the rapid change occurring. It also argues that the evidence suggests successful approaches to servitization requires a broader approach that benefits from inputs from customer liaison *and* liaison with supplier, distributor, and service organizations. Figure 7.7 considers serviceability as a significant value generating component within the overall the producibility infrastructure model.

Producibility: Managing Sustainability

What is sustainability? Is it corporate profitability or corporate longevity? Is it environmental sustainability, simply the sensible use of raw materials? Stakeholder value management requires a more comprehensive and descriptive statement: product-service design that identifies candidate components that may be designed to have multiple life cycles and manufacturing and distribution processes that are designed with recycling and remanufacturing (value renewal) in mind. It is important that sustainability is understood within the context of both the organization (individual companies and their network partnerships) *and* the resources environment if effective response capabilities are to be developed. The Ellen MacArthur Foundation proposes a circular value chain network: " one that is *restorative and regenerative by design and which aims to maintain products, components and other inputs at their optimal value to end-users at all times, distinguishing between technological and biological cycles"* (Ellen MacArthur Foundation website: www. ellenmacartherfoundation.org).

Of interest in this chapter is the circular value chain. The circular value chain differs from the traditional linear value chain which typically is a one-way linear flow of materials through industrial systems: raw materials are extracted (harvested, etc.) from the environment and transformed into products and eventually disposed of. In the linear system, such eco-efficient techniques that do exist seek only to minimize the volume, velocity, and toxicity of the material flow system but are incapable of altering its linear progression. Some materials are recycled but often as an "add-on decision"; no decisions to accommodate this activity have been made at the design stage and often as a response to regulatory decisions and mandates. This is in effect not recycling but downcycling as it results in a downgrade in its quality and limits its subsequent usability.

The work of the foundation has strong implications for the response capability model because much of what they advocate will require changes in current practices. Possibly the largest is a change in attitude towards moving away from the linear production processes of manufacture-sell-dispose, to the circular model of manufacture-sell-recover-remanufacture. This will require further physical changes to manufacturing and distribution facilities. Some examples will indicate the changes occurring and the results being realized:

Heineken is pursuing circular manufacturing practices across its entire value chain – thinking holistically; from water treatment – energy conservation –composting yeast – recycling bottles.

Philips introduced energy efficient technology (LED street lighting) by offering a product-SERVICE approach to enable municipal authorities to take advantage of energy savings without incurring up-front investment.

Renault (Choisy-le-Roi) re-manufactures some 200,000 components. It suggests these to be 30 – 50 per cent less expensive than new. Re-manufactured parts have the same warranty and QC process. The plant uses;

> 80% less energy
>   88% less water
>   92% less chemical products
>   70% less wasted production
>   No landfill services used

Parts being designed with re-manufacturing in mind – disassembly and recyclability. Product interchangeability is an objective.

Caterpillar (CatReman) suggests re-manufacturing is " … not about using less and less material but rather creating components that can be remanufactured a number of times. With fixed costs at 35% and variable costs (materials) 65% remanufacturing offers increased profits. For example, Caterpillar designs engine blocks with removable sleeves – reducing the time and cost of a reboring process". (Ellen MacArthur Foundation website: www.ellenmacartherfoundation.org)

Liam (Apple) a manufacturing process application that identifies and extracts serviceable components from returned Apple products.

Figure 7.8 considers briefly how the implications of managing for sustainability will impact the producibility model.

Convergence …

Industry Week (Hitch 2018) published research findings (based on a sample of 1000 manufacturers over a year) following up on a report by the World Economic Forum, suggesting seven out of ten manufacturers fail in pushing initiatives in data analytics, artificial intelligence, and additive manufacturing past the pilot phase. This was confirmed by their own research that indicated manufacturers are encountering difficulties finding "ways and means" of benefiting from the "consolidation." The findings are interesting. Organizations that have persevered have 20 to 50 per cent higher performance and create a competitive value advantage. This has been achieved by creating agile teams with intra- and inter-organizational sphere of interest, analytics, IIoT (Industrial Internet of Things), and software development expertise. They have deployed a common data/IIoT platform and have produced case study evidence of the efficacy of the approach. "They are thinking scale, acting agile, and resetting the benchmark" (Hitch 2018). Industrié 4.0 has introduced several technological and connected activities that are rapidly being adopted by these organizations. The holistic process activities of the value chain network are ideally structured to apply the relevant aspects of the "convergence" to increase the efficacy of the principles of collaborative leverage and are achieving impressive results as evidenced in the research reported above. The convergence of technologies has expanded the production and productivity options available from flow line mass production and batch manufacturing (mass customization) often to a batch size of one. Data management/analytics (collection, analysis, sharing, and collaborative decision-making) is increasing "organizational effectiveness and efficiencies"; component suppliers locate closer to their assembler customers, abandoning the one-time advantage of low costs and increasing quality reliability and time4response.

Figure 7.9 considers the impact of implications and concerns of producibility on stakeholder value-led management and the responses of the other response capabilities. Stakeholder value-led management requires the optimal performance that a seamless operational infrastructure can provide. Accordingly, the producibility response capability should reinforce an interactive approach:

Producibility: Comprises developing/changing a network, an operating infrastructure – the fusion of design and development, of manufacturing and distribution, and of serviceability and product-service renewal activities into a seamless and continuous process to contribute to a network value advantage.

> *Strategic activities involve:* Reviewing and restructuring activities and processes in the value chain network to improve network value engineering and/or delivery. And together with network partners evaluate major changes to the value proposition to reflect customer product-service requirements (for example transforming PRODUCT-service based products into product-SERVICE format); exploring collaboration opportunities with other procurement management teams within the industry that may result in buying exchanges being established (such as *Covisint*), resources substitution of non-compatible materials.

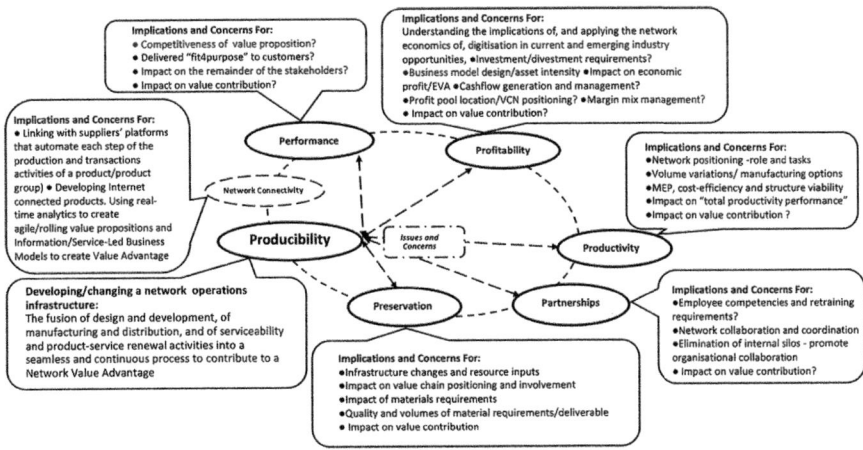

Fig. 7.9 Interfaces and interactions: Producibility

Profitability: Understanding the implications of and applying the network economics of digitization in current and emerging industry opportunities is critical as is adapting to the investment/divestment requirements a "convergent" industry environment requires. Changing business model design/asset intensity alternatives and the implications of CapEx/ OpEx alternatives may also be necessary. Exploring the impact of economic profit/EVA on strategic and operational activities and of the implications of cashflow generation and management. Finance and investment management will be exploring profit and productivity pools for "indicators" that point towards a change in location and/or VCN positioning. Overall implications of change for margin mix management and the Impact on value contribution.

Productivity: Strategic operations will require an "infrastructure" that optimizes utilization of network assets. The "utilization" of *relevant* assets, capabilities, capacities, processes, and relationships positioned in *relevant* locations *to* achieve the benefits of partners' organizations' network positioning – role and tasks. Productivity planning will be cognizant of maximum efficiency point (MEP), cost-efficiency, and structure viability. In the short term, operational management may work with a range of component, module, and assembly manufacturing options/services to cope with volume variations (low volume variants within the model range).

Partnerships: These are critical for both strategic operations and operational management. For the long-term designing network (intra- and interorganizational structures) and establishing network structure, roles and tasks are a main focus and the evaluation of advanced manufacturing, robotics/cobotics, AI, and machine learning and technology yet to appear. Operational management is likely to have to cope with ongoing changes such as identifying changing network employee expertise requirements, compliance with ethical employment policies, and agreeing remuneration equity within the network.

Preservation: Strategic issues will require management to understand and apply the developing social responsibility characteristics of Environmental Sustainability Corporate and Network Sustainability and the Circular Economy/Value Chain. This is accompanied by renewing the changing perspective of value and designing products and operational processes for repurposing. It is likely that many organizations will need to revisit and reimagine the GVCN Business Model. Operation concerns include infrastructure changes and resource inputs, impact on value chain positioning and involvement, and quality and volumes of material requirements/deliverable

Network Connectivity: producibility presents huge opportunities; however, to take advantage of these requires; an understanding of the implications, of and the application of, the network economics of digitization in current and emerging industry opportunities, understanding the investment/divestment requirements, business model design/asset intensity and the impact on economic profit/EVA, cashflow generation and management, profit and productivity pools, global location, and VCN positioning.

References

Agrawal, A., Dziersk, M., Subburay, D., & West, K. (2016). Design for value and growth in a new world. *McKinsey and Company.*

Barkai, J. & Manenti, P. (2011, January 12). The assembly plant of the future. *Industry Week.*

Boothroyd, G. (1994). Product design for manufacture and assembly. *Computer-Aided Design,* *26*(7), 505–520.

Boothroyd, G., & Dewhurst, P. (1990) *Product Design for Assembly.* Boothroyd Dewhurst, Inc., Wakefield, RI, USA, (First Edition 1983).

Collingridge, J. (2018, September 2). How the wheels came off Ford. *The Sunday Times.*

Dinges, V., Urmetzer, F., Martinez, V., Zaki, M., & Andy N. (2015, July 1). The future of servitization: Technologies that will make a difference. *Cambridge Service Alliance, University of Cambridge,* UK.

Evans, S. (2018, October 1). Bosch remakes itself after local car-making ends, *The Australian Financial Review.*

Fritzen, N., Heiko, N., & Wüllenweber, J. (2013). Capturing the full potential of design to value. *McKinsey and Company.*

Hitch, J. (2018, November 6). Nine smart factories lighting the way to a winning industry 4.0 strategy. *Industry Week.*

Leinwand, P., Mainadi, C., & Kleiner, A. (2016). *Strategy that works: How winning companies close the strategy-to-execution gap.* Boston: Harvard Business Review Press.

Murgia, M (2017, January 4). NHS to trial artificial intelligence app in place of 111 helpline. *Financial Times.*

Myerholtz, B., Tevelson, R., & Wood, E. (2016, January 19). The design-to-value advantage: Developing winning products with best economics. *Boston Consulting Group.*

Pratley, N. (2012, December 5). How Tesco's Fresh & Easy became stale & difficult. *The Guardian.*

PwC, (2015). Reinventing the wheel for the transformation.

Rocky road ahead (2018, August 25). *The Economist.*

Slywotzky, A. J., & Morrison, D.J. (1997). *The profit zone: How strategic design will lead you to tomorrow's profits.* New York: Wiley.

Ulrich, K., & Eppinger, S. (2008). *Product design and development.* New York: McGraw-Hill.

Vandermerwe, S., & Rada, J. (1988). Servitization of business: Adding value by adding services. *European Journal of Management,* Vol 6, Issue 4, Winter.

Waitrose unpacks groceries while you're out (2018, October, 5). *BBC Business News.*

Chapter 8
Partnerships: Managing Intra- and Interorganizational Relationships – The Global Value Chain Network

Abstract In recent years the pathway to revenue growth, profitability, and cash flow has been largely dominated by partnerships. These partnerships have varied depending upon the objectives of the organizations involved together with preferred routes to achieving their objectives, the time-span available to achieve the targets, the opportunities identified, the level and type of growth pursued, and the amount of risk they were prepared to accept. These criteria determine the scope of the required response capabilities. *Operational partnering* is typically cost-led and robust for as long as the advantage prevails; value builders, by definition, are long-term and involve *strategic partnering* in which individual organizations create networks of organizations, each of which contributes essential core capabilities, in order to create a value chain network that can compete with other networks successfully. The global financial crisis brought about a fiscal conservatism and a preference for minimizing investment. Large organizations are increasingly turning to "capital recycling" (whereby they are selling assets (businesses or divisions) of activities that no longer return a substantial proportion of the organization's income) and "financial engineering" repurposing tangible assets (e.g., production facilities). GM reinstated a retired plant to manufacture the Bolt EV, and intangible assets, and Nissan-Renault resurrected the Dacia brand for a low-priced vehicle. *Transformational/integration partnering* offers organizations the opportunity to change both strategy and structure in a cost-effective manner.

Keywords Partnerships · Operational partnering · Strategic partnering · Value engineering · Value delivery · Total Cost of Ownership (TCO) · Global Value Chain (GVC) · Smile curve

Introduction

In recent years the pathway to revenue growth, profitability, and cash flow has been largely dominated by partnerships. These partnerships have varied depending upon the objectives of the organizations involved together with preferred routes to achieving their objectives, the time-span available to achieve the targets, the opportunities

© Springer Nature Switzerland AG 2020
D. Walters, D. Helman, *Strategic Capability Response Analysis*,
https://doi.org/10.1007/978-3-030-22944-3_8

identified, the level and type of growth pursued, and the amount of risk they were prepared to accept. These criteria determine the scope of the required response capabilities. Figure 8.1 identifies the relationships and rationale behind the development of the type of relationship currently being used.

Operational partnering is typically cost-led and robust for as long as the advantage prevails; value builders, by definition, are long-term and involve *strategic partnering* in which individual organizations create networks of organizations, each of which contributes essential core capabilities, in order to create a value chain network that can compete with other networks successfully. The global financial crisis (GFC) brought about a fiscal conservatism and a preference for minimizing investment. Large organizations are increasingly turning to "capital recycling" (whereby they are selling assets (businesses or divisions) of activities that no longer return a substantial proportion of the organization's income) and "financial engineering" (repurposing tangible assets (production facilities – GM reinstated a retired plant to manufacture the Bolt EV) and intangible assets – Nissan-Renault resurrected the Dacia brand for a low-priced vehicle). *Transformational/integration partnering* offers organizations the opportunity to change both strategy and structure in a cost-effective manner: Lego sold its majority holding in four theme parks to provide cash to refocus the company on its core activity, education, and has launched a "Classroom of the Future" project with a US university to teach children about science and technology; it has also launched "LegoFactory.com" and "Lego Digital Designer" that offer an opportunity to design and order a unique Lego model. They have also entered a joint venture with the MIT Media Lab that introduces "Robotic Lego." As Fig. 8.1 suggests, the strength and duration of partnerships created depend very much upon motives such as reduced levels of cost, product-service differentiation, and product-service innovation, new PRODUCT-services, extending the value

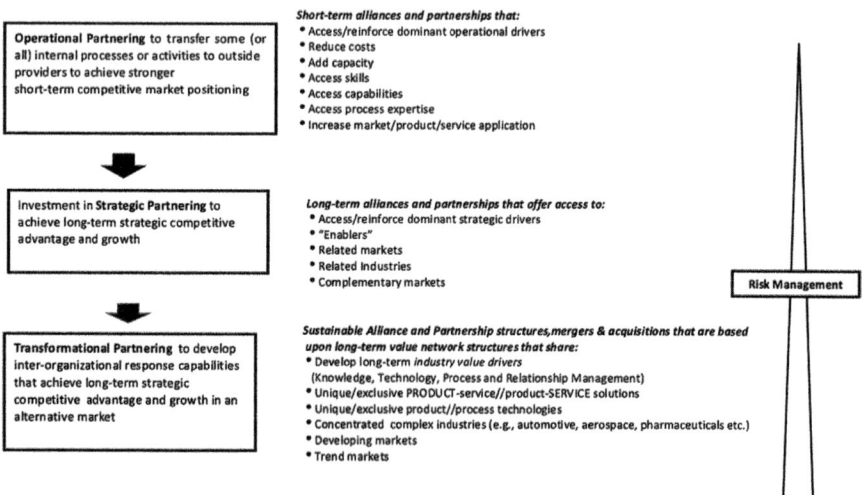

Fig. 8.1 Partnerships, alliances, mergers, and acquisitions

proposition to meet changing customer expectations for product-SERVICE "products." Figure 8.2 provides examples, and Fig. 8.3 provides examples of these relationships based upon observations of recent and ongoing partnering decisions.

Transaction Costs and the Boundaries of the Organization: The Logical Argument for Networks

Transaction costs and the boundaries of the organization. A paper published by Ronald Coase (1937) pointed out a glaring omission in the arguments of the "classical" economics "theory of the firm" of the late eighteenth and nineteenth centuries. The standard model of economics did not fit with what goes on within companies. The question posed by Coase was, why are some activities directed by market forces and others by firms? Coase's paper raised as many questions as it answered. If firms exist to reduce transaction costs, why have market transactions at all? Why not further extend the firm's boundaries?

Grossman and Hart (1986) focused thinking on this by distinguishing between two types of rights over a firm's assets (its plant, machinery, brands, client lists, and so on): "specific rights," which can be contracted out, and residual rights, which come with ownership. Where it becomes costly for a company to specify all that it wants from a supplier, it might make sense to acquire it to claim the residual rights (and the profits) from ownership. But, as Grossman and Hart (1986) noted, something is also lost through the merger. The supplier's incentive to innovate and to control costs vanishes, because he no longer owns the residual rights.

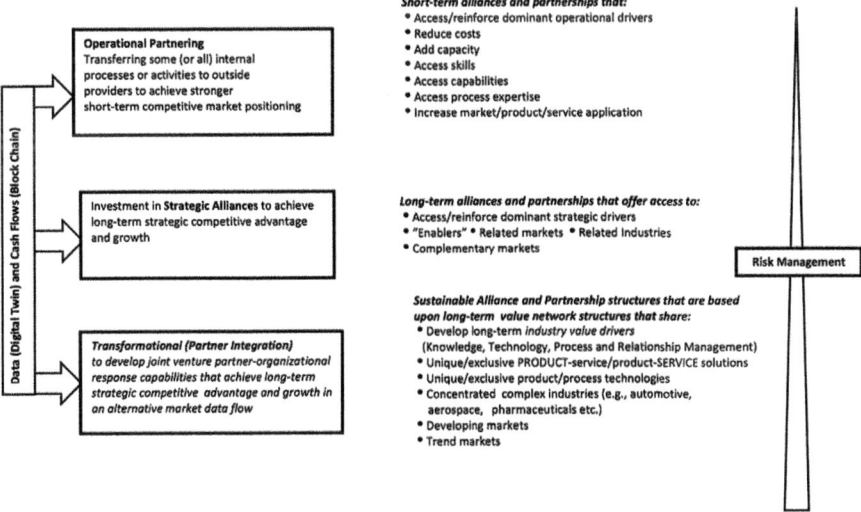

Fig. 8.2 Managing data flow and cash flow: Partnerships, alliances, mergers, and acquisitions

Partnering Typology	
Operational Partnering	**Lego** moved manufacturing to low wage centers in Eastern Europe; reduced color options; rationalized low performing 'secondary' product ranges; and reduced the number of suppliers
	Dyson UK manufacturer of consumer durable products transferred all manufacturing to Malaysia followed by physical distribution functions
	Pacific Brands outsourced their non-specialist apparel ranges to suppliers in the PRC (China) to reduce operating costs but retained manufacturing of bed linen etc.
Strategic Partnering	Leveraged technology – using expertise gained from another industry sector enables *Bishops Technology Group Sydney* to manufacture dedicated production equipment for automotive manufacturers
	Haier, the PRC white goods manufacturer has acquired twenty percent of **Fisher & Paykel** *(NZ based white goods manufacturer)* as part of its plan to access product R&D and to expand Asia/Pacific sales
Transformational Partner Integration	*Lego* sold its majority holding in four theme parks to provide cash to re-focus the company on its core activity education and has launched *"Classroom of the Future"* project with a US university to teach children about science & technology; it has also launched *"LegoFactory.com"* and *"Lego Digital Designer"* that offer an opportunity to design and order a unique Lego model. They have also entered a joint venture with the **MIT Media Lab** that introduces "Robotic Lego"
	General Electric sold its appliance division and is focusing on healthcare, oil field services, gas turbine engines, transportation equipment where it saw potential for growth. In 2017/2018 in response to shareholder pressures it reversed the program and has rationalised its activities

Fig. 8.3 Examples of network structures that address strategy and structural changes

Central to this was the idea that it is difficult to specify all that is required of a business relationship, so some contracts are necessarily "incomplete." Oliver Williamson (1979), added to the Coase concept by introducing "transaction costs" and Oliver Hart and Bengt Holmstrom, who shared the prize in 2016 for their work on the theory of contracts (Hart and Holmstrom 1987). Coase had noted in 1937 that the degree to which the mechanism of price is superseded by the firm varies with the circumstances; he suggested that the boundaries of the firm were not necessarily fixed, if a task could be undertaken by an external firm at a lower cost; this represented an extension of the firm's boundaries to include the activity of the contractor. Eighty years on, the boundary between the two might appear to be dissolving altogether. The share of self-employed contractors in the labor force has risen. The "gig economy" exemplified by Uber drivers is mushrooming. Products and services are delivered by a set of market-based transactions; the supply chains for complex goods, such as an iPhone or an Airbus A380 superjumbo, rely on long-term contracts that are often "incomplete." Coase was the first to spot an enduring truth. Successful economies need both the benign dictatorship of the firm and the invisible hand of the market.

Partnerships: Qualitative Relationships and Cost Considerations

Markets are not free; the market is a coordinating device, and transaction costs are a function of the organization's production/procurement processes. A transaction cost is any cost involved in making a business transaction. For example, when buying capital equipment, raw materials, components, or a service activity, there are transaction costs involved, in addition to the price of the good or service. The cost may be financial, extra time, or inconvenience. For management the decision concerns whether to "produce" the item or activity internally or to purchase it from an outside supplier. Assessing transaction costs are a function of the organization's production/procurement process:

- Does it have the capabilities internally to produce the product-service?
- Does it have the capacity to produce the product-service?
- Is there time (considering competitive time2market expectations) to produce the component and module and develop service functions?
- What would be the opportunity cost; i.e., if resources are allocated to this activity, which other activity will be under resourced?
- Is/are external sources of supply available? If they were used, could they present a threat to competitive positioning the organization is attempting to occupy?

Value Engineering: Strategic Network Effectiveness

Value engineering is concerned with new product services. It is becoming an integral activity in new product development. The focus is on reducing costs, improving function, or both, by way of teamwork-based product evaluation and analysis. Value engineering can be argued to be an optimizing activity; it enables management to optimize benefits delivered to customers *and* the acquisition costs involved at the time of purchasing and subsequently during installation, operations, retirement, and renewal. It is becoming increasingly important as it responds to customer expectations and, as we have established, they have become increasingly dynamic, as OpEx business models are favored rather than CapEx models in industrial markets as serviceability ("performance output" focused customer-centric transactions) becomes more significant. If it is to be effective, value engineering should be undertaken before any capital is invested in tooling, plant, or equipment. This is very significant, because according to many reports (see www.advice-manufacturing.com), up to 80 percent of product costs (that occur throughout its life cycle) are committed at the design development stage. This is understandable as the design of any product-service determines many factors, such as tooling, plant and equipment, labor and skills, training costs, materials, shipping, installation, maintenance, as well as decommissioning and remanufacturing (and repurposing) costs; its significance in the overall process increases as the PRODUCT-service becomes a product-SERVICE. The ownership of the "product hardware" remains with the vendor. Furthermore, the customer may well operate internationally, making a "responsibility cost" for the vendor. Producibility (a connected value ecosystem network or infrastructure created by the value chain network) is a total design activity that includes all relevant activities within the value chain network and, as such, creates a network of intra- and interorganizational partnerships to achieve stakeholder satisfaction; it is designed and empowered to maintain ongoing productivity and serviceability.

Clearly it is essential that value engineering performance is first estimated and then measured on an individual organizational basis and on a network basis; the two functions below use the EVA (economic value added; return to Chap. 5 to reread Kay's work on economic profit and its measurement as EVA (economic value added)) by each of the individual organizations comprising the producibility infrastructure network organizations and as an aggregate measure:

Organizational EVA	**ΣNetwork EVA**
Organizational AWACC	**ΣNetwork AWACC**
"Profitability"	"Profitability"
Effectiveness:	Effectiveness:
Value engineering	Network structure
	Value engineering

The analysis requires some questioning on future efficiencies that will accrue over time and may require the use of "experience curves" and "collaborative curves" to provide cost input estimates. Depending upon the target market and its "distance" from the existing market environment, use of beta factors (β) may prove helpful.

Value Delivery: Operational Network Efficiency

Value delivery is concerned with *existing* products; it includes the overall operational activity of distribution, support service (and serviceability), and remanufacturing/repurposing activities. It involves a current product being analyzed and evaluated against competitive product services (and against changing customer expectations), by a team, to reduce costs and improve product function or both. Value delivery activities use a plan which, step-by-step, methodically evaluates the product in a range of areas. These include costs, function, alternative components, and design aspects such as ease of manufacture and assembly. A significant part of value delivery is a technique called *functional analysis*, where the product is systematically subjected to "tear-down" analysis and reviewed as a number of assemblies. The criteria for the tear-down should reflect market change, for example, it may be based upon changes occurring in an *industry dynamic*, a *process* innovation, or a *regulatory change* concerning material inputs, packaging requirements, as well as changing consumer use practices. It may be the threat of a component failure suggesting modification of the product-service prior to a large incident resulting in a product-service recall. The use function is identified and defined for each product assembly and the new user expectations used to redesign the assembly or its components. The analysis considers the efficacy of the current product-service and of its current producibility network. Costs are also assigned to each component and to alternative manufacturing and servicing options. This is assisted by extending the analysis to consider an alternative producibility infrastructure should this appear to be a possible solution. Both value engineering and value delivery are group-led activities involving suppliers and customers that bring in wider experiences of product-service applications and of serviceability issues. *Value management* is often used to refer to the overall activity of value engineering and value delivery; this is becoming an ongoing activity facilitated by the convergence of Industrié 4.0, Value Chain 2.0, and stakeholder value management. Specifically, *data thread* analysis and *digital twinning* (a physical product-service supported by a digital replica) are being applied on an ongoing basis resulting in a "rolling" value proposition.

Performance can be measured on an ongoing basis. Digitization enables a cost/benefit approach as real-time performance data is available on an intra- and interorganizational basis. It will be recalled that Chap. 5 cited the practice of Volkswagen measuring activity and locational performance using an EVA-based metric. This is ideal for measuring ongoing activities. Data thread provides ongoing performance data; internal cost data can be captured and matched to provide measures on cost-effectiveness of operations. An aggregate period cost/benefit performance can be accessed from the digital twin. Network collaboration facilitates network-wide ongoing performance efficiencies.

Organizational EVA %	**$\sum$Network EVA %**
Organizational operating profit	**$\sum$Network operating income**
"Productivity"	"Productivity"
Organizational operating efficiency	$\sum$Network operating efficiency
Value delivery	Value delivery

Value Contribution and Network Partnerships

The convergence of Industrié 4.0, Value Chain 2.0, and stakeholder value management has increased the collaboration, the data transparency, and the understanding of value chain activities and decisions within and across the structures of market industries and value chains. It has also explained the logic of costs and their influence on value chain structures and relationships.

Partners in a value chain network have opportunities to create additional value for themselves and for other network members. See Fig. 8.4. The well-visited story of how Walmart profited from data sharing and how its improved logistics through better forecasting and inventory management is well understood. The collaboration with Procter & Gamble created a whole new world of data sharing, and collaboration not only improved forecasting and marketing but also provided major reductions in joint inventory holding. Walmart itself did not possess the resources to develop focused analyses on a given product because it carried hundreds of thousands of products. But its suppliers did because they had relatively small sets of products and significant vested interest in seeing those products' performances optimized. In addition, the suppliers were the experts on their categories and end consumers. Walmart was an expert on its stores and retail business (Waller and

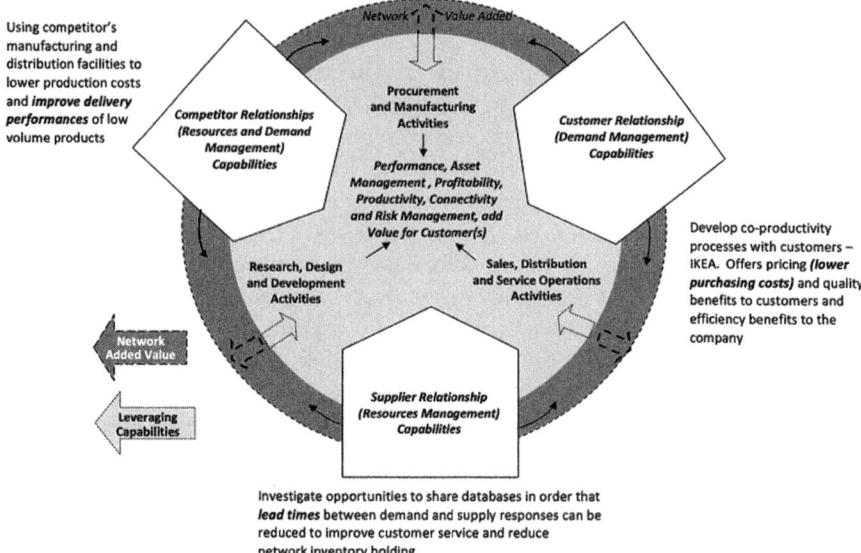

Fig. 8.4 Leveraging partners' capabilities to add value to the value delivery

Boccasam 2013). IKEA has leveraged customers' "resources" during past years by offering price/quality/design/aesthetics/value equivalence in return for customers undertaking order selection, assembly, delivery, and final assembly. More recently the company has offered connections with local service organizations who undertake delivery and assembly tasks. The company has also moved towards internet sales through a third-party company.

Coopetition and collaboration in manufacturing and physical distribution activities: innovative companies take collaboration beyond organizational borders. The Lego Group has created products by engaging customers for co-creation. Before the late 1990s, Lego operated in command-and-control mode by refusing to accept "unsolicited" ideas. Lego Mindstorms' value proposition included software and hardware to create small programmable and customizable robots. However, technologically aware users hacked the code and changed the products. Rather than sue these customers, Lego decided to collaborate with them. Since then, Lego users have collaborated with team members to create new product lines and distribution channels. Lego also uses revenue sharing to create incentives for customers (Rosen 2015). Figure 8.4 argues that coopetition and collaboration can increase the value added for customers, networks, and individual organizations within value chain networks.

For Volkswagen the economic return on investment serves as a consistent target in strategic and operational management. *If the economic return on investment exceeds the market cost of capital, there is an increase in the value of the invested capital and a positive value contribution.* The concept of value-based management (economic return on the annualized weighted average cost of capital) allows the success of the organizations and individual business units to be evaluated. It also enables the earning power of products, product lines, and projects – such as new plants and suppliers to be measured. Value contribution may be improved upon by leveraging network partnership capabilities. For example, a partner organization that is not utilizing production capacity may offer cost reductions to improve plant utilization, and a partner with a strong brand reputation for a specialist product-service characteristic can add differentiation in a competitive market.

For these reasons we consider how *operational partnering, strategic alliances,* and creating *partner integration* to pursue *transformational joint venture organizations* creates and increases value contribution throughout the network. Figure 8.5 suggests criteria to be used when considering partnership options. The diagram uses a two-by-two matrix in which the "y" axis identifies the values of value contribution/EVA percentage performances; the "x" axis identifies market segment/market growth rate. The purpose is to prescribe appropriate types of partnerships to meet the potential performance (value contribution/EVA/market segment/market growth rate). Clearly if both metrics demonstrate low growth, an *operational partnership,* cell one, would suffice; it requires monitoring activities to identify collaborative activities to enhance productivity. Cell 2 offers opportunity; the value contribution is high, and there is potential to expand the market with a collaborative *strategic alliance.* Cell 3 offers an opportunity for integrating capabilities and capital in order to "control" the direction of a market activity that may require restructuring the value chain network. A *strategic alliance* may be relevant for cell 4; however, a dominant partner is required to manage the performance continuity of both metric labels.

Value contribution has two factors that require consideration; the first is that the incremental increase in EVA (as a result of increases in revenues, decrease in costs, or possibly a change in funding structure) is greater than the incremental increase in the current value of EVA. For example, adding features to a product in order to create differentiation typically requires increases in both variable and fixed costs; the variable costs may be additional labor and material costs, or it may involve additional storage costs, such as maturing a product (whisky, cheeses, etc.). The second consideration concerns the impact of the added value to customers' attributes, by

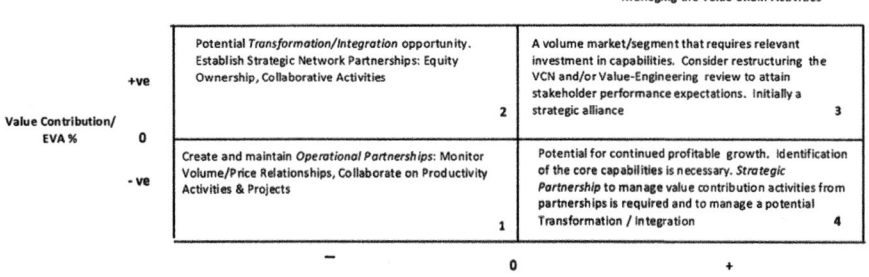

Fig. 8.5 Customer value contribution/market segment/market growth rate(s) %

strengthening their market or value chain network positioning and the time period over which the added value will exist (a consideration for both vendor and customer).

Procurement and Partnerships

The increasing importance of procurement since the events of the financial problems of 2008/2009 has led to the increasing involvement of procurement in planning in the strategic and operational activities of organizations and the role of suppliers as partners in the short and long term. The proactive procurement activity is one that understands it has more to offer the organization by looking at what it contributes to revenues than possibly it has by looking only at input costs. To make effective strategic and efficient operational contributions, procurement should take an *external* and an *internal* view of its role *and* short- and long-term perspectives of both. See Fig. 8.6.

The *short-term/internal* concerns of procurement are primarily to ensure that it contributes to short-term objectives by monitoring its own performance, within the business (and within the supplier network); the focus is on operational efficiencies such as ensuring supplier performance (quality, quantity, and timing reliabilities) is met. It is essential that procurement measures its contribution to customer satisfaction by monitoring overall performance contributions to meeting customer value driver expectations.

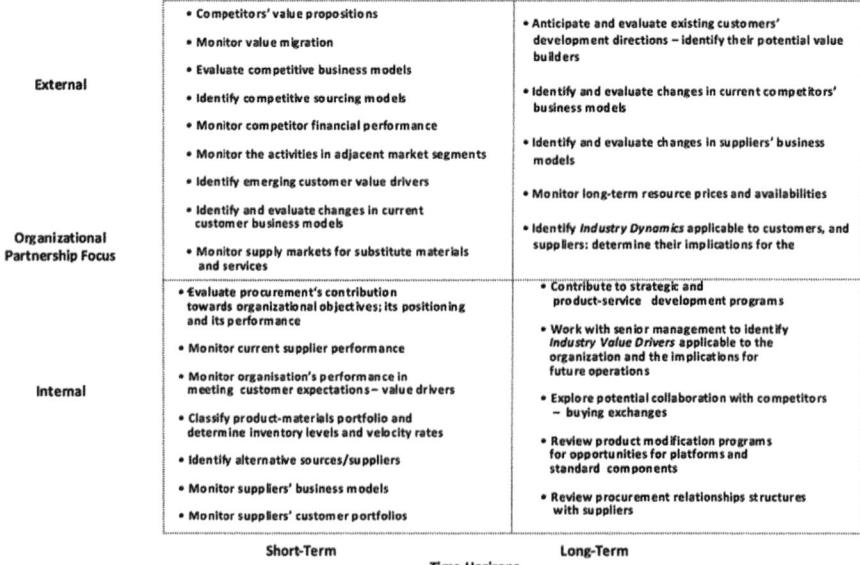

Fig. 8.6 Procurement has time and intra- and interorganizational partnership dimensions to consider

Short-term/external concerns involve procurement comparing competitive performance and structures with that of the organization. Even in the short term, changes can occur, and a close watch on changes in customer performance value drivers (and their business model structures, e.g., the introduction of digital connectivity) can be helpful for fine-tuning the value proposition. Procurement practice changes (digitization) in adjacent market structures can offer opportunities to increase revenues or to reduce costs and to be aware of resources; alternatives may have a similar impact.

Long-term/external factors will have an impact on long-term/internal strategy and structure decisions. In a dynamic business environment with often what appears to be ongoing change to the *industry dynamics* and with organizations divesting themselves of assets focused in one industry in order to focus on an emerging industry, it behoves all activities within individual and network organizations to be aware of potential "opportunities" and "threats" that may confront them. For this reason, procurement should contribute towards future strategic discussions and anticipate the implications of such change within existing customers (and supply networks) to identify these by considering existing customers' development directions – in strategy and structure – and to identify their potential value builders. Changes to competitor business models and supplier developments are equally important. In the long term, resource prices and availabilities will not only change – they may become irrelevant as entire product-service offers change to reflect environmental changes and customer acceptance of them; the EV (electric vehicle market) is a current example of this concern.

Long-term/internal perspectives are concerned more with procurement's contribution to strategic effectiveness. Working with senior management colleagues to identify and consider the organization's future value builders could be critical for its future relationships with both customer and supplier partnerships; clearly an awareness of developing resource markets can make a significant contribution to strategic and product-service development programs. Exploring collaborative opportunities with other procurement management teams within the industry may result in buying exchanges being established such as *Covisint* and *Elemica* (see earlier chapters). An opportunity for more effective product-service designs exists with earlier involvement with RD&D to explore the potential to design future product-service offers around standard components and product platforms. A review of procurement relationship structures with suppliers is essential if successful "fit" is to be found, and this may include ongoing and future changes to the *industry dynamics* influencing both customers and suppliers, and that determines the implications of any likely changes that may have for the organization or the network. For example, government action to restrict the sale of rare earth minerals (lithium) may suggest the need to find alternative suppliers or, perhaps, design products around substitute inputs and processes. Consumer pressure on packaging materials (e.g., replacing plastic bottles with recyclable glass) is another example.

It should be clear that the role of procurement within intra- and interorganizational decision-making is becoming increasingly important in a business environment where increasing revenues is difficult due to the rapidly changing and challenging business environment and when there is a focus on organizational efficiency by boards of directors. Figure 8.6 proposes that procurement decisions are made using similar criteria to those in other aspects of operations management, i.e., they (suppliers) share corporate approaches in the use of customer data (value builders and performance value drivers) and become more aware of the impact of interface implications on their decisions, resulting in an attempt to deal directly with an organization's customer.

Procurement has an increased responsibility in times of economic uncertainty; volatile currencies and input prices create problems for forecasting resource prices and cost structures Quirk 2011. A common solution, one that Caterpillar has used to advantage, has been to locate production units in large markets to avoid currency change problems. In the 1970s Caterpillar had two plants in China; currently it has about 20. Nissan-Renault now has a policy of locating manufacturing units in its major markets for this reason.

Centralizing procurement is another alternative solution. PepsiCo has implemented a procurement strategy for hedging forward commodity purchases; these amount to US$18 billion annually. Prices are effectively fixed for 9 months, thereby avoiding price increases (and not benefiting from unlikely decreases); local regional managers cannot deviate from the planned costs without very good reason. The chief purchasing officer (CPO) argues this gives the company the ability to forecast costs accurately, whereas forecasting the costs of inputs is becoming very difficult.

Procter & Gamble considers hedging to be an operational solution and therefore short-term; the CPO argues they are also expensive. P&G adopts a longer-term strategic solution: "innovative substitution." For example, the new packaging for its Pantene hair products uses biodegradable cornstarch rather than petrochemical resins. The CEO adds that such changes pass unnoticed by customer stakeholders but are applauded by shareholder stakeholders – particularly when they can reduce the cost of US$2 billion incremental commodity costs by 50 per cent.

Total Cost of Ownership: Evaluating Partnerships – The Recent Past

Total cost of ownership (TCO) is considered by organizations, when purchasing assets making investments in capital projects, purchasing significant material inputs for product manufacturing, and when purchasing "services." It is a quantitative, evaluative analysis of existing and potential partnerships. While most of these costs often are itemized separately on a company's financial statements, many are either ignored or unknown. Comprehensive analysis of the cost of ownership is a common practice for business dealings. Organizations use the total cost of ownership over the long term as a framework for analyzing business opportunities. The analysis includes the initial purchase price as well as all direct and indirect expenses. While the direct expenses can be reported easily, companies most often seek to analyze all potential indirect expenses that can be of significant influence in deciding whether to complete a purchase.

Offshoring manufacturing has focused the attention of organizations as not only do they encounter additional costs and constraints, not typically levied by domestic suppliers, but they also find a need to consider "opportunity costs." Usually offshore manufacturers insist on a down payment of some 25 percent of the value of the contract. Added to this is the necessity of increasing inventory to service demand over the "2- or 3-month" order cycle time and to cover inaccurate order responses (number and content), plus the possibility of rejecting the entire order because of quality problems. Given these circumstances customer loyalty quickly dissipates, and they move on to a more reliable supplier, and sales are lost.

J. Robins & Sons is an Australian footwear manufacturer that has been designing and producing branded footwear in Australia for over 100 years. Despite many businesses in the industry moving their manufacturing base offshore, Robins & Sons has taken a different approach. They are a footwear manufacturer with a particular focus on fashion and are aware they cannot compete on price or volume against countries like China or Indonesia; accordingly they "concentrate on being very quick to market," according to managing director, Butt (Roberts 2009). Robbins operates a multi-skilled workforce of some 250 employees who "move between work stations as part of a small-batch production process that gives Robins the flexibility to design, make and deliver a pair of shoes in only two weeks." With speed to market, one of the keys to success in the fashion industry, short lead times and quick supply are essential, and the structure of the company was changed in order to address these needs. "In the mid-80s we identified the importance of 'just in time' manufacturing, so we went overseas and examined the processes used by leading companies in Japan and the US and brought these back and modified it for Australian conditions" (Roberts 2009). As a result, the company structure is now an excellent example of vertical integration. Robins & Sons control everything from raw materials, design, and manufacture to when the shoe is packed in the box. Having a local manufacturing base and supply chain is one of the key advantages the company has over its international competitors. Butt added, We don't need to predict what the 'in' color will be, but overseas manufacturers have to know in up to six months in advance what the right color and style will be before each season. As a domestic quick response manufacturer with faster, smaller production runs, the result is a quick, flexible manufacturing process where from the time we get an order, it can be dispatched in five days. "Supply from places like China could be two or three months" (Roberts 2009). With "just-in-time" manufacturing, a major component of the company's strategy, this also led to changes to the company's team structure. With shoe manufacturing still a very labor-intensive process, Butt explains that the company not only increased expenditure in capital equipment but invested money into training and increasing the team skills of employees. "Our staff were literally trained to work in manufacturing teams, to multi task etc. and we enabled workers to have a far greater level of involvement in the company's business strategies and process development" (Roberts 2009).

To compete against the low-cost labor advantage of Asian manufacturers, Bonds, owner of Pacific Brands (an Australian-based manufacturer of branded apparel), has identified a market segment whose expectations could be met competitively.

Newman (2005) reports comments by Richard Abela, head of manufacturing and supply for the Bonds group, suggesting that provided a manufacturer is selective in the market segment they choose, there is scope for an organization that avoids commodity products that require significant labor input. The Bonds factory focuses on products "where the margins are not so critical, and customers are prepared to pay a premium." Bonds manufacturing process focuses on monitoring costs and speed of production, through a vertically integrated production process and in long-term agreements with its suppliers. Inventory holding is maintained at 3-day supply. Bonds is able to produce a finished garment "from yarn to a finished product in just four weeks, compared with the Chinese who require 12 weeks." Chinese manufacturers are increasingly unwilling to accept small garment runs as they are now more interested in the volume markets of Europe and the USA.

Bonds has invested in equipment specifically for small batch runs. In Australia 5000 items are considered medium sized, "…so small for Chinese manufacturers that it would be done in the sample room, with the attendant quality problems." Certainty of supply continuity is also an area where local manufacturers can offer advantage. Bonds also sees added-value product markets as an opportunity. Markets such as school uniforms that require specific crests and designs are of interest. Bonds has benefited from federal government funding. They used a grant to help fund a new packaging machine to reduce packaging costs by 50 cents an item. Abela, CEO, is conscious of the fact that the company needs to monitor and reassess the economics of manufacturing in the industry as innovation is an ongoing process.

Total Cost of Ownership of Activities: Recent and Future Developments

Understanding the "total cost of ownership is an important activity in understanding the role of partnerships and their impact on an individual organization. *Total cost of ownership (TCO)* has traditionally identified and analyzed the impact of all the lifetime costs that follow from owning certain kinds of assets. As a result, TCO is sometimes called *lifecycle cost analysis*. Knowing and understanding the total costs of ownership throughout a network provides an opportunity to optimize the cost of providing optimal value for stakeholders by identifying, and working with, the most capable network partners and therefore optimize the effectiveness of the total value added by the network. It has resulted in changes in product-service format value propositions; the PRODUCT-service becomes a product-SERVICE as Capex businesses move towards becoming Opex organizations by shifting their asset portfolios from tangible to intangible expenditures, for example, airlines purchasing hardware product outputs rather than the hardware product.

TCO analysis commences with a review of organizational and partnership objectives; the assumption being that by exploring stakeholder expectations, the organization as an entity and as a network partner, the optimal asset structure can be identified. It follows that the "ideal" solution may not currently exist; however, if the "ideal" solution is likely to be superior to the status quo and has substantial, attractive, benefits for primary stakeholders and does not impact the concerns of secondary stakeholders in a negative context, it is likely that it will be pursued further. Essentially an effective TCO analysis is asking a number of "what if?" questions concerning the organization and the network's current operating capabilities against two sets of criteria: its objectives and its value proposition. Increasingly large organizations are using internal resources (such as data collection and analysis systems and processes) to obtain effectiveness and efficiency improvements.

The "convergence" of Industrié 4.0, Value Chain 2.0, and the stakeholder value management perspective offers a new approach to total cost of ownership; rather than appraise alternative individual input capability sources on traditional criteria of cost, time, and quality, organizations are seeking supplier partners that have a synergistic response capability of adding additional added value. It extends the TCO analysis and uses it to identify cost-effective (strategic formats for creating producibility operating structures) and cost-efficient (operational methods of implementation and operations). This shift in supplier appraisal can be seen in the examples of organizations that are leveraging elements of the convergence to create product-service value propositions that offer cost-effective and cost-efficient implementation and operations throughout their networked activities.

An IndustryWeek publication (Hitch 2018) published research findings (based on 1000 manufacturers over a year) following a report by the World Economic Forum, suggesting seven out of ten manufacturers fail in pushing initiatives in data analytics, artificial intelligence, and additive manufacturing past the pilot phase. This was confirmed by their own research that indicated manufacturers are encountering difficulties finding "'ways and means' of benefiting from the "consolidation." The findings are interesting. Organizations that have persevered have 20 to 50 per cent higher performance and create a competitive value advantage. This has been achieved by creating agile teams with intra- and interorganizational sphere of interest, analytics, IIoT (Industrial Internet of Things), and software development expertise. They have deployed a common data/IIoT platform and have produced case study evidence of the efficacy of the approach. It argues "They are thinking scale, acting agile, and resetting the benchmark." Some examples demonstrate the effectiveness of this:

Bayer Biopharmaceutical, Italy: By intensifying data analytics, Bayer has increased operating efficiencies by 30–40 percent and decreased maintenance costs by 25 percent.

Bosch Automotive, China: Data analytics eliminated output losses, stimulate and optimize process settings, and predict machine interruptions before they occur.

Haier, China: An application of artificial intelligence used data analysis to create a user-centric mass customization model to manufacture electronic products on demand. Maintenance requirements are predicted before downtime is incurred using an "intelligent" cloud platform.

Phoenix Contact, Germany: Uses digital twinning to produce customer-specific clones that have reduced production time for repairs or replacements by 30 percent.

Proctor & Gamble, Czech Republic: Digital connectivity reduces product changes during manufacturing costs by 20 percent and output by 160 percent.

Siemens, China: Leveraging augmented reality to create 3D simulations has optimized its production lines and resulted in reducing cycle times and a 300 percent increase in output.

Total cost of ownership has in fact become "total cost of ownership of activities" and is to be implemented in the value chain at relevant locations to improve VCN 2.0 structures and costs that have overall network benefits. UPS utilizes 3D printing capabilities by putting printers in their UPS stores, creating a network for 3D printing on demand for manufacturers across the USA. This development is changing the supply chain response to service parts management. Hitherto, manufacturers of capital equipment and consumer durable (automotive, white goods, etc.) maintained large field stocks of service parts in a large number of locations. Additive manufacturing in strategic locations can manufacture replacement parts to meet demand thereby reducing the cost of field inventories significantly.

Identifying and Evaluating the Impact of Procurement Partnership Decisions: A Qualitative Approach to Evaluating Total Cost of Ownership

Clearly any organization must identify both opportunities and threats within its planned response capabilities and consider the impact of the decision alternatives on its competitive market and value chain positioning, across relevant time horizons. The analysis commences with an examination of the qualitative characteristics of the organization's current capabilities and the constraints they confront in the market place when making decisions concerning a structured network-based response. The decision is influenced by a number of considerations in making strategic and operational decisions that concern whether or not to internalize the production of product/process or enter into a contractual arrangement with an external supplier and to establish what investment returns and free cash flow will result from various alternatives. Qualitative characteristics currently considered include:

Investment constraints: Concerns such as the levels of financial and operational gearing, the impact on profitability (return on annualized WACC, ROI, etc.), cash flow, and share price of further investment. Access to sources of finance and expected constraints such as amount, interest rate, and concerns of major shareholders.

Search and information costs: Time and management expertise are required when identifying and assessing alternative external sources of supply. Legal costs are involved in developing and managing contracts. Again, there is a need to identify potential problems concerning enforcement issues when contracting with offshore organizations.

Acquisition costs: These can include many types of expenditure due to searching, identifying, selecting, ordering, receiving, installing, and training staff to use equipment. Typical costs are setup and deployment costs, outside services, operating costs (labor, materials, services), service/maintenance costs (labor, materials, energy, infrastructure), specific support costs (heating, lighting, cooling), and environmental impact costs (expenses for waste disposal, cleanup, and pollution control). These may also include insurance costs.

Complexity: Managing the design, manufacturing, serviceability management, and the possibility of remanufacturing comparisons of full transaction costs of alternatives. Process and procedural complexities are also important. The Brexit decision is one example and the "tariff war" is a concern.

"Time" constraints: Frequency of purchase is the requirement for an "occasional one off" (where potential suppliers may be concerned over an investment with doubtful profit potential) or is the need for a supplier who can meet a volume order with frequent and reliable deliveries (where the delivery schedules may be based upon JIT scheduling or may be for service parts inventories that have some flexibility).

Intellectual property: Typically, patents are protection of the intellectual property and rights of the organization (or individual); however, they may be licensed for an agreed fee and used by the licensee for revenue-generating purposes. Patents are often "neutralized" (e.g., biosimilar pharmaceutical products); offshore locations may have quite different patenting process and laws that could be a threat.

Asset specificity: An external supplier may be cautious about making an investment (size and sunk costs) and may require a contribution to the capital requirements or request permission to supply variants to other customers (the organization competitors). Again, intellectual property and competitive value advantage are a problem.

Sustainability (corporate and environmental sustainability) costs: It has three considerations: the first concerns long-term profitability and cash flow generation from operations (a need to be concerned about resource inputs (materials and human resources)); the second is the need to respect the environment and adopt design strategies that use substitute inputs or, if currently unavailable, make improvements in the use of existing inputs (such as recycling/remanufacturing); and the third is to identify the expectations of primary and secondary stakeholders and to make a positive response. Cost may include charges for "environmental compliance" reporting.

Indirect contract costs: Many organizations are surprised by the requirements of offshore suppliers for a significant amount of the value of the contract ahead of work commencing. Quality control and agreed delivery schedules are very difficult. Furthermore, the overall order management time should consider the cost implications of longer delivery times and the potential of unusable items (due to failure to meet specifications or inadequate labor or material content). The additional cost of poor customer serviceability (lost business and no repeat business) is to be computed in the overall transaction costs.

Economies of scale and scope: Are important considerations, where large consumer durable producers often outsource the total manufacturing of low-volume products within the product range of otherwise high-volume product-services. Often the solution is to contract out the manufacturing of an entire product. *Magna Steyr*, an international organization, has developed the capabilities required for manufacturing low-volume automotive products profitably. The company has an extensive range of services covering product group engineering services from systems and modules to complete vehicle engineering and complete vehicle manufacturing, where they offer world-class flexible solutions from niche to volume production. Magna Steyr has produced more than 3.3 million vehicles – of 24 different model variants. Recently, March 2018, it was announced that two new sports cars jointly developed by BMW and Toyota are to be manufactured by Magna Steyr. Economies of scope are important; the use of an existing tangible asset (a manufacturing facility operating at less than optimal capacity) or an existing intangible asset (a successful brand, an existing market with unsatisfied demand) potentially offer growth opportunities.

Collaborative contracting/alliance (aka integrated project delivery (IPD) in the USA): Owners, contractors, and engineers are integrated into a single contract which has been heralded as the cure for the problems dogging contracting in the collaborative approach, parties work within the boundaries of traditional contracts, and their agreements rest on a fundamental belief that both owners and contractors want the best possible outcome and that each party contributes *unique strengths and capabilities* (Abidi et al. 2018. McKinsey (Abramowicz et al. 2018) argues that if participants hold these beliefs and implement a number of simple but important collaborative practices, collaborative contracting leads to better project outcomes. Stringent adherence to planned outcome, budgets, and behavior focused on an agreed vision of the end result is essential. See also Hardt et al. (2007).

Social responsibility: The global value chain has enabled the manufacturing capabilities of developing countries to contribute input into international markets AND to generate income for a large number of semi- and unskilled workers. This clearly has two benefits; it offers multinational organizations a pool of labor that is less costly than its domestic locations, therefore to remain cost-competitive in all of its markets; it also generates markets in the developing countries as the spending power of the labor force accumulates wealth.

Clearly the convergence of Industrié 4.0, Value Chain 2.0, and stakeholder value-led management offers response capabilities that will facilitate decision-making concerning the outsourcing decision through connectivity and collaboration that improve relationship management, particularly the offshore relationships; it can be argued that the "coalition" will decrease internal cost options. For example, Rolls-Royce engines and General Electric are using additive manufacturing in the production of aeroengines; the process can manage the manufacturing of complex shapes within components, thereby reducing the number of components required, resulting in simplifying manufacturing and maintenance costs, and realizing weight reductions that increase performance. Stakeholder value-led management will influence "people" concerns, for example, the impact of robotics on skilled, semi-skilled, and unskilled employment as well as the environmental issues of major concerns: global warming and environmental impact of operations processes on wildlife and habitat (Great Barrier Reef, Arctic regions, etc.). These last paragraphs suggest rigorous analysis is required when evaluating partnerships; this is practiced using total cost of ownership (TCO) analysis. See Fig. 8.7 for a detailed example of the process.

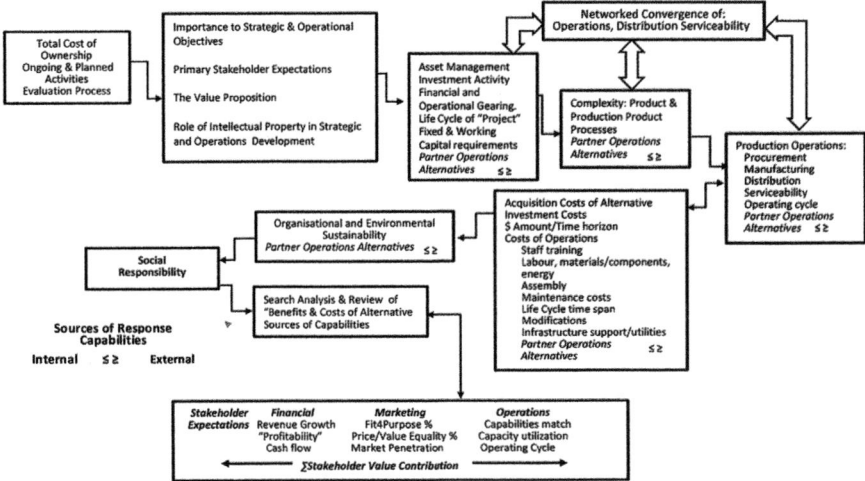

Fig. 8.7 TCO analysis procedure exploring sources of response capabilities

The Global Value Chain

By definition operating within global value chains involves working with offshore partners. One view of this activity is the outsourcing approach, whereby offshoring reduced variable costs, typically labor, and perhaps material costs. A strategic perspective expands this view to include situations (e.g., Dyson, the UK consumer durable manufacturer) in which the entire operations activity is located to achieve an optimal cost structure. The global value chain (GVC) with its capability to coordinate sequential production activities on an interorganizational basis, across international borders, has been enhanced by the convergence of Industrié 4.0, Value Chain 2.0, and stakeholder value management. The global value chain's digital connectivity, real-time collaboration, and data transfer offer opportunities to increase its effectiveness and efficiency response capabilities to opportunities to add value throughout the value chain.

Dollar (contributor to the 2017 "Global Value Chain Development Report") classifies GVCs into "pure domestic production," those in which production and consumption occur within one country, "simple" GVCs in which the value added crosses an international border just once, and "complex" structures in which value added crosses national borders at least twice (Wang et al. 2017). Dollar (2017) gives the example of the GVC structure of the ubiquitous mobile phone (and most other electronic products) as an example of a complex GVC; the RD&D is based in North America or Europe, for example, semiconductors and processors are produced in Asia (Korea, China, and Taiwan), and the product is assembled in China. Physical distribution is international and after-sales servicing conducted in the end-user country.

Dollar (2017) reviewed the patterns of development of the GVC from the early 1990s and during the 2008/2009 global financial crisis, during which there were some setbacks, to the current period. He identifies China as being accountable for the ongoing expansion of the GVC growth; the acceleration of the expansion occurred soon after China joined the World Trade Organization (WTO) and the fact that the rate of value-added growth was also very high. However, since 2011, slowing rates of GDP growth appear to have had a disproportionate impact on GVC channels, particularly for complex GVCs, which were the primary influence of growth in preceding economic cycles.

Using the "Smile Curve" to Plot Added Value by Global Value Chain Partners

Dollar (2017) discusses the distribution of value adding activities (and the rewards that accrue) throughout the GVC using the *smile curve* (Meng et al. 2019). Taylor (2017) provides an "edited version" of Dollar's more detailed explanation of the smile curve. Figure 8.8 simplifies the explanation (but not the importance to management), the vertical axis plots labor compensation ($ per hour) in the participating countries, and this, it is argued, reflects the value added by each country, indicating

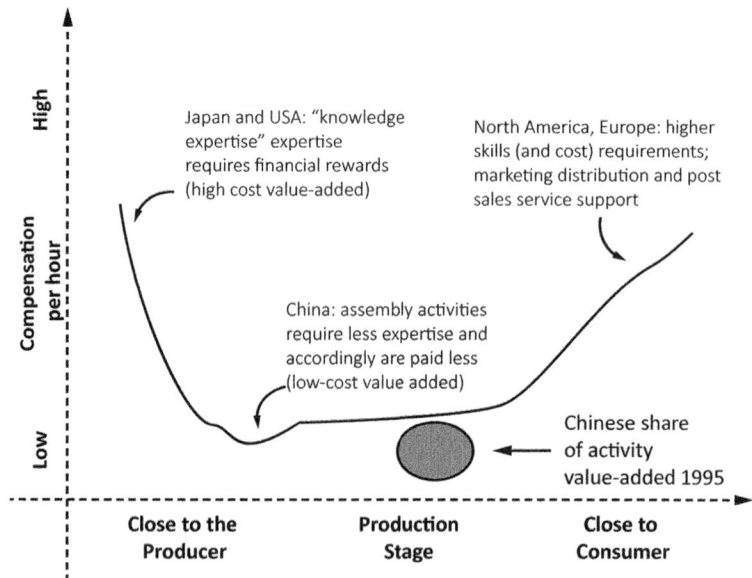

Fig. 8.8 The "smile curve": Identifying added-value activities in the value chain network 1995

high versus low value-added activities. The horizontal axis plots the aggregate production cycle length from RD&D activities through component manufacturing and assembly to end-user consumers and after-sales support activities; the example used was electrical and optical equipment.

It is argued that as RD&D, knowledge management benefits are high value-added activities; occurring early in the overall production process, they are carried out in more advanced activities (e.g., in the Meng et al.) in Japan and the USA. See Fig. 8.8, representing the smile curve in 1995.

By 2005 the smile curve had deepened as sector labor rates in the USA had increased considerably (from $25.00 per hour to $60.00 per hour); Chinese wage rates remained low; this resulted in a large increase in employment (probably the intention of the Chinese government) and a tenfold increase in the Chinese share of the value-added at that point of the global value chain network.

We suggest the convergence of Industrié 4.0, Value Chain 2.0, and stakeholder value management has had an impact on the GVC, and this is illustrated by Fig. 8.9. The impact of digital connectivity, real-time collaboration, and data transfer is becoming apparent with decreasing order cycle times, reductions of inventory holding throughout value chain networks, and increasing levels of customer satisfaction levels such as fit4purpose, time2market, service completion time, and customized product service formats.

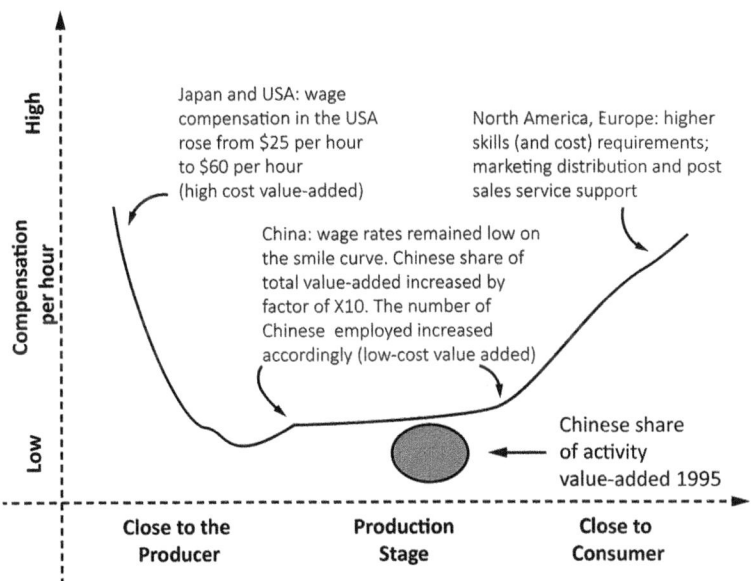

The Global Value Chain Network: Production/Consumption Cycle Time

Fig. 8.9 The "smile curve": Identifying added-value activities in the value chain network

Considerable change is underway, and we suggest that the smile curve should reflect these. The overall measurement of value added using labor rates (hourly compensation) does not reflect the technology investment in operations in the GVC. An alternative measure is to use EVA data (economic value added); EVA performance is accessible and reflects an increasing interest in economic profit (see earlier paragraphs in this chapter). The operating result shows the economic performance of the organization. A "smile" GVC network performance metric comprises:

The three activities comprising the horizontal axis of the Smile Curve (i.e. Close to the Producer, Production, Close to the Consumer) also require modification to reflect the impact of servitization/serviceability as a dominant feature in the industrial-capital markets where the PRODUCT-service has become a product-SERVICE. In addition, it is necessary to consider the impact of concepts such as the growing support for the circular value chain/circular economy that adds; recycling – returning to the producer remanufacturing/repurposing to the Smile Curve (Chap. 9, *Preservation*, deals with this activity in detail).

Figure 8.10 includes a matching of value engineering and delivery cycle with the "Smile" activities and with a detailed "producibility infrastructure activity sequence." There are a number of interests that can be considered and addressed by constructing a smile curve for existing network activities to evaluate opportunities. Dollar (2017) suggests the identification of the value-added benefits exposes "winners and losers." His analysis of the Ye et al. (2015) work identifies changes in socioeconomic structures responsible for large and/or increasing shares of the total value added produced. He suggests the Chinese reforms in the 1990s/2000s resulting in large wage increases as rural agricultural labor transferred into manufacturing employment, suggesting these benefits were being available to developing countries; however the "very large" benefits accrued to a small number of highly skilled employees and Chinese and offshore investors. This contrasts with the developed countries in which the overall increases in share of total market value-added share benefited the high-skilled workers and multinational corporations.

Dollar (2017) suggests the initial findings do not permit drawing strong causal conclusions. Clearly additional information is required, for example, government activities in investment in education as well as production facilities with the purpose of increasing the share of value added. From the GVC investors' perspectives, evidence of a potential market in participating countries would provide encouragement for including (and expanding) activities within these countries. Dollar (2017) argues for additional steps towards redistribution of some of the income towards policies they would benefit to others in these countries. A review of alternative producibility infrastructure options could help. It would require a review of both the product-service design and the design of the manufacturing processes involved. The example of General Electric's problems (and solutions) with healthcare equipment in Asia demonstrates how a better understanding of market user needs, product design that reflects these needs, and a restructure of production operations can lead to success.

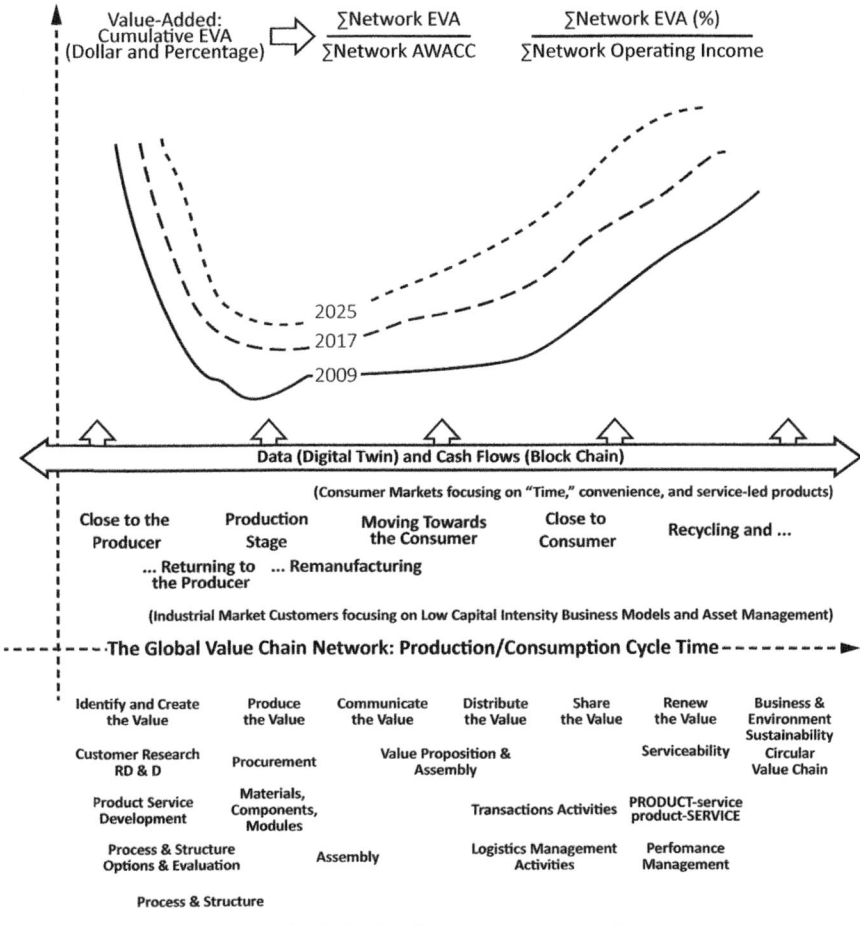

Fig. 8.10 Modifying the "smile curve" to reflect the growth of the circular value chain network

Convergence …

The primary importance of partnership management is to build and maintain value-generating partnerships.

Partnerships: To do this it effectively requires a network stakeholder focus on satisfying the expectations of the prime source of income, the end use, and concurrently ensuring the remaining network partners' objectives (revenues, "profitability," and cash flow) are met. For this to be effective, contributions from the other response capabilities are necessary. An adaptable strategy and structure model that ensures current (and future) management and operations staff possess relevant capabilities (skills and experience) to meet current and future challenges' requirements' responses – to meet growth opportunity requirements. Ethical policies (equitable rewards for network partners, ethical working hours and conditions for network employees, etc.), if a positive value advantage contribution is to result.

Performance: A primary concern is for strategic synergy among partners concerning shared resources (RD&D results, data collection and analysis, etc.) and identifying and accessing additional/specialist resources. Operational implications for network convergence on customer-centricity concerning fit4purpose, cost/price/service/value equivalence, time-2market time4service response, and locational/demand availability.

Profitability: Developing Capex and Opex business model structures that optimize asset management investment performance decisions, network-based product-service, margin management, and margin mix decisions. Adoption of economic profit and financial (EVA) as realistic performance metrics – sustainable strategic and operating cash flow. Network convergence on fit4purpose, cost/price/service/value equivalence management, time2market Time4Service Responses, and locational/demand availability (suppliers, distributors, and customers).

Productivity: Optimizing capabilities, processes, and capacities across a connected, collaborative network. Data transfer and transparency are essential requirements for success. Comparative productivity (EVA/$\sum$EVA) metric for network comparative productivity monitoring "total productivity." Network agreed productivity drivers, collaborative relationship connectivity, and shared process innovation.

Producibility: An integrated, seamless, connected infrastructure comprising all value-producing activities from RD&D to value renewal. Data collection/analysis transparency. Shared capability and capacity resources and access to additional/specialist resources. Network convergence on role of specialist outsourcing.

Preservation: A network approach to incorporate collective social responsibility concerning relevant and applicable issues (the UN's 17 Sustainable Development Goals (SDGs) shared sustainability governance indicators). Adoption of the GRI (Global Reporting Initiative) expectations of business. A network agreement on environmental protection and on corporate and stakeholder sustainability/longevity. Network connected producibility infrastructure to enhance overall productivity.

Network connectivity: Transforming the network enterprise from a closed system to an open system with free-flowing communication and connections with network partners. Linking with suppliers and customers' platforms. Using real-time analytics to create agile/ rolling value propositions and information/service-led business models to *create value advantage.*

References

Abidi. A, Russo, F., Sommerer, M., & Streif, A. (2018). Digital procurement: For lasting value, go broad and deep. *McKinsey & Company.* November.

Abramowicz, L. J. Banaszak, T. Jaynath, T., & Zarrinkoub, T. (2018), Collaborative contracting: Making it happen, *McKinsey and Company.* July.

Coase, R. H. (1937), The nature of the firm. *Economica,* 4. Retrieved from https://onlinelibrary. wiley.com/action/showCitFormats?doi=10.1111%2Fj.1468-0335.1937.tb00002.x

Dollar, D. (2017). Global value chain development report. *International Bank for Reconstruction and Development, The World Bank.*

Grossman, S., & Hart, O. (1986). The costs and benefits of ownership: A theory of vertical and lateral integration. *Journal of Political Economy.* Retrieved from https://www.investopedia. com/terms/t/totalcostofownership.asp#ixzz5VFdB735M

Hart, O., & Holmstrom, B. (1987) The theory of contracts, in Advances in Economic Theory: Fifth World Congress (T. Bewley, Ed.). UK: Cambridge University Press.

Hardt C, Reinecke, N., & Spiller, P. (2007). Inventing the 21st century purchasing organization, *McKinsey Quarterly,* November.

Hitch, J. (2018, November 6). Nine smart factories lighting the way to a winning industry 4.0 strategy. *Industry Week.* Retrieved from https://*www.*industryweek.com/technology-and-iiot/nine-smart-factories-lighting-way-winning-industry-40-strategy

Meng, B., Xiao, H., Ye, J., & Li, S. (2019). Are global value chains truly global? A new perspective based on the measure of trade in value-added. *IDE Discussion Papers 736, Institute of Developing Economies, Japan External Trade Organization (JETRO).*

Newman, G. (2005, October 22). Flexible Bonds stretched, but working to sew up niche markets. *The Australian.*

Quirk, B. (2011, July 4). How smart companies avoid getting burned by wild dollar swings. Fortune.

Roberts. P, (2009, May 26). Shoemaker puts best foot forward. *The Australian Financial Review.*

Rosen, E. (2015, May). Collaborative manufacturing creates value: The bounty effect: 7 Steps to the culture of collaboration. *Industry Week.* Retrieved from http://www.industryweek.com/leadership/collaborative-manufacturing-creates-value

Taylor, T. (2017, August 25). The smile curve: The distribution of benefits from global value chains. *Conversable Economist.*

Waller. M, & Boccasam, P. (2013), How sharing data drives supply chain innovation. *Industry Week.* Retrieved from http://www.industryweek.com/supplier-relationships/how-sharing-data-drives-supply-chain-innovation

Wang, Z., Wei, S., Yu, X., & Zhu, K. (2017). Measures of participation in global value chains and global business cycles. *Working paper 23222, National Bureau of Economic Research.* April.

Williamson, E. W. (1979). Transaction-cost economics: The governance of contractual relations. *Journal of Law and Economics.*

Ye, M., Meng, B., & Wei, S. (2015). Measuring smile curves in global value chains. *Institute of Development Economics. Discussion Paper No. 530,* August.

Chapter 9
Preservation: Corporate Sustainability

Abstract What is *sustainability*? Is it long-term corporate profitability or corporate longevity, in other words, *commercial sustainability*, a measure of continued financial and market success that provides employment, taxes, and a contribution to GDP and a measure of philanthropy? Or is it *environmental sustainability*, the sensible use of raw materials and recycling and remanufacturing (value renewal) and product-service design that identifies candidate components that may be designed to have multiple life cycles? The answer is that it is both. Both corporate and environmental sustainability are required if the organization and the network it works with are to succeed in creating satisfied stakeholders. Clearly primary stakeholders lean towards commercial success, while secondary stakeholders will favor policy and strategies that support long-term environmental sustainability. They interact with each other, and managements' task is to ensure the interaction is planned to ensure longevity. *Environmental sustainability* is the rates of renewable resource harvest, pollution creation, and nonrenewable resource depletion that can be continued indefinitely. If they cannot be continued indefinitely, then they are not sustainable. It is defined as the responsible interaction with the *environment* to avoid depletion or degradation of natural resources and aims for long-term sustainable "quality." Environmental sustainability is essentially resource based. The debate concerning what some might call "the reckless use of natural resources" has become stronger in the past few years. So much so that it has persuaded "government" to become involved. It has to be admitted that not all politicians are convinced by the "climate change debate," but most national governments are acknowledging the need to consider current use of basic resources and manufacturing processes.

Keywords Sustainability · Governance · Social responsibility · Value renewal · Sustainability equilibrium · Circular economy · B Corporations · Corporate citizenship

© Springer Nature Switzerland AG 2020
D. Walters, D. Helman, *Strategic Capability Response Analysis*,
https://doi.org/10.1007/978-3-030-22944-3_9

Introduction

What is *sustainability*? Is it long-term corporate profitability or corporate longevity, in other words, *commercial sustainability*, a measure of continued financial and market success that provides employment, taxes, and a contribution to GDP and a measure of philanthropy? Or is it *environmental sustainability*, the sensible use of raw materials and recycling and remanufacturing (value renewal) and product-service design that identifies candidate components that may be designed to have multiple life cycles? The answer is that it is both. Both corporate and environmental sustainability are required if the organization and the network it works with are to succeed in creating satisfied stakeholders. Clearly primary stakeholders lean towards commercial success, while secondary stakeholders will favor policy and strategies that support long-term environmental sustainability. They interact with each other, and managements' task is to ensure the interaction is planned to ensure longevity.

Environmental sustainability is the rates of renewable resource harvest, pollution creation, and nonrenewable resource depletion that can be continued indefinitely. If they cannot be continued indefinitely, then they are not sustainable. It is defined as the responsible interaction with the environment to avoid depletion or degradation of natural resources and aims for long-term sustainable "quality." Environmental sustainability is essentially resource based. The debate concerning what some might call "the reckless use of natural resources" has become stronger in the past few years. So much so that it has persuaded "government" to become involved. It has to be admitted that not all politicians are convinced by the "climate change debate," but most national governments are acknowledging the need to consider current use of basic resources and manufacturing processes: some have introduced legislation that was enacted within weeks. The European Community, following extensive research, responded quickly, and EU members were obliged to adopt its requirements rapidly. Extracts from the EU's Official Journal are reproduced in precise form here in order to emphasize the importance placed upon environmental sustainability (European Commission COM (2018) 656 Final):

> The EU's Circular Economy Package (CEP) will see new targets set for the recycling of municipal waste, which the EU estimates accounts for between 7 percent and 10 percent of the total waste generated in the EU. Member states will now be expected to reach a recycling rate of 55 percent by 2025, 60 percent by 2030 and 65 percent by 2035.

> Recognizing the importance of waste management and the circular economy to the EU, the text state: Improving the efficiency of resource use and ensuring that waste is valued as a resource can contribute to reducing the Union's dependence on the import of raw materials and facilitate the transition to more sustainable material management and to a circular economy model.

> That transition should contribute to the smart, sustainable and inclusive growth goals set out in the Europe 2020 strategy and create important opportunities for local economies and stakeholders, while helping to increase synergies between the circular economy and energy, climate, agriculture, industry and research policies as well as bringing benefits to the environment in terms of greenhouse gas emission savings and to the economy.

The CEP also includes specific targets for packaging and separate requirements for biowaste and landfill. EU member states will be expected to achieve stated recycling rates by 2030 for all packaging (70 percent), plastic (55 percent), wood (30 percent), ferrous metals (80 percent), aluminum (60 percent), glass (75 percent) and paper and cardboard (85 percent).

In the UK Parliament, the House of Commons Environmental Audit Committee called an enquiry into the fashion industry as it is a major source of the greenhouse gases that are overheating the planet: discarded clothes are also piling up in landfill sites, and fiber fragments are flowing into the sea when clothes are washed. The committee is investigating the social and environmental impact of disposable "fast fashion" and the wider clothing industry. The inquiry examines the carbon, resource use, and water footprint of clothing throughout its life cycle and looks at how clothes can be recycled and waste and pollution reduced. The committee provided statistics that caused them some concern over the impact of the "fast fashion" industry:

- UK consumption per head in 2017 amounted to 276 kg of clothing product.
- 235 million items were sent to landfill.
- 70,000 fibers are released in a single domestic wash.
- 1.2 bn. tons of carbon emissions were produced by the global fashion industry in 2015.
- 3781 liters of water are used in the full lifetime of a pair of Levi's 501 jeans.

On 27 November 2018, the BBC World News reported CO2 emissions were rising for the first time in 4 years – a definitive United Nations report has found that the world is well off course on its promises to cut greenhouse gas emissions and may have even further to go than previously thought. Seven major countries, including the USA, are well behind achieving the pledges they made in Paris 3 years ago, the report finds, with little time left to adopt much more ambitious policy measures to curb their emissions. "We have new evidence that countries are not doing enough," said Philip Drost, head of the steering committee for the UN Environment Program's (UNEP) annual emissions gap report released in Paris on 27 November (Mooney 2018).

Or, is "sustainability" something more expansive – Is the "Triple Bottom Line" an expansion of "stakeholder value" expectations. A phrase coined in 1994 by John Elkington and later used in his 1997 book *Cannibals with Forks: The Triple Bottom Line of 21st Century Business* describes the separate financial, social, and environmental "bottom lines" of companies. A triple bottom line measures the "company's economic value" (its return on capital used), the "people account" (which measures the company's degree of social responsibility), and the company's "planet account" (which measures the company's environmental responsibility). Elkington argued that companies should prepare three bottom lines – the triple bottom line – instead of focusing solely on its finances, thereby considering the company's social, economic, and environmental impact.

What is the role for *social responsibility*? Social responsibility is the idea that businesses should balance profit-making activities with activities that benefit society. It involves developing businesses with a positive relationship to the society in which they operate; their view is stakeholder based. The International Organization for Standardization (ISO) emphasizes that business' relationships to society and the environment are a critical factor in operating efficiently and effectively. Therefore, social responsibility means that individuals and companies have a duty to act in the best interests of both their environments and their society as a whole. Social responsibility, as it applies to business, is known as *corporate social responsibility* (CSR). Many companies, such as those with environmental sustainability policies, have made social responsibility an integral part of their business models. https://www. investopedia.com/terms/s/socialresponsibility.asp#ixzz5Y6rrOH7h

Corporate social responsibility is a self-regulating business model that helps a company be socially accountable to itself, its stakeholders, and the public. By practicing corporate social responsibility, they can be conscious of the kind of impact they are having called *corporate citizenship*. Corporate citizenship refers to a company's responsibilities towards society. The goal is to produce higher standards of living and quality of life for the communities that surround them and still maintain profitability for *stakeholders*. The demand for socially responsible organizations, individual and as networks, continues to grow, encouraging investors, consumers, and employees to use their individual power to negatively affect companies that do not share their values by not purchasing shares in them. (See https://www.investopedia.com/terms/c/corporatecitizenship.asp#ixzz5Y7eSK1IO.)

To engage in corporate social responsibility means that, in the normal course of business, a company is operating in ways that enhance society and the environment, instead of contributing negatively to it. (For more details see www.investopedia. com/terms/c/corp-social-responsibilty.asp#ixzz5Y7aQMMyj.)

Common examples of corporate social responsibility include:

- Reducing carbon footprints to mitigate climate change
- Improving labor policies and embracing fair trade
- Engaging in charitable giving and volunteering within the community
- Changing corporate policies to benefit the environment
- Making socially and environmentally conscious investments

Some examples of large organizations activities include:

Coca-Cola continues to make strides toward aiding in the alleviation of *environmental issues.* (carbon emissions) After realizing that its fleet of delivery trucks accounted for 3.7 million metric tons of greenhouse gases (GHGs) in 2014, Coca-Cola made significant changes to its supply chain, investing in trucks powered by alternative fuels. Those changes should support the company's goal of reducing its carbon footprint by 25%, by 2020. Ford Motor Company is attempting to improve their *environmental performance*. In an effort to reduce its GHG emissions, an EcoBoost engine was developed to increase fuel efficiency and the company hopes to offer 13 new electric vehicle models by 2020. In addition, American Ford dealerships now use wind sail and solar PV systems as their primary power source.

Improving employment policies and embracing fair trade. For example, Netflix offers their employees 52 weeks of paid parental leave, which applies to both parents. Within that time, employees have the option of going back to work and then resuming their paid leave as it suits them. No matter how they choose to take their leave of absence, they receive their full salary for the entirety of its duration. Netflix also offers unlimited vacation time to their workers, as does LinkedIn. Spotify's benefits include 24 weeks of paid leave for both parents, which can be divvied up however they choose over the 3 years following a child's birth.

Engaging in charitable giving and volunteering. Pfizer makes the list of corporations that "give". The pharmaceutical company raises awareness for non-infectious diseases and works to provide healthcare to women and children who struggle with adequate care. The company's philanthropic activities also include supplying multi-dose vials of its vaccine Prevenar 13 at the price of just a few dollars for one dose. Wells Fargo sets an annual goal to donate up to 1.5% of its revenue to charitable causes such as NeighborWorks. Besides financial donations, Wells Fargo employees are given two paid days off a year to volunteer, which is indicative of a growing trend among corporations. Over half of Fortune's 100 Best Companies to Work For, offer paid leave specifically to encourage their workers to give back to their local community. Motivating employees to get involved in CSR activities is an easy way to strengthen your corporate citizenship.

(See Reputation Management website for more detail: www.reputationmanagement.com/blog/corporate-social-responsibility-examples/.)

Coca-Cola Corporation makes clear their corporate social responsibility policy.

Coca-Cola Europe's Corporate Responsibility and Sustainability Report

People are core to enacting change and must be engaged, well-trained, flexible and skilled.

Big data Products and outputs will begin to be increasingly tailored to customer preferences, which can be tracked through the effective capturing and use of consumer data.

Technology The rise of big data and a shift away from globalization could lead to some radical innovations. Technology will play a vital part, but it must support new ways of working.

Collaboration will develop into a more symbiotic relationship between a company and its supply chain, with more involvement with local communities.

Knowledge sharing and collaboration between competitors to reduce resource use and waste will become key.

Value will have a fundamental impact on what resources are used – encompassing high standards, convenience, trust and "doing good" for human kind as well as the environment as consumers become increasingly concerned about the traceability of products and the value of waste. Developing circularity in how a business controls its resources will become prominent. The "servitization" and the adaptability of products that can be used beyond their intended purpose to deliver 'value beyond profit' will grow.

Resilience – the ability to adapt to change and to do so at speed will be key to future decision-making, as businesses seek to maintain a supply of quality, ethically-sourced raw materials. Flexibility and transparency in both sustainable manufacturing and the supply chain will be vital in delivering this.

(See: https://www.coca-cola.com)

However, there is an argument to be made that environmental sustainability is an important aspect of social responsibility which, together with commercial sustainability, represents aspects of *corporate sustainability*, an "aggregate measure" of both viability (practicability) and (economic) feasibility. *Corporate sustainability* is suggested by Lozano (2013) to have internal and external components. Internal components comprise leadership and the business case; external components are reputation, customer demands, and expectations. This further suggests it to be a prerequisite for doing business. Within the context of this text, it suggests the necessity of ensuring stakeholder satisfaction: meeting the *expectations* of primary stakeholders and the *concerns* of secondary stakeholders. He defines corporate sustainability as: "Corporate activities that proactively seek to contribute to sustainability equilibra, including the economic, environmental, and social dimensions of today, as well as their inter-relations within and throughout the time dimension (i.e. the short, long, and longer term) while addressing; the company's systems i.e. operations and production, management and strategy, organisational systems, procurement and marketing, and assessment and communication; as well as its stakeholders." Given the rapid development and adoption of Industrié 4.0, we would add *technology* to the definition. This is proposed as Fig. 9.1, a perspective of the components,

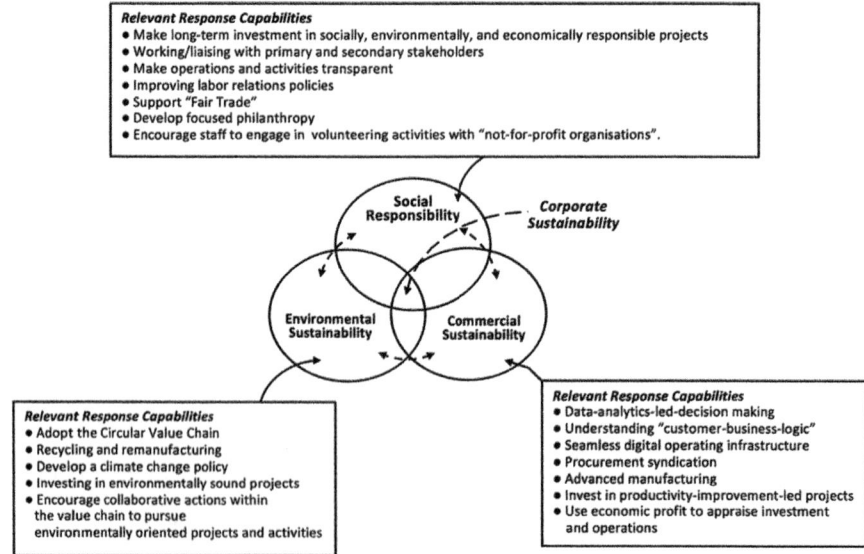

Fig. 9.1 Identifying relevant response capabilities as components of corporate sustainability

their activities, and the fact that interact with each other creating *corporate sustainability* (see: www.reputationmanagement.com/blog/corporate-social-responsibility-examples/).

Herrara (2011) conducted qualitative research among senior executives of large US companies asking them how they saw their corporate sustainability drivers for the future. Their responses suggest a range of considerations very much along the lines of Lozano's respondents:

> Investments that improve efficiency and workflow while allowing the company to meet customers' needs in an environmentally responsible way will drive sustainability efforts, *US Foods*.

> Companies will continue to move beyond sustainability as an obligation driven by outside forces. They will uncover more tangible economic value that drives both top and bottom line growth. But they'll need to get ahead of the curve in order to do that. They'll need to become savvy life cycle practitioners, innovators, and collaborators. The recognition that economic value drives sustainable development should influence decision-making at every turn, from capital investments, to hiring, to marketing products and services, *Dow Chemicals*.

> Across all sectors, corporations have been setting 5–10-year sustainability goals and putting their organizations on a flight plan to be more socially, economically and environmentally responsible. As financial institutions start investing in companies that are setting and making progress toward long-term goals, and companies work closer with stakeholders and increase transparency in reporting, the company will make significant progress toward the best, long-term sustainable impact clients, *PepsiCo*

> A key driver for 2012 and beyond will be the increase in demand for food, feed, fiber and fuels caused by population growth, expansion of the middle class, and urbanization. The impacts of climate change on agriculture may motivate companies to invest in production systems that ensure long-term crop yields along with the healthy soils, watersheds and pollinators essential for food security. This will require harnessing corporate supply chains in ways that go beyond certified product sourcing to stimulate the development of sustainable agricultural landscapes that promote healthy, resilient economies. Leaders in the private sector are ahead of most governments in understanding the urgent need for an integrated strategy that delivers funding and technical support to enable farmers to adapt to climate change, and to maintain productivity and profitability while contributing to reduced greenhouse gas emissions, *Conservation International.*

> No. 1: Our own people. In our annual employee survey, almost nine out of 10 employees believe that we are environmentally responsible and responsible in the community. And these beliefs are one of the top drivers of their engagement. Our employees reflect the wider community: Consumers care about the company behind the brand.
>   No. 2: Money. Being more efficient and reducing risk is good business, particularly in unsettled times, *Molson Coors*

It is argued that sustainability is an important component of corporate social responsibility as shown in Fig. 9.1. A "preservation response capability" should include *all resources,* not to include consideration of such topics as employment benefits, fair trading, development needs of subsidiaries, and supplier organizations in developing countries overlook the opportunities available to fine tune response capabilities within the value proposition.

Figure 9.1 suggests *corporate sustainability* is a confluence (or a convergence) of social responsibility, environmental sustainability, and commercial responsibility. Each has both specific and interacting characteristics, and the strength of an individual/network organization depends upon the depth of analysis applied in research in each of the three components and the subtilty with which the overlap/interacting elements are identified and combined in its value proposition.

B Corporations

B Corporations make a commitment to corporate social responsibility by agreeing a charter of social and environmental development that is purpose-driven and creates benefit for all stakeholders, not just shareholders. B Corporations argue that most challenging problems cannot be solved by government and not-for-profit organizations alone. The B Corp community works towards reduced inequality, lower levels of poverty, a healthier environment, stronger communities, and the creation of more high-quality jobs with dignity and purpose. By harnessing the power of business, B Corps use profits and growth as a means to a greater end: positive impact for their employees, communities, and the environment. B Corps form a community of leaders and drive a global movement of people using business as a force for good. The values and aspirations of the B Corp community are embedded in the B Corp Declaration of Interdependence. Certified B Companies are a new kind of business that balances purpose and profit. They are legally required to consider the impact of their decisions on their workers, customers, suppliers, community, and the environment. This is a community of leaders, driving a global movement of people using businesses as a force for good.

Examples of B Corporations: Drawn from Mazzoni (2016)

Ben & Jerry's overall mission is to make the best product they can, be economically sustainable, and at the same time, create positive social change – specifically to advance new models of economic justice that are both sustainable and replicable.

Ben & Jerry's has a progressive, nonpartisan social mission that seeks to meet human needs and eliminate injustices in our local, national and international communities. They have long supported nonviolent initiatives that seek to achieve peace. In all of their dealings, they are guided by a mission statement which makes the community's quality of life integral to and inseparable from our product and financial goals.

Patagonia grew out of a small company that made tools for climbers. Alpinism remains at the heart of a worldwide business that still makes clothes for climbing – as well as for skiing, snowboarding, surfing, fly fishing, paddling and trail running. These are all silent sports. None requires a motor; none delivers the cheers of a crowd. In each sport, reward comes in the form of hard-won grace and moments of connection between us and nature.

Patagonia's values reflect those of a business started by a band of climbers and surfers, and the minimalist style they promoted. The approach taken towards product design demonstrates a bias for simplicity and utility. For Patagonia, a love of wild and beautiful places demands participation in the fight to save them, and to help reverse the steep decline in the overall environmental health of our planet. They donate time, services and at least 1% of sales to hundreds of grassroots environmental groups all over the world who work to help reverse the tide. Patagonia aims to build the best product, cause no unnecessary harm, use business to inspire and implement solutions to the environmental crisis.

The B Corporation is described on the website (https://bcorporation.net):

Certifying as a B Corporation goes beyond a product or service certification. It is the only certification that measures a company's entire social and environmental performance.

The B Impact Assessment is rigorous. It evaluates how your company's operations and business model impact your workers, community, environment, and customers. From your supply chain and input materials to your charitable giving and employee benefits, B Corp Certification proves your business is meeting the highest standards of verified performance.

Positive impact is supported by transparency and accountability requirements. B Corp Certification doesn't just prove where your company excels now – it commits you to consider your impact on stakeholders now and in the future by building it into your company's legal structure.

The Circular Economy/Value Chain

The Ellen MacArthur Foundation was launched in 2010 to accelerate the transition to a circular economy. Since its creation, the charity has emerged as a global leader, establishing the circular economy on the agenda of decision-makers across business, government, and academia. With the support of its global, philanthropic, and knowledge partners, the Ellen MacArthur Foundation works across five areas, business, institutions, governments and cities, insight and analysis, and learning and systems, to bring positive economic change on a global level. The foundation aims to inspire and transmit circular economy thinking through insights and analysis, learning activities, and collaborative business opportunities.

Companies need to build core competencies in circular design to facilitate product reuse, recycling, and cascading. Circular product (and processes) design requires new skills, information sets, and working methods. Areas that are important for economically successful circular design include material selection, standardized components, designed-to-last products, design for easy end-of-life sorting, separation or reuse of products and materials, and design-for-manufacturing and assembly (DFMA see Chap. 7 *Producibility* Criteria) that consider possible reuse and other useful applications of by-products and wastes and the remanufacturing and repurposing of the products once their planned life cycle expires.

The foundation's research identifies opportunities to extend the concept into "circular communities and cities." Of particular interest to this chapter is the circular value chain, as the components of the convergence of Industrié 4.0, Value Chain Network 2.0, and the acceptance of stakeholder value satisfaction have to incorporate the logic of the circular economy; we see that stakeholder value management requires a more comprehensive and descriptive statement: product-service design that identifies candidate components that may be designed to have multiple life cycles and manufacturing and distribution processes that are designed with recycling and remanufacturing (value renewal) in mind. It is important that sustainability is understood within the context of both the organization (individual companies and their network partnerships) *and* the resources environment if effective response capabilities are to be developed. The Ellen MacArthur Foundation proposes a circular value chain network: one that is *restorative and regenerative by design and which aims to maintain products, components, and other inputs at their optimal value to end-users at all times, distinguishing between technological and biological cycles.* (www.ellenmacarthurfoundation.org)

The work of the foundation has strong implications for the response capability model in this regard because much of what they advocate will require changes in current practices. Possibly the largest is a change in attitude towards moving away from the linear production processes of manufacture-sell-dispose to the circular model of manufacture-sell-recover-remanufacture.

This will require further physical changes to manufacturing and distribution facilities. Some examples will indicate the changes occurring and the results being realized:

Heineken is pursuing circular manufacturing practices across its entire value chain – thinking holistically; from water treatment, to energy conservation, composting yeast and recycling bottles.

Philips introduced energy efficient technology (LED street lighting) by offering a product-SERVICE approach to enable municipal authorities to take advantage of energy savings without incurring up-front investment.

Renault (Choisy-le-Roi) re-manufactures some 200,000 components. It suggests these to be 30–50 percent less expensive than new. Re-manufactured parts have the same warranty and QC (quality control) process. The plant uses;

- 80% less energy
- 88% less water
- 92% less chemical products
- 70% less wasted production
- No landfill services are used

Parts are being designed with re-manufacturing in mind; disassembly and recyclability. Product interchangeability is an objective.

Caterpillar (CatReman) suggests re-manufacturing is " … not about using less and less material but rather creating components that can be remanufactured a number of times. With fixed costs at 35% and variable costs (materials) 65%, remanufacturing offers increased profits. For example, Caterpillar designs engine blocks with removable sleeves – reducing the time and cost of a reboring process.

Apple (Liam) a manufacturing process application that identifies and extracts serviceable components from returned Apple products.

The shift to a circular economy requires innovative business models that either replace existing ones or seize new opportunities. Companies with significant market share and capabilities along the linear value chain can play a major role in developing the circular value chain business model by leveraging their collaborative experiences in vertical and horizontal integration.

Clearly there are problems with the existing business model. It is arguable that up until very recently the generic industry business model has been linear, with scant evidence of anything to suggest an alternative to the dominant linear model; raw materials are extracted (harvested, etc.) from the environment and transformed into products and eventually disposed of. In the linear system such eco-efficient techniques that do exist seek only to minimize the volume, velocity, and toxicity of the material flow system but are incapable of altering its linear progression. Some materials are recycled but often as an "add-on decision process"; no decisions to accommodate this activity have been made at the design stage and often as a response to regulatory decisions and mandates. This is in effect not recycling but downcycling as it results in a downgrade in its quality and limits its subsequent usability.

Beyond the Circular Economy

Hannon et al. (2016) of McKinsey offered an approach to developing a strategic circular operating model, with "developing products for a circular economy." The authors suggested a staged approach: The *first* is to devise a highly collaborative product-development process with network partners that help to determine sourcing requirements, production methods, marketing, sales, serviceability, and how the product-service is handled at the end of its life. The second is to use, *design thinking*, which can help identify, ways of meeting customers' needs with much greater resource efficiency than in the past and add more value throughout the network.

The authors suggest the product-development process begin with a premise, based upon the notion of a circular economy. Instead of considering only functionality and cost and assuming that products will be consigned to waste management, the network partnership would consider how it might manage the entire life cycle of its products in order to maximize the total value added by them and their component materials. For example, a mechanical components manufacturer might give customers rebates for returning end-of-life parts, so the manufacturer can refurbish them for resale at a lower price or dismantle them for recycling or possibly (as Caterpillar has with *Cat Reman*) creating components that can be remanufactured a number of times.

The McKinsey article provides examples of the *circular* model working as a collaborative network model. Product development is becoming a collaborative process involving the whole value chain; value chain partnerships develop effective investment profiles and efficient scale-based production units, thereby offering competitive low-cost and high-price/value product-service offers.

A medical equipment company, for example, gave its sales department ambitious targets in emerging markets and a portfolio of high-priced products. Bringing together sales, product development, and other teams revealed this problem and gave product developers a chance to help solve it. They realized that by refurbishing used medical equipment from developed countries, the company could offer a lineup that would be appealing and affordable in emerging markets.

The medical equipment company's experience illustrates another benefit of making the product-development process more collaborative: it helps companies center the process on customers' needs rather than product specifications. It is interesting to note that UK earth-moving equipment manufacturers used this principle in the 1960s to dispose of obsolescent equipment and to open markets in Africa.

Customer centricity, creating products with customers first in mind, becomes very powerful when design thinking is taken beyond a "product-service-in-use" approach and is based on what it is the customer organization's business is about. For example, airlines are in a service business; recent moves by aerospace manufacturers to focus on selling the "output" of their product offer result in vendors taking an asset management perspective of the customers' needs. Selling "power-by-the-hour," not product hardware, allows the customer to invest in intangible assets (such as brands and differentiated routes and inflight service facilities), thereby finding the best way to meet customers' needs, rather than the best way to design products.
The PRODUCT-service becomes a product-SERVICE
The authors provide an excellent example of a product-SERVICE:

> A floor covering company, Desso, have introduced a carpet-leasing service. Instead of buying carpet, customers now have the option to lease carpet from the company, which takes care of installation, maintenance, and removal. This arrangement gives Desso an incentive to manage materials efficiently. Indeed, Desso has cut waste and reduced its consumption of virgin material by treating old carpet as a valued commodity. The company collects carpet from its customers and other sources, including its competitors, and removes the fibers from the backing. The old fibers are recycled into new fibers; the backing is used as an ingredient in roads and roofs.

Clearly the traditional linear business model requires reimagining as it is failing to consider the long-term aspects of *corporate sustainability* as it is becoming to be accepted. Beal et al. (2017), Boston Consulting Group (2018), pursue a similar path to Lozano (2013) concerning the "width and depth" of stakeholder value expectations. Their contribution, *Total Societal Impact: A New Lens for Strategy*, is very similar to *corporate sustainability* and explores the potential for TSI (total societal impact). Their argument replicates that of Lozano (2013) (and these authors) that have been identified in this chapter. We all argue that enlightened organizations have moved on (or are moving on) from the notion of a business purpose focused solely on creating shareholder wealth.

> A company's TSI (aka Corporate Sustainability) includes the impact of its products and services, its operations, and its corporate social responsibility initiatives. It also includes the

result of explicit decisions the company makes to adjust its core business to create positive societal benefits. Activities related to TSI often have a material impact on total shareholder return (TSR) - but not always. (Beal et al., 2017)

Beal et al. (2017) provide some hypothetical "examples" of corporate approaches:

The intrinsic benefit to society of a product or service (a drug that saves lives, for instance, or a bank loan that enables a farmer to buy a plough).

Strict adherence to ethical business rules and inclusive hiring policies, that directly or indirectly impact societies in the countries and communities where the company operates.

The jobs created as a result of the materials a company purchases and services associated with the company's supply chain.

The impact on the environment—both negative (such as the environmental footprint of operations) and positive (such as innovations that reduce pollution).

Towards a Structured Approach to Using Corporate Sustainability in Developing Response Capabilities

If organizations are to address the changes and challenges of the "convergence" of these game-changing developments, an organized approach is required. The initiators of the changes are, in the most part, educated and career minded. They are persuading networked value chains to be mindful of the "total environment"; not just that part of it that, hitherto, provided a one-way stream of resource inputs but of the component of the environment that influences consumer convenience, the use of time, the changing attitudes towards ownership (short-term availabilities rather than long-term of inflexible, tangible assets), the changes in interactions at work and in leisure situations, and of the impact of gender at work, in politics and casual (or not so casual) relationships. Given the network structure of business, maybe we should be labelling corporate sustainability and organizational sustainability, to reflect the interdependencies that are being created at all levels of activity in the value chain *and* the collaborative integration that the digitization of products and processes currently ongoing, the expectations prompted by data analytics, and the algorithms being constructed.

Chapter 1 introduced the notion that stakeholder-led value was eroding the older view of the purpose of the business; its primary duty is/was to create shareholder value, principally through dividends, generated from profitable activities, and increasing the value of its shares by growing the business organically or by astute mergers and/or acquisitions. In a review of changing perspectives of value, the chapter identified the declining need for ownership that is being replaced by a need to pay for the use or utility value an asset has and the convenience of access to its availability.

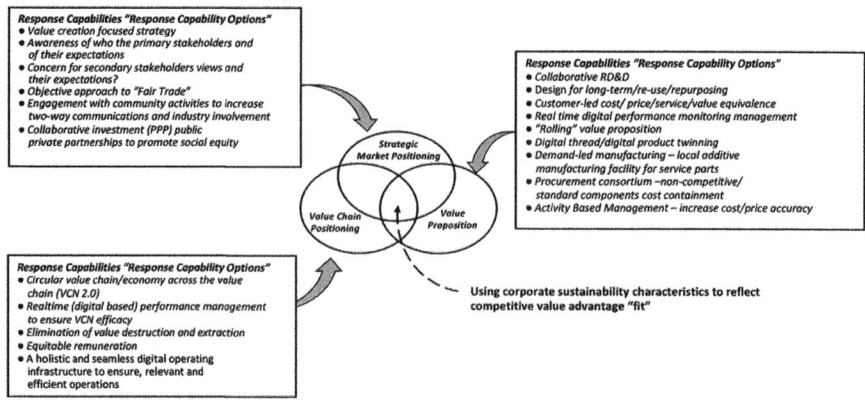

Fig. 9.2 Identifying relevant response capabilities that create corporate competitive positioning advantage from focused corporate sustainability decisions

The chapter also reviewed the role of the value proposition in a business environment undergoing rapid changes sparked off by the convergence of Industrié 4.0, Value Chain 2.0, and the growing importance of stakeholder value management; the chapter identified one important change as being the "Rolling Value Proposition" – the benefit of connectivity providing customer and supplier with real-time data with which to monitor performance and to use the performance data to make operating changes *and* for the supplier to make ongoing changes to the value proposition and the product-service.

The value proposition offers a means by which *response* capabilities that are significant to customers may be identified and how they may then be used to communicate their relevance to customers in the value offer to customers. Figure 9.2 aligns the response capabilities with the value proposition. Emphasis is added by including *strategic market positioning* and *value chain positioning*. By including strategic market positioning, a message is passed to stakeholders demonstrating the relevance of the product-service to the strategy portfolio; value chain positioning communicates with suppliers (components, services, distributors, and serviceability providers). Figures 9.1 and 9.2 suggest an approach. Figure 9.1 identifies relevant response capabilities as components of corporate sustainability; Fig. 9.2 extends the process by identifying relevant response capabilities that create corporate competitive value advantage positioning advantage that emerge from focused corporate sustainability decisions.

Building Corporate (Organizational) Sustainability into the Business Model Review

The analysis required to reach decisions concerning the response capabilities that will support the organization's future strategic direction is explored in Fig. 9.3.

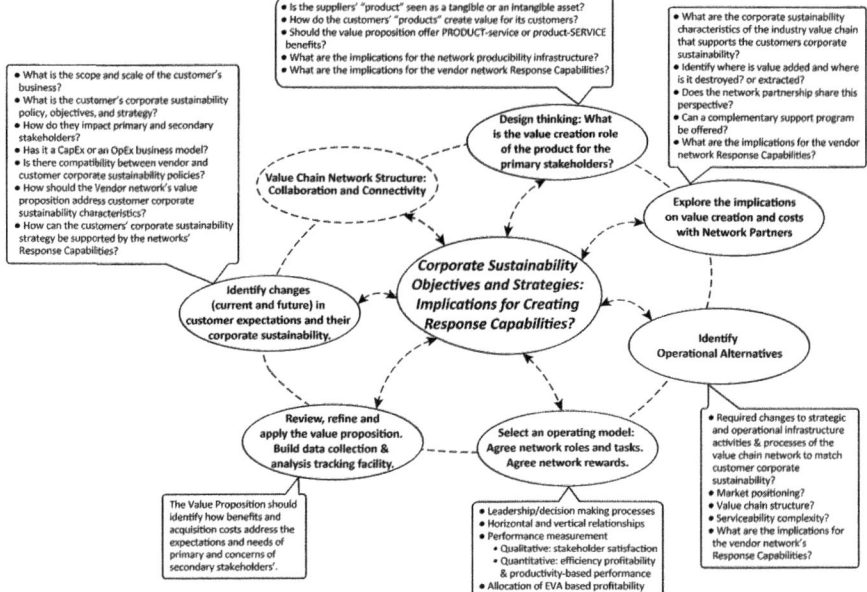

Fig. 9.3 Corporate sustainability: Determining response capability requirements

Corporate Sustainability Objectives and Strategies: Implications for Creating Response Capabilities

Given the benefit of observing ongoing corporate decisions and practicing managers' contributions, Fig. 9.3 describes a process that helps identify stakeholder expectations. Primary stakeholders are customers, employees, management, shareholders and investors, and partners' suppliers and distributors. Secondary stakeholders comprise competitors, the community, and government. Primary stakeholders welcome open, close, and ongoing relationships with organizations, involving the exchange of creative ideas, innovation, management approaches to change, and performance improvements. The interests of secondary stakeholders have specific focus: competitors monitor progress and have interests in opportunities for collaboration (typically co-opetition, licensing patents, and using any excess capacity they have to manufacture and distribute competitors product-services); the community is interested in employment opportunities and expansion of revenues (and therefore, in local taxes); and government has interests that include employment, innovation, exporting activities, and a contribution to GDP. A clear understanding of both stakeholder groups' expectations of the organization, its product-services and corporate sustainability policy, and its implementation will enable the organization to identify relevant response capabilities and can invest in them and the future.

Identify changes (current and future) in customer expectations and corporate sustainability. Several questions need to be asked and answered concerning the customers' businesses. What is the scope and scale of the customer's business?

Customers' corporate sustainability policies, objectives, and strategies and their impact on primary and secondary stakeholders? Is it a CapEx or an OpEx business model (which provides an insight into its financial strategy and its ability to react to stakeholder expectations of corporate sustainability. The compatibility between vendor and customer corporate sustainability policies. How the vendor network's value proposition should address customer corporate sustainability characteristics. Identify options for supporting the customers' corporate sustainability strategy by the networks' response capabilities

Design thinking: What is the value creation role of the product for the primary stakeholders? It is important to identify how the customer creates value for its customers and how this is incorporated as a value delivery activity. The customer PRODUCT-service versus the product-SERVICE decision is a major input into its suppliers strategy and structure decisions. Sheppard et al. (2018), at McKinsey, comment concerning the customer's interpretation of the value it offers downstream is important. Value chain partnerships develop effective investment profiles and efficient scale-based production units, thereby offering competitive low-cost and high-price/value product-service offers. However, it should be remembered that stakeholder value-led management implies that the total value set may differ in some way, for example, the industry may have competitors that profit from "patent hopping"; they monitor competitors' product development activities (by patents granted); if the patent suggests that it will give its owner some advantage in the market, they negotiate a price and access to the license where upon they will focus their product development activities on the license, thereby saving RD&D investment and time.

"Similars" (biosimilars in the pharmaceutical industry) are product-services that are designed to produce an end result that is the same as a patent-protected product or process. Similars depend upon a range of market characteristics, high price, volume markets, and insufficient supply, for success. There are examples other than the pharmaceutical drugs; often low-cost manufacturers identify successful high-price/high-quality products; they "commoditize" the product and produce low-price "similars," typically the product is not branded; and they offer no service. Hilti (a leading hand-power tool manufacturer) realized that its value proposition with its emphasis on being a premium brand had lost its visibility with major customers. The product was becoming "commoditized"; users began to regard it as disposable, and after-use care was ignored. The hand-power tool market has been "devalued" as a flood of "look-alike" products from Asian manufacturers gained market penetration; this introduces another problem for "premium brand" manufacturers – imitation – as many of these products often infringe international copyright agreements. Hilti responded with a change in its value proposition; the PRODUCT-service became a product-SERVICE; Hilti undertook to maintain client productivity by offering a full-time replacement service.

Explore the implications on value creation and costs with network partners. Creating a value proposition and structuring its network operations to identify strategic and operating cost options consider more than just identifying network resource requirements of managing and implementing a tangible value proposition.

It is clearly insufficient only to consider *operations response management* as simply delivering and servicing the value proposition, monitoring value delivery and performance, and implementing changes to achieve customer expectations. Furthermore, it is apparent that the *circular economy/value chain model* on its own is insufficient. The concerns involved in value delivery are shared by secondary stakeholders (as well as the primary stakeholders) and may have long-term perspectives. For example, compressing time for mass producers of automobiles and consumer durables with just-in-time operations has implications for infrastructure serviceability. There have been observations made concerning the impact of the increase in delivery frequencies on road maintenance – a concern for local governments who welcome the increase in employment on the one hand but are not too enthusiastic concerning the road maintenance costs, on the other. Clearly there are other issues to be considered.

Identify operational alternatives. The connectivity and flexibility facilities of Industrié 4.0 combined with the more collaborative approach of Value Chain 2.0 have made the notion of customization and demand-led manufacturing almost possible. The *producibility* model of an integrated operations infrastructure, supported by low-volume producers (e.g., Magna Steyr), has increased the product variety available in the automotive industry; the Li and Fung model for textiles/apparel products fulfils the same facility. An interesting and new consideration for manufacturing companies is that depending upon their interpretation of the response capability requirements and new aspects of data availability, they can adopt relevant market, value chain, and serviceability positioning depending upon the customer. Not so new, but of relevance given the data availability facility of Industrié 4.0 is a contribution by Leszinski and Marn (1997) who argued that customers do not buy solely on low price, they buy according to *customer value,* the difference between the benefits a company gives customers, and the price it charges. Therefore, customer value equals *customer perceived* benefits minus *customer perceived* price. Within this equation the acquisition costs are costs involved in obtaining and owning the product-service (i.e., operating costs, modifications, and decommissioning costs) that may be considered to be "negative" benefits or a component of the price. See Fig. 9.4a.

The authors propose a "value map" to explore customer value and price/benefit trade relationships. They suggest that with constant market shares, the market's offers will align themselves along a *value equivalence line* (VEL) identifying the range of choice available to customers. Any changes in market share will be indicated by horizontal movement (left suggesting share loss as the customers perceive less value – right suggesting an increase in relative value).

As the authors contend, the value equivalence line may shift downwards and/or towards the right, reflecting either a decrease in market prices or perhaps an increase in the benefits that are available. This may occur because of production efficiencies developed by one or more competitors or due to design or technology advances that individual R&D processes have developed (see Fig. 9.4b). Essentially the model identified customer groups (segments) who share a specific product-service, or brand, that offers their required price/quality expectations. For example,

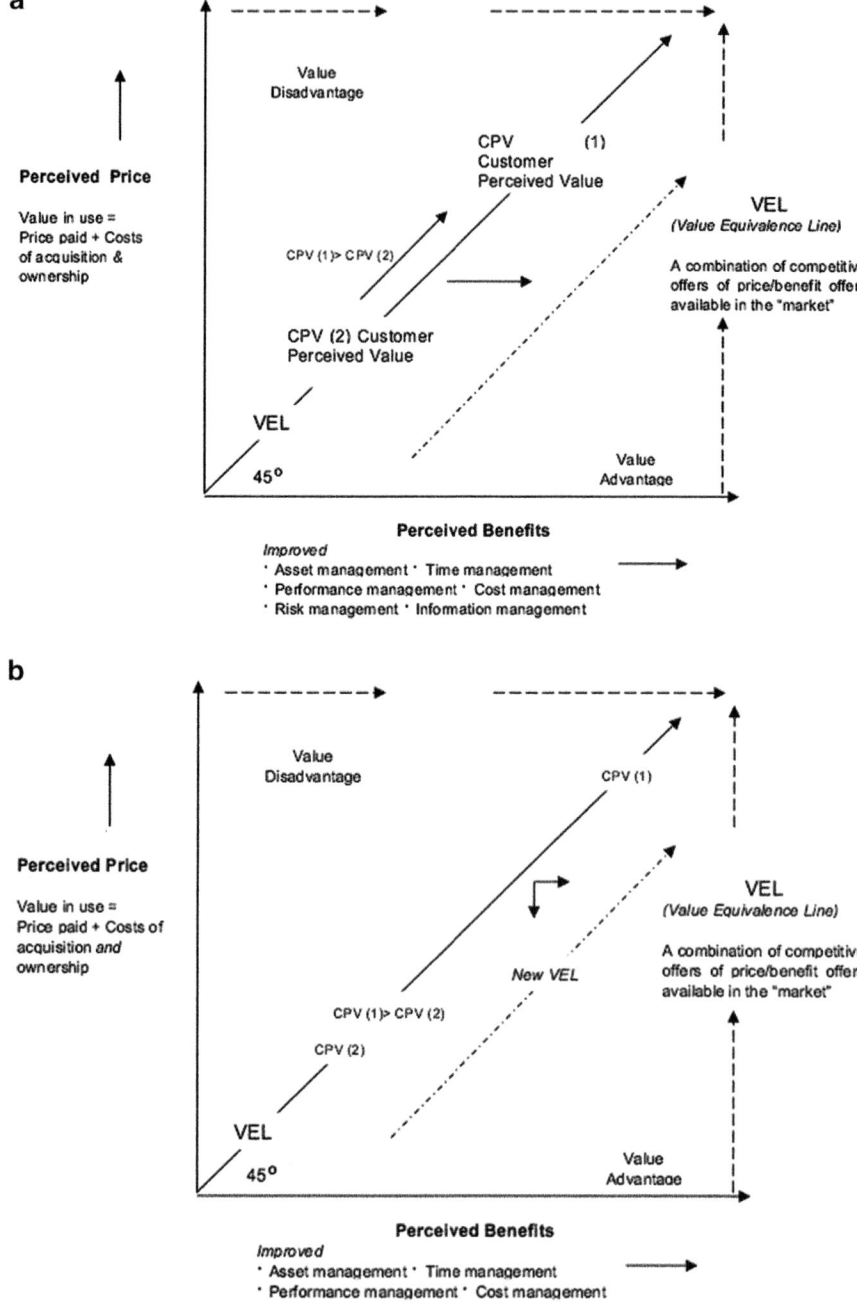

Fig. 9.4 (**a**) Value, price, and value equivalence. (**b**) Value, price, and value equivalence

the automotive market comprises a wide range of vehicles at differing quality (equipment) levels at an equally wide range of prices; a small vehicle with manual transmission and a basic radio/CD player (CP 1) is all some buyers need; by contrast a top of the range Jaguar will offer automatic transmission plus, satellite navigation, parking assistance, road safety warning sensors, etc. (CP 2) on Fig. 9.4a. If both customers are on the diagonal line (VEL), then their price/value equivalence expectations are balanced, and the manufacturers need do nothing to improve their value propositions. Figure 9.4b suggests that competition and changes in consumer expectations can occur (value migration) which disturbs the price/value equivalence relationships and vendors then need to reconsider their offers.

The relevance here is that the value equivalence line identifies different price/value perceptions (preferences) and each offers different combinations (market segments) of PRODUCT-service and eventually product-SERVICE and different response capability expectations each of which offering different opportunities (operational alternatives) and returns. The "convergence" offers opportunities to explore alternatives in more detail than before.

Select an operating model: agree network roles and tasks and agree the allocation of network rewards. Given agreement on the structure and membership of the value delivery network performance parameters (ideal and minimum values for volumes, margins, etc.) are to be agreed and set. Clearly, customer perceptions are important if the role of the product-service fails to meet customer expectations at a price the customer is prepared to pay the price/value equivalence model requires investigation. Real-time data monitoring enables managers to get a clear picture of their entire process and to make timely decisions to avoid unnecessary down/change over time. Once the data is collected in a database, statistical process control (SPC) or machine-learning technologies carry out predictive analysis of that data providing insights into data that, hitherto, could not be seen just by looking at it, because of the huge volume of data available. Moreover, there is evidence to suggest that the "convergence" offers an opportunity for greater flexibility with fit4purpose (quality, cost, serviceability, delivery options, etc.). In the value chain network structure, *collaboration and connectivity* are critical in creating and managing value chain network partnerships in delivering value to stakeholders and are an essential feature of the model. The "convergence" of Industrié 4.0, Value Chain 2.0, and stakeholder value management as interpreted by the IIOT (Industrial Internet of Things) has enabled the connectivity of production processes *and* connectivity between products, thereby increasing the value delivered. Both strategic and operational performance require monitoring; Fig. 9.5a, b suggests headings and topics for this purpose. Figure 9.5a, b suggests a range of performance measures for strategic and operational activities.

Review, refine, and apply the value proposition by building data collection and analysis tracking facilities. The value proposition has been refined throughout the overall process, and the relevant response capabilities are identified and being developed or obtained by partnership leverage agreements. Figure 9.6 provides a hypothetical example of the overall process together with examples of the response

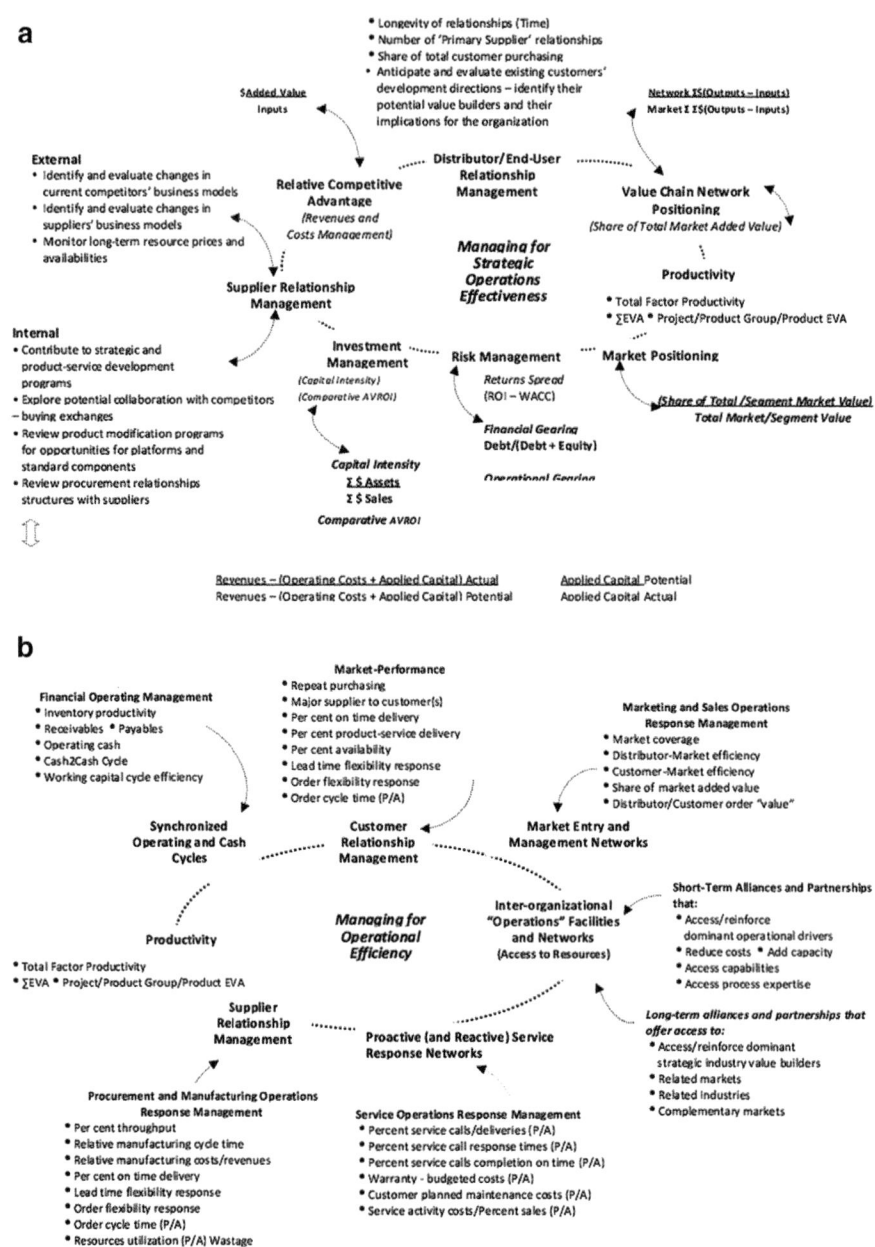

Fig. 9.5 (**a**) Managing strategic performance. (**b**) Managing operational performance

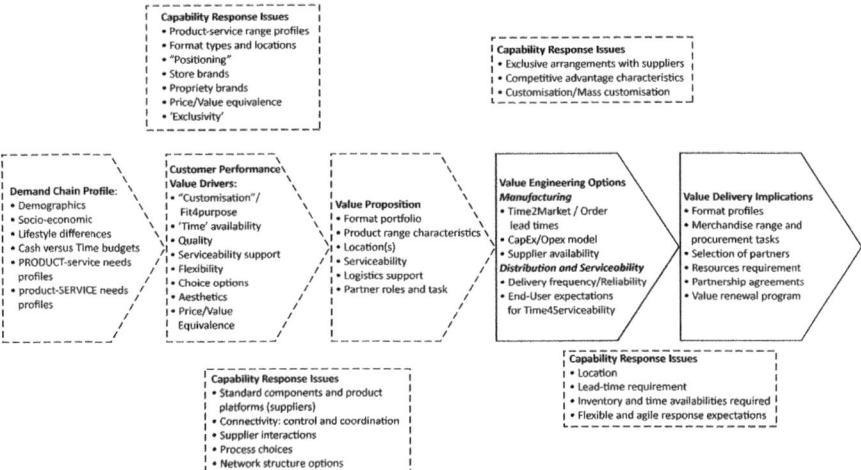

Fig. 9.6 Using demand chain analysis to develop and apply the value proposition

capabilities. The data capture output can be designed to identify the potential for value renewal of the product-service and design capability together with customer acceptance of remanufacturing.

To provide data feedback on the ability of the value proposition in meeting stakeholders' expectations, a data collection and analysis tracking facility is the purpose of Fig. 9.7. The tracking model addresses the interests of stakeholders and details marketing and financial performance expectations and identifies relevant response capabilities. The data capture output can be designed to identify the potential for value renewal of the product-service and design capability together with customer acceptance of remanufacturing.

Convergence

For the contemporary business model to be successful it is necessary to take a broader definition of preservation. The remit of corporate or organizational sustainability extends well beyond the environment of natural input resources and takes on an ecosystem approach. Value Chain Networks 2.0 consider themselves to be integrated, connected and collaborative business units. They are stakeholder driven which implies a responsibility portfolio that has increased beyond the traditional concerns of shareholder value management to include; maintenance of employment, "fair' wage rates, staff development; customer interests and expectations are important as are liaison with consumer reference groups and government sponsored consumer protection associations. It also extends to a consideration of the concerns of secondary stakeholders concerning involvement with communities and "interest" groups' concerns. Involvement in education is becoming important particularly

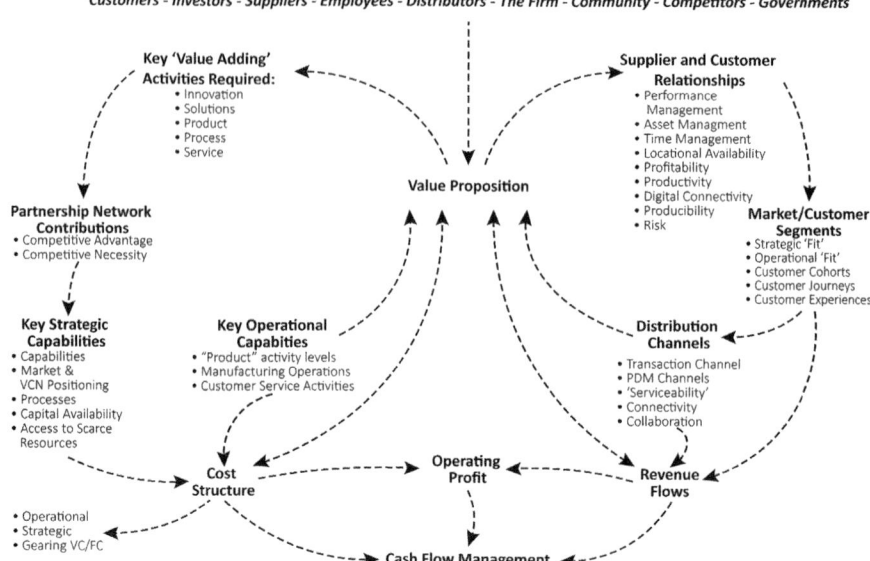

Fig. 9.7 The integrating role of the value proposition, response capabilities, relationships, and performance metrics

with curriculum changes being made to meet the STEM (science, technology, engineering, and mathematics) requirements for high school level students. The Toyota Trust is actively with developing a better understanding of STEM in schools; robotics kits and "scienceworks" excursions for disadvantaged primary schools, and a scholarship program to alleviate financial barriers experienced by talented and motivated students wanting to pursue STEM related education and careers. A review of Individual and network relevance of corporate sustainability is apparent. Toyota's corporate sustainability demonstrates the statement that preservation extends well beyond organizations developing corporate sustainability programs. Toyota's program is based upon corporate governance (Toyota regards sustainable growth and the stable, long-term enhancement of corporate value as essential management priorities. Building positive relationships with all stakeholders, including shareholders, customers, business partners, local communities and employees, and consistently providing products that satisfy customers are key to addressing these priorities), risk management (Toyota has been working to reinforce its risk management structure since the recall issues in 2010 and has appointed risk managers globally and at each level of its operations to prevent and mitigate comprehensively the impact of risks that could arise in business activities) and compliance (The Guiding Principles at Toyota state that Toyota shall honor the language and spirit of the law of every nation and undertake open and fair business activities to be a good corporate citizen of the world. Toyota believes that adhering to this principle is to fulfil corporate

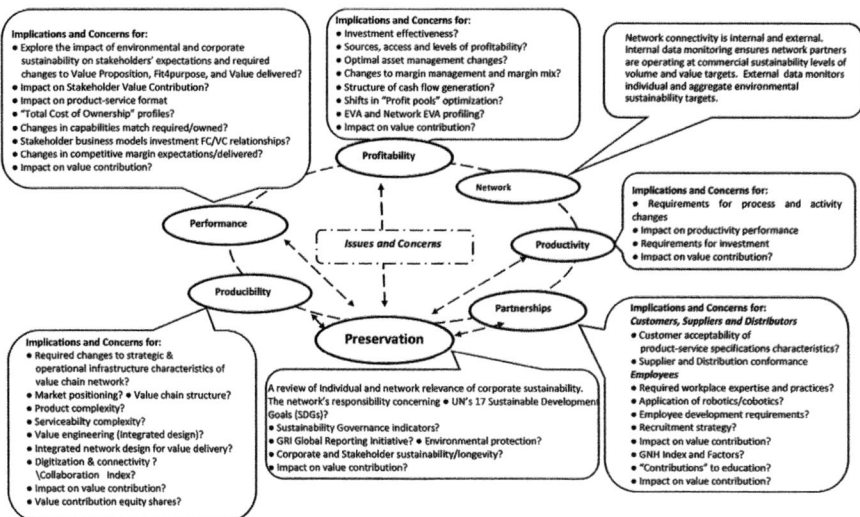

Fig. 9.8 Issues and concerns: Preservation

social responsibility and ensure compliance. In accordance with its basic internal control policies). (Fig. 9.8).

Performance: A major concern for performance can be the impact of environmental and corporate sustainability policies that may impact on current stakeholders' expectations and the required changes to value proposition fit4purpose and value delivery they impose. The adoption of the circular economy/value chain will impact on product-service format decisions and on "total cost of ownership" profiles.

Changes in capabilities match required/owned are likely to differ given a possible change in input materials and in changes in activities and processes if significant regulatory compliance requirements appear to be unavoidable.

Profitability: Business model investment fixed cost/variable cost profiles and relationships may change if organizations need to change elements of their corporate/network sustainability policies. The result could involve rethinking the extent of manufacturing and serviceability activities that are marginal in terms of investment effectiveness, profitability, and risk. Changes in competitive margin expectations/delivered should be monitored as will be the impact on value contribution. Other aspects of profitability that may be affected are sources, access and levels of profitability, optimal asset management changes, changes to margin management and margin mix, structure of cash flow generation, shifts in "profit pools" optimization, and organizational EVA and network EVA profiling.

Productivity: There are implications and concerns for requirements for process and activity changes, impact on productivity performance and the requirements for investment, and impact on value contribution. The use of EVA monitoring could prove to be valuable when comparing the implications of change and the options that appear available.

Partnerships: The primary importance of partnership management is to build and maintain value-generating partnerships. There are some aspects of corporate sustainability that require review as organizations implement (or make amendments to) corporate sustainability programs. *B2B customer* acceptability of changes to product-service specifications characteristics are essential if changes to product-service efficacy are likely to occur (either positive or negative) and should be explored with customers. This is essential as the changes may have a major impact on downstream value delivery processes. For example, the introduction of sell by/best by dates had major implications for manufacturer/retailer supply changes as order cycles became mandatory performance metrics. *B2C customers* of generic and biosimilar versions of drugs require assurance that the product-service impact is the same as the heavily promoted branded original version. *Suppliers and distributors* in established value chains are invariably impacted by product-service innovations that may be imposed by regulators; there has been intervention in a number of markets prompted by government concerns on the impact on health of tobacco and alcohol. Climate change is likely to become a topic that attracts legislation in the shape of renewable energy systems and emission controls. These changes will have impact on both value engineering and value delivery of power supply and "total vehicle mobility" (public and private transportation markets), much of which is currently being discussed. Supplier, distributor, and end-user change in the automotive industry can be seen to be occurring. *Employees* are important partners and are impacted by sustainability changes in a number of ways; required workplace expertise and practices, the application of robotics and cobotics, changing employee skills development requirements, and organizational recruitment strategy.

Producibility: the "convergence" of Industrié 4.0, Value Chain Network 2.0, and the emphasis on stakeholder-led-value management has made the cost-effective and efficient integrated operations infrastructure attainable. Corporate sustainability as a *preservation response capability* may require changes to the producibility model. The automotive industry is one example; the EV (electric vehicle) and automatic/driverless vehicles are no longer "what-if" concepts but are "when and where" issues. Bloomberg (6 December 2018) reported Volkswagen's intension to "accelerate a push to lower costs and lift profits at its biggest car brand to add financial muscle as the world's largest automaker pursues a $50 billion technology shift to counter Tesla Inc". And "over the next 5 years, the VW brand alone plans to spend 11 billion euros on next-generation technology, including 9 billion euros for electric cars as it increases the number of battery-powered models to 20 by 2025 from two now." The article also cites the similar intentions of PSA Group's Opel unit will add three pure-electric variants over the next 2 years to meet stricter emission rules in Europe. Aerospace is also responding to change. The Boeing 787 marked a change in materials inputs as carbon composites replace conventional metals. Changes in the assembly process were made so that supplier-built component module assemblies and then shipped these to Boeing for final assembly. This took some time to become effective.

Network Connectivity: An internal and external network connectivity is essential to ensure network partners are operating at commercial sustainability levels of volume and value targets. External data monitors individual and aggregate environmental sustainability targets. The external data network also connects the network to allied research centers and activities such as the MacArthur Foundation to RD&D activities of "similar" industries to identify "best practice" technologies and identify research activity in materials.

References

Beal, D,. Eccles, R., Hansell, G., Lesser, R., Unnikrishnan, S., Woods, W. & Young, D. (2017, October 25). Total societal impact: A new lens for strategy. *BCG (Ideas and Inspirations)*.
European commission (2018). Report from the commission to the European Parliament, the council, the Economic and Social Committee and the Committee of the Regions. Com. Brussels.
Hannon, E., Kuhlmann, M., & Thaidigsmann, B. (2016, November). *Developing products for a circular economy*. Berlin: McKinsey Quarterly.

Herrara, T. (2011, December 30). What will be the biggest driver for corporate sustainability in 2012. *GreenBiz.*

Leszinski, R., & Marn, M. (1997). Setting value, not price. *The McKinsey Quarterly, Number One, 1*, 98–115.

Lozano, R. (2013). A holistic perspective on corporate sustainability drivers. *Corporate Social Responsibility and Environmental Management, 22*, 32–44. Wiley Online Library.

Mazzoni, M. (2016, December 10). 3P weekend: 12 B corps leading their industries. *Triple Pundit (The Business of Doing Better).*

Mooney, C. (2018). Countries vowed to cut carbon emissions. They aren't even close to their goals, U.N. *The Washington Post*, November 27.

Sheppard, B., Sarrazm, H., Kouyoumjian, G. & Dore, F. (2018, October). The business of value design. *McKinsey Quarterly.*

Part III
Introduction

Part I of the text reviewed recent topics that have increasing importance and influence on the business environment in which progressive organizations are working. Part one (A Changing and Challenging Future: How the Future Is Developing) considered changing perspectives of value, industry dynamics, and capabilities as strategic response resources – an interactive model. We suggested the changes ongoing across these topics identified the need for agile thinking and actions when considering future direction and specifically when exploring the capabilities required for successful responses.

Part II undertook a detailed exploration of each of the response capabilities suggested (from ongoing research) of choice options and decisions being taken by progressive organizations and networked structures to which they belong. Part II also offered a review of current practice and application of management thinking, and application within the disciplines is outlined as interacting capabilities seen as requirements for success in an increasingly changing business environment. The disciplines identified are *performance* (strategic and operational marketing, finance, and operations), *profitability* (economic profit, economic value added (EVA), asset management and gearing financial and operating, profit pool analysis as a method of reviewing past and future investment activities, and risk assessment), *productivity* (the problems of current measurement methodology and the use of EVA as an alternative), *partnerships* (the increasing relevance of networks (Value Chain Networks 2.0) increased integration, coordination, collaboration, and transparency brought about by the richness and reach of data availability and analysis), *producibility* (the concept identified by the creation of production infrastructures that link the activities required to create an end-to-end "connected" series of activities from RD&D, manufacturing, distribution, consumption, and, increasingly, remanufacturing and repurposing), and *preservation* (the notion that sustainability concerns organizational longevity providing stakeholder satisfaction, as well as longevity of environmental resources and limitation of damage caused by production process pollution).

Part III commences with a review of the impact the "convergence" of Industrié 4.0, Value Chain Network 2.0, and the shift towards the broader concept of stakeholder value management (away from shareholder value management) as the focus (or alignment) of profit and not-for-profit organizations. Following discussion of the impact of the "convergence" on production, two chapters, *building the capability-based business model* and *managing the business model,* pursue the application of "the convergence" to current strategic and operational management.

1.1   Creating, Managing, and Capturing Value Using the "Convergence of Industrié 4.0, Value Chain Network 2.0, and Stakeholder Value Management

The increasing number of references to the role of demand management in the value chain (i.e., the ability to decrease the response time to customer orders as well as the increasing ability to customize their order) requires a concise perspective of a value-based strategic and operational business model, described by Fig. P3.1. The activities comprise strategic and operational value-led choices, creating value, constructing and managing the value network, and capturing value. Essentially it is "connected" sequence of activities involving value engineering and value delivery.

The emphasis here is on connectivity; the accuracy and relevance of the response are based upon, in turn, the relevance of the response capabilities and the application of the facilities of the "connectivity-led" operating model shown as Fig. P3.2 that supports the extended process of identifying strategic opportunity, assessing the relevant set of response capabilities, and effecting efficient value delivery.

There are numerous features that a connectivity-led operating model is capable of delivering. An example will help. We define *serviceability* as "an overarching

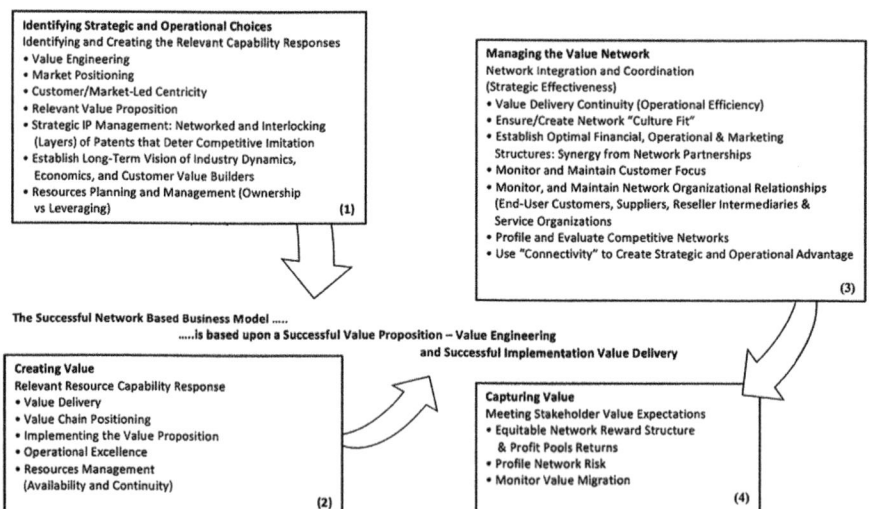

Fig. P3.1 Value creation, network management, and value capture the basis of successful strategic planning

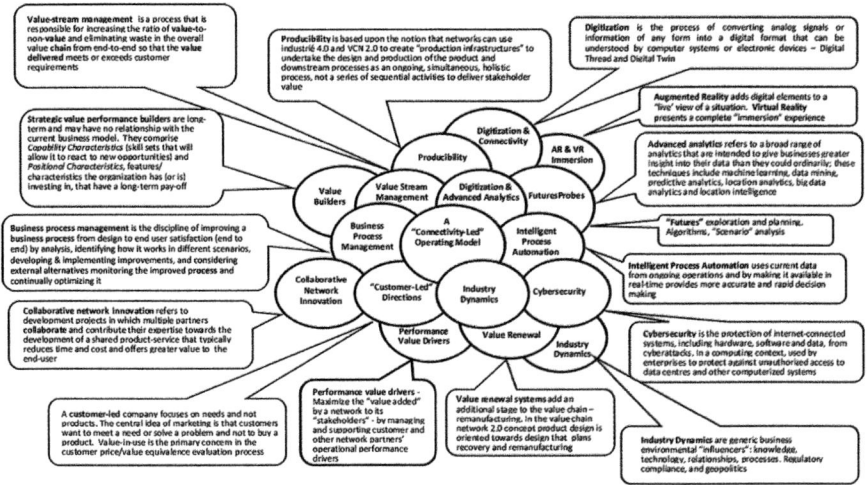

Fig. P3.2 A "connected" operating model enhances strategic and operations activities

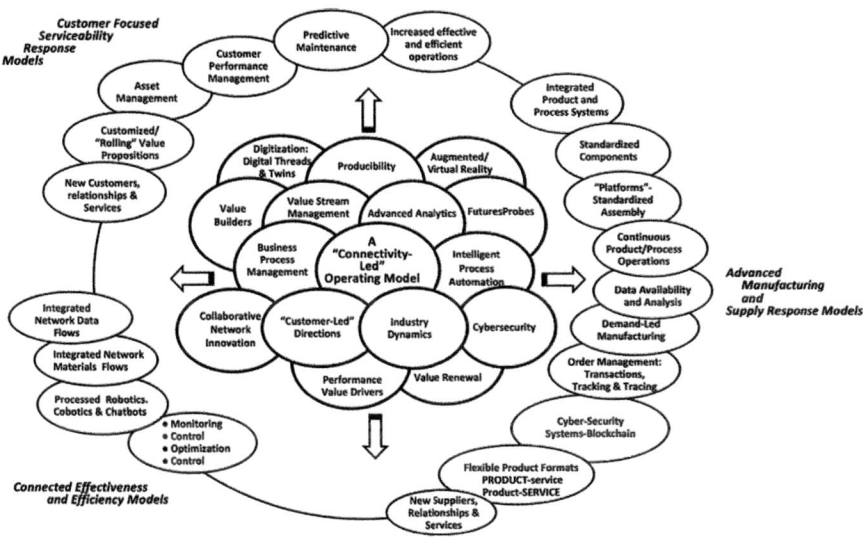

Fig. P3.3 A "connected" operations model facilitates the implementation of strategic and operations decisions

comprehensive term that extends beyond conventional terms such as 'servitization' which current parlance considers to be the adding characteristics such as planned maintenance, operator training, and the flexibility of product format (product-service outputs as products – not the hardware that produces the outputs), see Fig. P3.3. *Serviceability* is longer-term; it works with customers once an order is placed and is regarded as a strategic alliance driven by a mutual regard for both strategic effectiveness (selecting and succeeding in markets and industries by using share of market/industry value added as a primary metric rather than the somewhat, short-term, market share). *Serviceability* also has an operational efficiency consideration such as the use of real-time product performance data to modify operating procedures to increase performance, or decrease operating costs, to digitally modify an installed product to improve on planned performance, and to be incorporated in product development. It also includes aspects of product-life cycle management and working with customers to identify opportunities to modify value proposition positioning to attract alternative market segments. *Serviceability* also includes aspects of supply chain response management such as inventory management characteristics that include inventory profiling management (customizing inventory holding to reflect local demand characteristics) and ensuring vendor inventory management works in a collaborative context.

1.2   Expanding Networks and Convergence Connectivity

A characteristic of Value Chain Network 2.0 is the increase in collaborative activities by network partners. The availability of data and its transparency, "reach," and "richness" through innovative software packages and systems (notably digital thread and twin developments) have enabled RD&D collaboration (product and process innovation) and optimal operational activities resulting in compressing activity time and reduced network system inventories. The net result has been one of the operational and financial efficiency improvements of profitability, economic value added (EVA), and cash flow management. Figure P3.4 modifies an approach taken by Slack et al. (2009). Figure P3.4 profiles a cost curve that reflects the impact of customer performance value driver expectations on the investment in operations resources. The curve identifies the benefits of *producibility* by identifying the fact

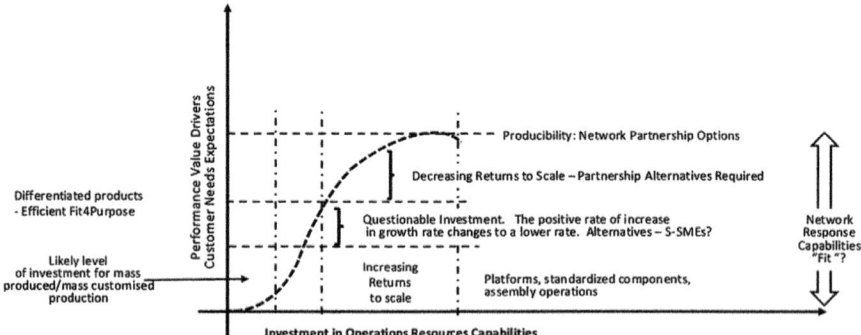

Fig. P3.4 The impact of value migration (changing consumer demand and competitive entries require flexible supply chain responses if competitive cost/price profiles are to be maintained. (Adapted from: Slack et al. (2009) Operations and process management second edition)

that there are high-volume aspects of product-service manufacturing (platforms, standardized components, and assembly operations) where economies of scale operate. There are also low-volume areas where the production of differentiation products and services takes place. Finally, there are the effects of life cycle age, customers' changing preferences, and where competitive market entrants capture value and the volume is unsustainable in terms of costs. The analysis can be used to predict likely economic returns on investment and to identify where external partners should be sought. Earlier discussions on organizations such as Magna International are examples of how this "model" could prove to be beneficial concerning network partners and structures.

An effective value strategy approach takes an organization beyond its own boundaries. It involves identifying the resources necessary to compete and to produce and deliver customer value expectations and integrate and coordinate the resources used in the value production process. Examples of creative resources management in value production are readily available. The well-known examples, such as Dell and Nike, have established models that have been replicated and implemented by a number of organizations and industries using a value chain approach. Recognition of the fact that not all of the necessary capabilities and/or capacities are internally available leads the progressive business towards identifying where in the value chain its resources are most effectively applied and who would make ideal network partners. Examples were given in Chapter 1 of how companies identify opportunities and structure value production and coordination to meet customer value expectations. Economists suggest that production is constrained by resources and support the argument with a two-input model that demonstrates the limited choice available to the organization. While the economists' model has been simplified, it serves the purpose of focusing managements' concerns by identifying relevant and appropriate market-based opportunities and the need to match these with "relevant and appropriate" resources. This identifies another issue that while an organization may not have all of the "relevant and appropriate" resources, they may be available within another organization – this is the basis of collaborative networks – and it is a primary task for operations management to identify not only what resources are required but where and who owns them and how they may be leveraged (accessed).

1.3   The Production Possibility/Efficiency Frontier Curve and Convergence Connectivity

A favorite model of the organization for micro-economists has been the *production possibility curve* and is illustrated as Fig. P3.5. It identifies how the limited sum of an organization's resources may be deployed to meet specific customer price/value equivalence requirements and is a useful approach if we wish to identify specific value-based opportunities, to identify effective responses and the

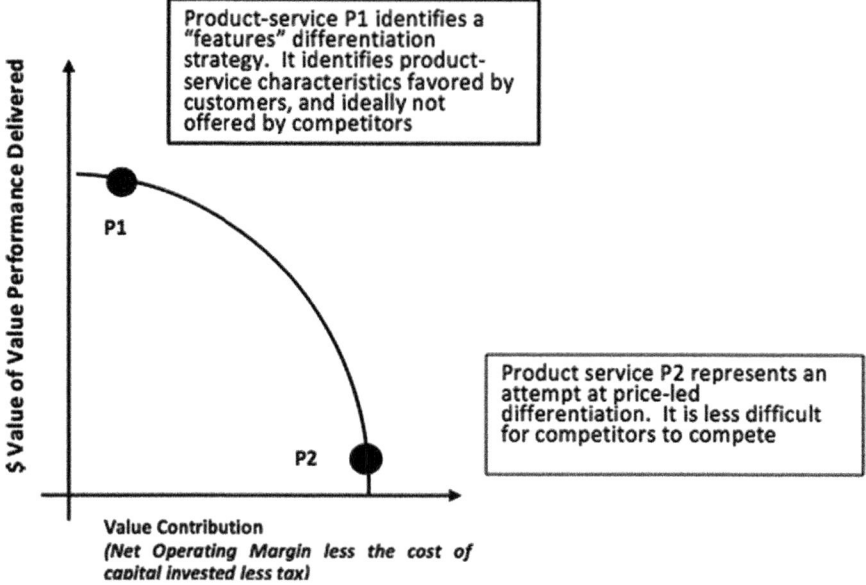

Fig. P3.5 Limited resource budgets constrain customer segment choice

possible constraints. Chopra and Meindl (2010) identify this as the cost-respon-
sive efficient frontier suggesting the organization has a choice either focusing on
price-led differentiation or on non-price-led differentiation, both options being
based upon conclusions made from an analysis of customer performance value
drivers. Price-led differentiation is based upon achieving a leading position on
costs among its competitors and offering a very basic PRODUCT-service (typically the service element barely covers the legal consumer protection requirements of the law) at a competitive low price. Non-price-led differentiation seeks to
compete on product-SERVICE characteristics that it has identified as being important to the target customer (the performance value drivers). Chopra and Meindl
(2010) argue this to be the essence of "strategic fit." An organization is only operating efficiently if it is operating on the *production possibility curve*; it is here
where the concept of network structures becomes significant because it offers all
members of the network to "leverage" the resources of other members and extend
the boundaries of their existing businesses.

Skinner (1986) and Porter (1996) addressed the productivity/profitability issue.
Skinner's comments were concerned with the problem that "The very way managers define productivity improvement and the tools they use to achieve it push their
goal further out of reach." Skinner argued that the pursuit of productivity to restore
competitive positioning does more harm than it does help. He focused on the fact
that simply applying efforts to reducing labor costs did not necessarily give the
results required because quite often labor costs accounted for a small proportion of
total costs; furthermore, such efforts are damaging to staff morale and customer

service and eventually result in alienation of both groups. Skinner also argued that innovation suffers: the preoccupation with cost performance shifts management focus towards cost reductions rather than product and/or process innovation and eventually has a negative impact on customer service. More importantly an exclusive focus on productivity ignores other ways to compete; ways that use manufacturing as a strategic resource, quality, reliability, flexibility/agility, short lead times, attention to service detail, rapid time-to-market, and effective implementation of economies of specialization and integration are the primary sources of competitive response in a dynamically competitive environment. And finally, Skinner contends that the productivity emphasis "…fails to provide or support a coherent manufacturing strategy." Figure P3.5 illustrates the problems of this approach.

A manufacturing operations strategy is essential if an effective value strategy is to be developed and implemented. Skinner's perspective is dated inasmuch that it does not give sufficient emphasis to the potential competitive responses now available from strategic partnering. It also predates the view that "production," logistics, and service integrate into an operations strategy and structure that is required in an era of global activities. Nevertheless, Skinner's manufacturing strategy components of what to make and what to buy, capacity levels to be provided, the number and location of plants, choices of equipment and process technology production and inventory systems, cost management systems, information systems, workforce management policies, and organizational structure are all relevant when considering value strategy and its management. Increasingly as the organization extends beyond its boundaries and, as a consequence, operations management adapts and considers not only the intraorganizational aspects of managerial planning and control decisions, but also the inter-organizational and international activities, it will discover that Industrié 4.0 with Value Chain 2.0 offers increased connectivity, collaboration, and transparency, thus making Skinner's manufacturing strategy realistic.

It follows that a value-based strategy is one that identifies customer expectations and addresses them with innovative responses, thereby creating a strong customer-supplier relationship. It also requires a similar approach to supplier relationships. An innovative response is also one that will require creative approaches to both product-service and process technology, i.e., the overall management of technology. Neither of these can be successful without a strong knowledge and relationship management base. Figure P3.6a, b provides a more detailed review of these choices.

Developing sophisticated approaches to platform technology has been a feature developed in the automotive and aerospace industries with the purpose of maintaining quality and containing costs. An example of this activity can be seen in work on platform technology by Volkswagen. Volkswagen's MQB platform is the company's strategy for shared modular design construction of its transverse, front-engine, front-wheel-drive format. The platform underpins a wide range of cars from the small vehicle class to the mid-size SUV class. MQB allows Volkswagen to assemble any of its cars based on this platform across all of its MQB ready factories. This allows the Volkswagen group flexibility to shift production as needed between its different factories. MQB stands for *Modularer Querbaukasten*, (Modular Transversal Toolkit) or "Modular Transverse Matrix". MQB is one strategy within

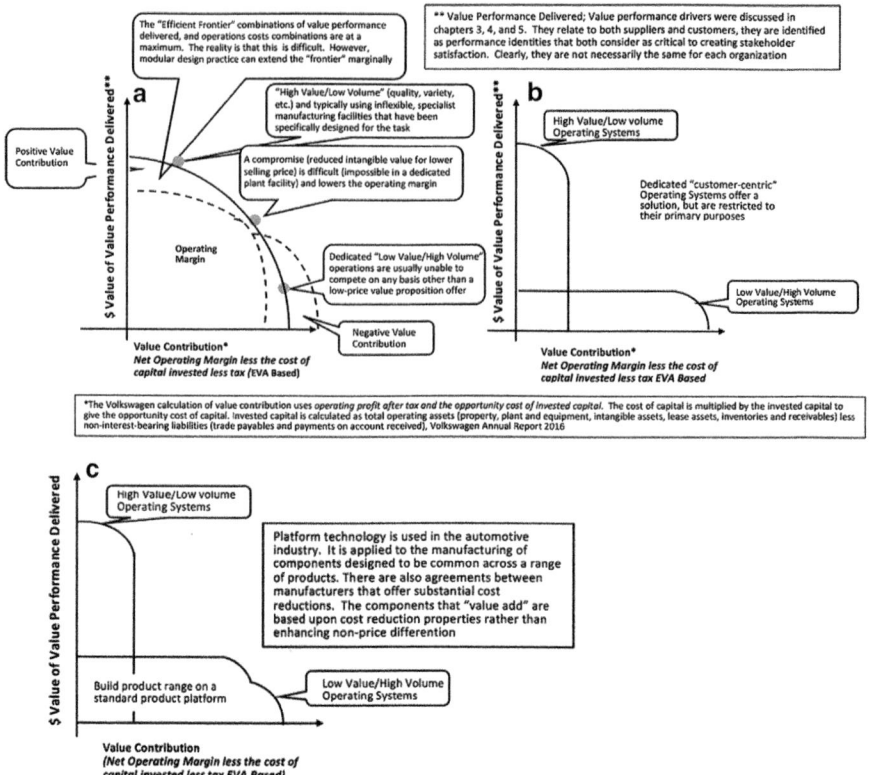

Fig. P3.6 (a, b) Recent (and current) operating system practice has constraints. (**c**) Platform technology reduces product build complexity and costs

VW's overall MB, *Modularer Baukasten* (modular matrix), program which also includes a similar MLB strategy for vehicles with alternative longitudinal engine orientation. The result (illustrated as Fig. P3.6c) has been a successful exercise in contained cost without impacting on the characteristics of its numerous brands and model variants within the brands.

1.4   Convergence…

The examples of the application of the "Convergence" of Industrié 4.0, Value Chain Network 2.0, and stakeholder value management suggest the availability of increased cost-effective (strategic) and cost-efficient (operational) solutions to existing activities and processes together with a suggestion of more in the future. However, there is a strong indication that to benefit from the current and future developments, the *response capabilities* required will require a very specific fit4purpose. Furthermore, the speed and characteristics of product-service and production

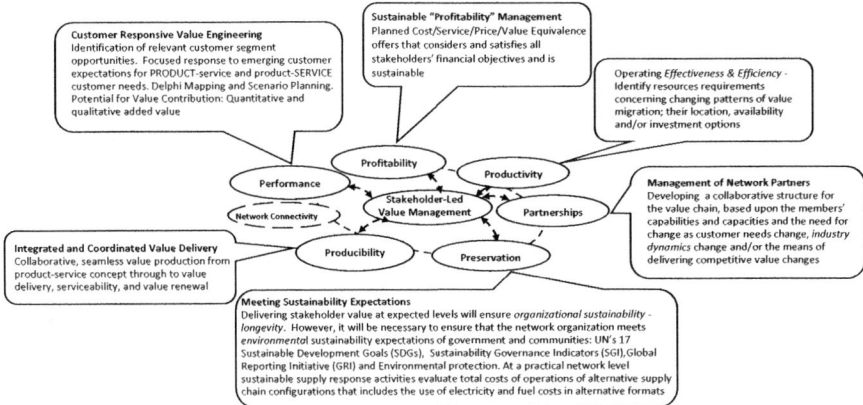

Fig. P3.7 Developing operating model capability responses

process development suggest these (*response capabilities*) may not be incremental changes; rather, they are likely to incorporate new, specific, technological developments. Figure P3.7 provides a framework for managing strategic future development. It suggests:

> Successful organizations require six interlocking and interrelated *response capabilities* that create value for *all* organizational stakeholders: product-service *performance*; economic *profitability*; mutual *productivity*; a strong but agile *partnership* value-creating network; an integrated and coordinated production infrastructure, *producibility*; and a robust organizational/environmental sustainability strategy, *preservation*.

> The "model" is holistic at all levels of its application; if this does not occur, its primary purpose, performance (stakeholder satisfaction), will not be optimal, and the network will be demonstrating this is happening with exit and entry activities of partners.

> To be successful it is essential that decision outputs are reached by ensuring that the outputs meet the expectations of each of the stakeholders.

References

Chopra, S. & Meindl, P. (2010). *Supply chain management*. New Jersey, NJ: Pearson.

Porter, M. (1996, November/December). What is strategy?. *Harvard Business Review*.

Skinner, W. (1986, July/August). The productivity paradox. *Harvard Business Review*.

Slack, N., Chambers, S., Johnston, R. & Betts, A. (2009). *Operations and process management* (2nd Ed). Harlow, UK: , FT-Prentice Hall.

Chapter 10
Building the "Connected" Business Model: Identifying Capability Requirements

Abstract *A business model describes the method or means by which a company tries to capture value from its business.* A business model may be based on many different aspects of a company, such as how it makes, distributes, prices, or advertises its products. *The business model concentrates on value creation.* It describes a company's or organization's core strategy to generate economic value, normally in the form of revenue. It identifies the capabilities required for successful responses to potential strategic opportunities. The model *provides the basic template for a business to compete in the marketplace*; it provides a template on how the firm is going to make money and how the firm will work with internal players (firm's employees and managers) and external players (stakeholders such as customers, suppliers, and investors). Customer satisfaction has become more focused. In the 1960s a strong brand promise targeted towards broad socioeconomic segments sufficed; the power and effectiveness of television advertising supported by trade marketing through efficient distribution (increasingly being offered by multiple retailing intermediaries in almost all product categories) was a successful formula. The value chain and its supply chain response are undergoing considerable change. The older measures – contribution to profitability, productivity, and added value – continue to be applicable, but we now see digital transformation is poised to change almost everything, increasing collaboration within the value chain and a broader view of "returns" to stakeholders.

Keywords Connected business model · Capability audit · Convergence management · Convergence of technology management · Product life cycle management (PLM) · Scenario planning

Introduction

A business model describes the method or means by which a company tries to capture value from its business. A business model may be based on many different aspects of a company, such as how it makes, distributes, prices, or advertises its products. *The business model concentrates on value creation.* It describes a company's or organization's core strategy to generate economic value, normally in the form of revenue. It identifies the capabilities required for successful responses to potential strategic opportunities. *The model provides the basic template for a business to compete in the marketplace*; it provides a template on how the firm is going to make money and how the firm will work with internal players (firm's employees and managers) and external players (stakeholders such as customers, suppliers, and investors).

Business models are an essential part of strategy – they provide the fundamental link between product markets, within the industry, the markets for the factors of production such as labor and capital, and the required response capabilities. Any resilient business model must be able to create and sustain returns for its investors over time and to generate positive cash flow; otherwise, it is likely to go out of business. Interactive processes have matched structural change. The growth of ICT and latterly IOT applications has moved swiftly from bar codes that made inventory management and availability more reliable for the end-user towards using the Internet supported by these systems, to facilitate online shopping as well as customer relationship interactions and management (the success of the Tesco "Loyalty Card" is testimonial to this success).

A McKinsey report, "The challenges ahead for supply chains", Gyorey et al. (2010), suggests that the organizations they had surveyed expressed concern over three customer related issues; the increasing volatility of customer demand, increasing consumer expectations about customer services/product quality and, increasingly complex patterns of consumer demand. It would appear that a different perspective on the role of product and service categories for marketing to consider is required, with the importance of the role of serviceability being identified here. As competition intensifies the service content of the package can become a critical factor in vendor/customer relationships and at a particular time (or situation) in the relationship the PRODUCT-service becomes a product-SERVICE, Johnson (2010). Hilti (a leading hand-power tool manufacturer) realized that its value proposition with its emphasis on being a premium brand had lost its visibility with major customers. The product was becoming "commoditized"; users began to regard it as disposable, and after-use care was ignored. The hand-power tool market has been "devalued" as a flood of "look-alike" products from Asian manufacturers has gained market penetration; this introduces another problem for "premium brand" manufacturers – imitation – as many of these products often infringe international copyright agreements. Clearly events such as these have changed the nature of market opportunities and of marketing responses, and we suggest a change in the way in which marketing should approach a business environment in which end-user attitudes and behavior towards products and service are changing.

> Manufacturers and retailers are rethinking and reimagining products, services and processes because of the new capabilities that a digitally enabled, thinking supply chain can offer. But we must emphasize that it isn't enough just to have technology for technology's sake. Manufacturers must continue to innovate and create value from their investments in solving business problems or enabling new offerings. The coming years will greatly alter the technology landscape for business functions in the manufacturing and retail industries.

Digital transformation is now the overriding priority for most manufacturers and retailers, with the adoption of digital technologies aimed at improving efficiency and effectiveness as well as providing the opportunity to either disrupt their market segment or be resilient to others that may try. *Simon Ellis* (2019), *Program vice president IDC Manufacturing Insights.*

The value chain and its supply chain response are undergoing considerable change. The older measures – contribution to profitability, productivity, and added value – continue to be applicable, but we now see digital transformation is poised to change almost everything, increasing collaboration within the value chain and a broader view of "returns" to stakeholders. This reality is the overriding theme suggested by IDC's top ten predictions for 2019 (Ellis 2019).

IDC suggest that now more than ever business leaders who recognize the real impact of digital technologies and other changes on their industry, customers, partners, suppliers, functions, and business practices stand to gain substantial advantage over their competitors. And the value chain network is increasingly viewed by manufacturers and retailers as a critical function for their future success.

Value chain transformation is both about driving efficiency and effectiveness today for ROI and building resilience to disruption tomorrow (almost 60% believed their business would be disrupted in 2018: data is not yet available to validate this prediction). If companies are going to embrace new business models, or digitally enable older ones, they are going to need a modern, capable supply chain to do it. IDC offer ten predictions to help guide technology leaders and their operations colleagues in their strategic and operational planning efforts for 2019 and much of the next decade beyond. These predictions cover the near term to the medium term, and the impact will likely be felt for years to come:

Prediction 1: By 2024, over 60% of 2000 manufacturing organizations will rely on artificial intelligence platforms to drive digital transformation across the supply chain, leading to productivity gains of over 20%.

Prediction 2: By 2022, over 40% of manufacturers worldwide will be integrating data from product life cycle apps into their supply chain data to improve overall after-sales service levels, achieving increases of 60%.

Prediction 3: By 2020, 65% of e-commerce operations will make use of autonomous mobile robots within their order fulfillment processes, thus helping increase productivity by over 100%.

Prediction 4: By 2021, smart supplier life cycle management solutions will automate 50% of suppliers' enterprise activities, from onboarding to exit, thus improving both performance and relationships.

Prediction 5: By 2024, one-third of large manufacturers will have moved to using actual demand data instead of short-term forecasts, resulting in an on time and in full OTIF (on time in full) delivery improvement of 2 percentage points on average.

Prediction 6: By 2020, half of the large manufacturers will have begun shifting their supply chain applications from enterprise-centric to network-centric, driving productivity gains of 2 percentage points.

Prediction 7: By 2022, digital technologies will have enabled the automation of repetitive operational tasks, leading to 50% less planner intervention and "touchless" sales and operations planning.

Prediction 8: By 2023, talent shortages in the supply chain for 75% of the top 500 manufacturers worldwide will largely have been mitigated by the use of supply chain digital assistants.

Prediction 9: By 2019, 25% of manufacturers will have doubled investments in distribution automation in reaction to dramatic increases in single-item orders originating from the growth of online marketplaces.

Prediction 10: By 2020, track-and-trace investments will have increased by 30% to improve forecast accuracy and customer experience metrics, and real-time order visibility will have become the norm for consumers.

A conclusion that emerges from Ellis' (2019) projections is there will be a continuum of development responses as stakeholder expectations, competitive activities, and other influences on value migration occur. This suggests the *response capability* approach being advocated by the authors will help refine the inputs into the business model. Figure 10.1 is a reminder of the model developed in earlier chapters.

The Changing Characteristics of Network Economics and of Operational Complexity

The introduction to this part of the text suggested that a concise perspective of a value-based strategic and operational business model is described by the activities comprising strategic and operational value-led choices, creating value, constructing

Fig. 10.1 Developing responses capability for building the business model strategic operating model

and managing the value network, and capturing value. Essentially it is a "connected" sequence of activities involving value engineering and value delivery. Before undertaking the task of designing and building the business model, some consideration of the influences responsible for changing the value creation, delivery, and capture processes should be considered, and they comprise:

Market Trends Direction

End-users' expectations of customer centricity, immediacy, price/service competitiveness, and price/value equivalence have led to greater differentiation and increased costs. There has also seen a concern in industrial markets towards investment and asset management: asset life cycles are becoming shorter, and, consequently, a preference for purchasing output and serviceability of hardware products rather than investment in the hardware has been witnessed. Accordingly, the PRODUCT-service has become a product-SERVICE option, reducing investment in tangible assets and releasing funds for focused RD&D and marketing-related activities such as product-service development, brand building, and developing customer loyalty programs.

Convergence: "Connectivity" has linked activities such as customer searching, selection, purchasing, transactions management, manufacturing, distribution, and "serviceability" functions.

Maintaining Profit Margins

Competitive marketing, for example, in automotive and aerospace, has impacted *value capture,* in these markets. An analysis of profit pools suggests a shift in profitability away from assemblers towards component and module suppliers and to "service parts and after-markets accessories" and supplier managed serviceability activities. It is noticeable that in the automotive industry, assembler/OEMs have delegated/abdicated RD&D to SSMEs (specialist SMEs) such as Magna Steyr (Magna International), Delphi, and other large suppliers. This together with overcapacity in the industry and the uncertainty of the timeline for electric vehicle development accounts for reported long-term decreasing margins.

Convergence: The digitization and connectivity characteristics of Industrié 4.0 have clearly reinforced technological developments in aerospace, automobiles, and "generic" consumer durable industries so too has the structure of product life cycle management. See Siemens' software organization (Teamcenter) below.

Product Life Cycle Management (Marketing)

PLCM refers to the commercial management of the life of a product in the business market with respect to costs and sales measures: the traditional product life cycle model was useful for considering alternative strategic (and operational) marketing opportunities and responses by specific organizations, perhaps brands. Ansoff (1984) updated the PLC model, offering a more realistic and current view (see Fig. 10.2). Ansoff argued that demand for a product-service is ongoing – in other words end-user product needs remain unchanged – we continue to "count and calculate" and many other activities; what has changed is the ways and means of achieving the outcomes. Ansoff argues that it is technology that has provided faster, more accurate means for conducting these tasks; calculating "technology" has moved along from the 4000 BC "digital technology" of using fingers, to the abacus, and onto various mechanical applications, electronic dedicated equipment, to dedicated computer software applications. Technological development has accelerated, and Fig. 10.2 indicates this by the decreasing time periods between the *demand technology life cycles (DTLC)*.

Demand technology life cycles' life spans are decreasing as sequential life span replacement (more efficient solutions and/or more cost-efficient solutions) increases. Figure 10.3 adds to the Ansoff concept. Within each DTLC *applications life cycles* can be observed. The technology of the DTLC provides a basic, underlying technological capability that can be applied to a range of product-services, for example, mobile cell phones, audiovisual products, and the miniaturization of the processes introduced by the "application technology" to expand its uses (e.g., the Sony Walkman and mobile entertainment). Another *application* is the adaption of military

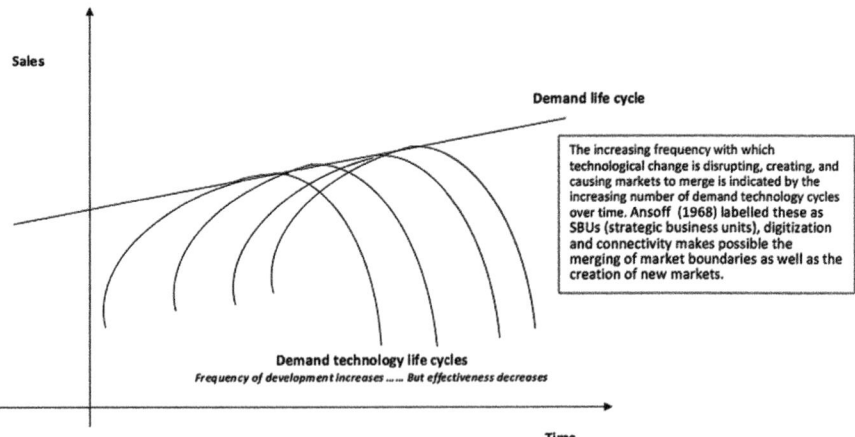

Fig. 10.2 The demand life cycle and demand technology life cycles respond to ongoing technological development

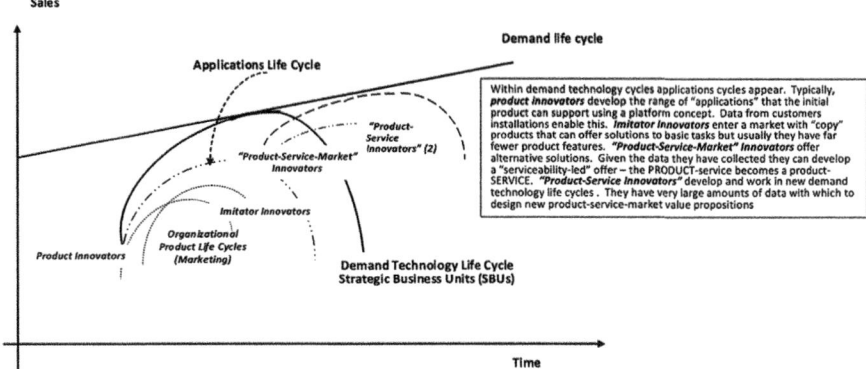

Fig. 10.3 Applications life cycles and the organization's product-service life cycle

technological hardware for consumer markets; ABS automobile braking systems and global positioning systems are two of the high-volume applications.

Within each *applications cycle, organization cycles,* identify opportunities within market segments using the application technology; the benefit that value chain networks offer is that to compete no single organization necessarily needs to have the specific resources; these can be identified within other companies and successful partnerships forged. *Minelab,* a subsidiary company of *Codan,* an Australian technology-based organization, is an example of how the technology of metal detection for military purposes can be converted into a viable consumer product using the resources of partner organizations to manufacture the product. Application of "convergence technology" driven by real-time data analysis (digital threads) is extending the effective life span of DTLC application cycles by monitoring and improving performance.

Convergence: Digital twins can provide the high-quality inputs needed to create models of operations. Artificial intelligence (AI), machine learning, and other advanced technologies use these models to suggest improvements to operations. Once operations are improved, the digital twin also improves and provides higher-quality inputs to models, which are then used by advanced technologies to suggest further improvement.

Product Life Cycle Management (Operations)

PLM (product life cycle management) is the process of managing the entire life cycle of a product from inception, through engineering design and manufacture, to service and disposal of manufactured products.

The stages are traditional sequential engineering workflow. The exact order of event and tasks will vary according to the product and industry. "Convergence technology" models alternative scenarios, making engineering product management processes and the resulting decisions proactive instead of reactive. They use digital twins to create and run simulations and ask "what-if" scenarios of the digital twin and using the results to improve operations and employee skills. Siemens (with Teamcenter) and General Electric with (Predix) have well publicized examples of software applications, (e.g. www.siemens.com and www.generalelectric.com) that offer a digital sequence of the operations aspects of product lifecycle management.

The main processes in the development of data-led operations product life cycle models are: Identify a current user need activity/process that could be improved or an alternative, new approach, is used to create a solution and a product/process specification; develop a concept design (with a detailed design in mind); conduct validation and analysis (simulation using a computer based prototype, tool design for manufacturing) - this may be additive manufacturing initially; realize viability; plan manufacturing; manufacture, build/assemble; test (establish quality control criteria metrics); plan service activities and schedule (the PLM program may address this and implement predetermined servicing using real time performance data); sell, deliver and install; monitor in use performance; maintain and support; determine remanufacture and eventual disposal timing and activities. During the design phase of the product life cycle, many iterative designs are created and prototyped before the version that goes to market is finalized. Often, those who are *designing* the prototypes and those manufacturing them are in different teams. A PLM system with built-in project management can facilitate the design process and reduce the amount of iterative designs because miscommunication is reduced and a structured place to store product information is offered. PLM accelerates the product life cycle by automating and managing the processes and people involved with the design, production, and distribution of a new product. It achieves this largely by maximizing information visibility. PLM software is essential for remaining competitive and producing new articles on budget and on time.

Convergence: "Total connectivity" is becoming an integral part of the manufacturing process and is becoming more accessible for every size of business. It will continue to challenge and change the way product-services are designed, manufactured, and marketed by reducing time to market cycles and enable a new dimension of personalization for customers. Growth of lean methodologies currently employed by modern software developers in traditional industries such as the apparel or footwear market can be expected.

Network Structure Economics

The "investment fixed costs" of integration and coordination of manufacturing systems have increased with system complexity. Risk is also a factor, measured by the "returns spread" (EROI *less* AWACC), economic operating profit *less* the annualized weighted average cost of capital, and by increases in operating complexity as are the increasing expectations of customers.

Chandler's (1977) analysis of the development of industry in the USA identifies the role of the vertically integrated business of the early twentieth century and the emphasis on *economies of scale*. However, Chandler did suggest that it was the *economies of speed* (the ability to coordinate the production sequences) that were more critical for manufacturing efficiency. Subsequently *economies of scope* (these exist when the cost of jointly producing a range of products (or managing a process) within one organization is less than the cost of producing the products separately across independent organizations: tangible and intangible aspects (e.g., manufacturing and distribution and brands, R&D, and service) shared importance with economies of scale in manufacturing.

Economies of Scale: Create price advantages through *"managed" volume of production*. Early business models (e.g., the Ford production and sales model was driven by one product, the Model T, in one color) enabling high manufacturing volumes to be achieved. The costs of the twenty-first-century model of automotive production even now have MEP (minimum economic production) quantities. While the contemporary automotive production plants operate on a multi-model basis, they are unable to cope with low-volume (specialist/luxury) vehicles selling some 30,000 units annually and it is for this reason that Magna International is in business. Note: Economies of scope involve cost savings from joint production utilizing one set of assets, whereas economies of scale involve efficiencies from the production of higher volumes of a given product.

> Convergence: The benefits of Industrié 4.0 and Value Chain Network 2.0 are becoming apparent. The digital connectivity of automotive production lines using programmed robotics to perform a range of assembly tasks enables assemblers to manufacture vehicles with a range of different specifications in a serial production system.

> *Economies of Scope:* Exist when the cost of jointly producing a range of products (or managing a process) within one organization is less than the cost of producing the products separately across independent organizations: tangible and intangible aspects (e.g., manufacturing and distribution and brands, R&D, and service). They offer variety and choice through *"product platform"* design management.

> Convergence: The application of platform technology offers a capability of product-service differentiation at a competitive price by designing product-services on a principle that assumes end-user customers make purchases based on intangible value (aesthetics) and tangible value (the acquisition cost of the product and the assurance of reliability and serviceability guarantees and its performance).

Scalability: Is one of the most important factors to be considered when building the business model or perhaps for taking the existing business to the next level. Successful business growth depends on a scalable business model that will increase profits over time, by growing revenue while avoiding cost increases. Scalable businesses grow exponentially, not being weighed down by the same sales-cost growth relationship as linear models. Instead, as sales increase, costs stay flat, allowing for higher levels of profit over time (Housh 2015). Scalability differs from *economies of scale* because while costs do increase, they do reach a minimum point; given this low-cost point may persist over volume increases eventually, it reaches a point at which working additional shifts maximize plant and labor capacities and begin to increase. *Economies of scope* can reduce a sales/cost ratio, but eventually they require additional capital (an increase in intangible asset investment). Technology companies, Google, Apple, and Microsoft, have perfected this type of scalable business model. Their initial costs for developing an advertising platform or operating system are high, but once it is on the market, they can sign up users or sell many copies of the related software with relatively minimal cost increases. It is suggested that using economic profit (EVA) provides more analytical accuracy as it is able to identify required changes in capital, labor, and services.

Convergence: Each of the components of convergence Industrié 4.0 and Value Chain Network 2.0 and the shift of focus away from shareholder value management towards stakeholder value management have an input. Product-service performance management remains constant, and customer, vendor, and other stakeholders' expectations will be met (it does raise issues concerning the impact of value migration and agility capability); both customer and vendor profitability margins will remain constant. Productivity may improve. Partnership collaboration will pay a large part in the success of a scalability strategy. The producibility infrastructure may provide profitability and productivity increases due to an "experience effect" and to an impact from "collaborative curves" activities. The model suggests an optimal use of material inputs, and, therefore, preservation objectives can be planned and met.

Economics of Value Added: A term used in a quantitative context representing "margin" – the difference between a selling price and input costs. We have referred to this in the sections on EVA. In a qualitative context, it is used to identify a "value" or "utility" benefit that is a reference to the monetary benefit received or an ability to use it to increase the "value" added by additional activity or resource input before being "sold on" within a value chain as a component, finished product, or service. For example, physical distribution adds value by delivering a product-service where and when it is required.

Convergence: The convergence of Industrié 4.0 and Value Chain Network 2.0 and the shift of focus away from shareholder value management towards stakeholder value management have had a significant impact, and it is foreseeable this will continue. Industrié 4.0 has impacted the value engineering and value delivery processes by using digitization and connectivity to change the response capability requirements of customer response in significant directions. "Time-to-market and serviceability-response-time" have been significantly reduced; in some examples, it has been eliminated by "connecting" capital equipment that is in service to digital twin replicas of the hardware product; performance management can be optimized; many product-service upgrades and modifications can be digitally installed and the performance data aggregated, stored, and analyzed for use in product development and other-related activities. Value Chain Network 2.0 has evolved with an emphasis on integration, cooperation, and collaboration resulting in strongly competitive sustainable networks. The shift of focus away from shareholder value management towards stakeholder value management has introduced the notion of a wider set of responsibilities to secondary as well as primary interests in organizational activities.

Recent (2018/2019) shareholder impact on executive remuneration is an example. The availability, analysis, and application of data to identify the what, who, where, when, how, and why of customer expectations and the use of the data outputs as algorithms in manufacturing (product-service formats) and demand satisfaction (availability schedules).

The Economics of Learning (Experience Effect): A characteristic of the experience curve concept is that the unit cost of added value to a standard product declines by a constant percentage each time cumulative output doubles. This may result in an uncompetitive and unacceptable cost/price/value/equivalence relationship to a potential customer. The value chain network structure has the capability of reducing the cost context of the relationship by combining the experience curves of members of the network to reduce the aggregate cost, making the offer more appealing to that customer. ***Collaboration curves*** (Hagel et al. (2009) **hold the potential to mobilize larger and more diverse groups of participants to innovate and create new value.** In so doing they may also reverse the diminishing returns dynamics of the experience curve and deliver increasing returns to performance instead. Hagel et al. (2009) (see Fig. 10.4)

Convergence: Digitally connected processes evaluate the learning benefits of experience; data analysis identifies the correlation coefficients between process activities and the "value " of their collective (collaborative) application.

Economies of Integration: Cost-effective management of overhead through *strategic part-nering.* The principle of business system integration; is that the linkage of isolated activities *into a single, integrated system (producibility infrastructure)* is fast, responsive, flexible and relatively low cost, and results in a situation in which unit costs decrease as output increases (because the activities of the *entire* operation) and are increased by optimizing (and "leveraging"

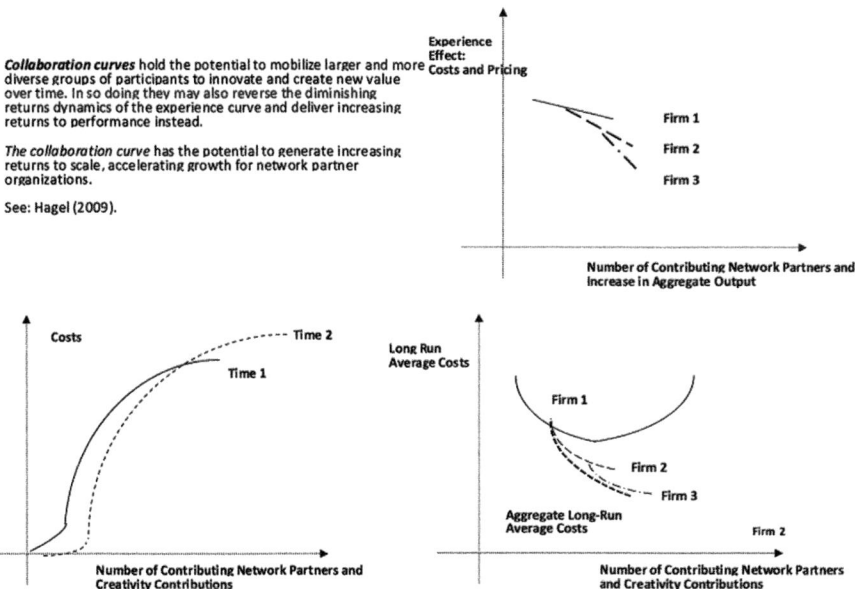

Fig. 10.4 Collaboration curves – aggregate and cumulative economies of scale

the ownership), the distribution and location of assets throughout a value creation system, and result in aggregate, investment and operating costs that are decreased.

Convergence: From digital machine performance management towards connected digital manufacturing process management by connecting relevant machines to create an overall single, production process.

Economics of Coordination: Optimize the costs of transformation, transactions, and interactions throughout the value creation network (its fixed and variable cost structures) by establishing a strategy (and a structure) that ensures the tasks required for success are identified and planned and that they are implemented and managed efficiently; coordination requires that the organization "does the right things right"!

Convergence: Logistics services companies have built platforms to match current and short-term excess capacity to demand.

Economics of Interaction: The searching, coordinating, and monitoring undertaken by organizations for effective and efficient means to exchange products, services, and ideas. They occur on an intraorganizational basis as well as an inter-organizational basis. IOT developments continue to enhance the interactive capacity of industries and individual consumers such that it will create new ways to configure businesses and organizational structures and to service customers. Accordingly, it will have a major impact on the strategy, structure, and competitive dynamics of entire industries.

Convergence: Workflow data is being moved into the "Cloud" and are increasingly digitized and capable of interacting with other digitized workflow data and identifying opportunities for restructuring process management to realize cost-efficiency opportunities.

Economics of Motivation (to commit to network membership and membership continuity): Network structures optimize the use of resources. Adam Smith suggested that markets are one very prominent mechanism for solving the problems that arise with the interdependencies of specialization and division of labor. Smith was referring to end-user markets, but the comment is applicable to resource markets within networks; a well-functioning network leads to "internalization" as interdependence implies that one organization's choices and actions have an impact on other network members but that it is end-user satisfaction that is the final arbiter, and this is a major influence in the price mechanism *within* the network structure.

Convergence: Coase (1937) focused attention on the notion of optimizing resources. His Nobel Award is seen as the initial move towards "partnerships" in networks. He argued that the firm's boundaries moved on the basis of costs: any task that could be undertaken at a lower cost by an external supplier would, if accepted, move the boundary of the firm to include the external supplier (i.e., it would be outsourced). Value Chain 2.0 is becoming a sophisticated extension of the Coase (1937 paper. "Profitability" and "Productivity" both benefit from an optimal use of resources; the principle driver of value chain networks is to maximize aggregate resources to achieve optimal utilization of fixed and current assets of individual members.

The Economics of Transaction: Searching, coordinating, and monitoring for effective and efficient means/media to exchange products, services, and ideas that maximize added value during developing specifications, searching/sourcing products and services, evaluating alternatives, negotiating with suppliers, coordinating performance commitments, monitoring performance, and supplier and customer relationship management

Convergence: Online, real-time "searching" and transactions data has passed the assumed levels of activity and consumer involvement. Data analysis provides inputs for customer segmentation and customer journey profiling. It also has input into product-service inventory range profiling and availability levels.

The Economics of Relationships Management: Adding value by working with suppliers, intermediaries, and end-user customers to identify cost-efficient value delivery alternatives using collaboration through co-productivity, coopetition, and prosumerism.

Convergence: The customer journey is the complete sum of experiences that customers undertake when interacting with an organization or with a specific brand. Customer journeys can be mapped, instead of looking at just a part of a transaction or experience; the customer journey map documents the full experience of being a customer. Data analysis builds aggregate data banks by customer segments (behavior and actions) into algorithms that are used for marketing planning.

The Economics of Differentiation/Specialization: Adding customer value by offering "product-service exclusivity" through customization and "mass customization" using specialization, process, and capability collaboration.

Convergence: Creating important "difference or distinction" is being achieved by using co-creativity (e.g., automobile websites that encourage potential customers to "build" their vehicle from a range of optional packages) and co-productivity (the IKEA approach of "trading-off" design and price reduction for undertaking delivery and manufacturing (assembly) processes.

Economies of Time Response: Product and service customization through design- to-order (DTO), build-to-order (BTO), and quick-response (QR) operations process management.

Convergence: Time and location availability are important responses capability sought by various partners within the value chain and by the end-user customer. Response time can be reduced by working with customers to create a fit4purpose specification (co-creativity). Additive manufacturing is being used to address time constraints and location availability. International logistics companies offer track-and-trace monitoring and some, e.g., are working with localized additive manufacturing (3D printing) to add time and product-service customization to their value propositions.

The Economics of Complexity: The success of reverse innovation (General Electric's response to the needs of the Asian healthcare equipment market) has identified the benefits of identifying the *appropriate* level of product-service sophistication on production costs *and* in market success. Connected product performance data to manufacture databases is enabling performance management and predictive maintenance.

Convergence: *Digital threads (real-time monitoring of product/component performance) and digital twinning (reviewing performance data with predicted data from a digital twin replication of the hardware/tangible product) is providing data to modify operational activities and to identify predictive maintenance requirements as well as identify product upgrade requirements and to provide design input for product development.*

Network structures expand the nature of organizational economics from a limited perspective, based upon economies of scale (within which the firm became volume-oriented striving to achieve its minimum cost/volume position on its long-run average cost curve (Chandler: 1962)), to a "collective" perspective based upon a notion of dispersed operations (i.e., the complete range of value creation, production, delivery, and service provision is digitized, and the assets and activities are "connected"). It is no longer sufficient to be the *lowest cost provider* in a market, but rather it is now essential to be the most effective and efficient *solution provider*: end-user markets are product-service dominated. These may be PRODUCT-service markets; however, in the New Economy, many industrial markets are product-SERVICE markets: the customers are aware of product application performance but are often more influenced by *service maintenance availability* rather than low prices, hence the approach by major manufacturers of such products as aero engines which are priced by the hour of serviceable use. The cost/price/service value equivalence relationship is being refined by the application of the "convergence" capabilities. The "big data" facility of Industrié 4.0 has the ability to track, merge, and consolidate purchasing behavior and purchasing group motives, suggesting that price/value equivalence data is becoming a useful segmentation factor.

Control of Network Relationships and Organizational Complexity

Chapter 8 dealt with partnership development and management. The primary importance of partnership network management is to build and maintain value-generating partnerships. To do this effectively requires a network stakeholder focus on satisfying the expectations of the prime source of income, the end-user, and concurrently ensuring the remaining network partners' objectives (revenues, "profitability," and cash flow) are met. An adaptable strategy and structure model that ensures current (and future) management and operations staff possess relevant capabilities (skills and experience) to meet current and future challenges and the capability requirement responses to meet growth opportunity requirements is required. In this chapter the concerns that should be addressed and researched and decisions reached are dealt with prior to *firm* decisions being made.

Relationships: With suppliers and customers that identify required responses capability that are supported by developing synergistic relationships with and between suppliers and distributor network partners in order to reinforce critical performance characteristics. Some considerations include:

Effectiveness: Value engineering is a systematic and organized approach to providing the necessary functions in a project at the lowest cost. Value engineering promotes the substitution of materials and methods with less expensive alternatives, without sacrificing functionality. It is focused solely on the functions of various components and materials, rather than their physical attributes. Value engineering processes comprise a comprehensive review of a product-service and the purpose for which it is used, the raw materials used, the processes of transformation, the equipment needed, the distribution activities and facilities required, and the serviceability structures (organizations) and activities (roles and tasks). It also questions whether what is being used is the most appropriate and economical alternative. This applies to all aspects of the product.

Convergence: Digitization connectivity and collaborative transparent data exchange agreements are working towards more rapid agreement on product-service specifications and the use of platforms to optimize investment in value chain assets, for example, payment systems, physical distribution services, and serviceability.

Manufacturing Capabilities: Are the technical and physical limitations of a manufacturing firm and each of its plants. Categories of "capability" include innovative product-services and production processes, application of technological processing methods, and flexible production capacity.

Convergence: Digitizing processes offers the opportunity to "extend" the scope of activities on an intra- and inter-organizational basis. For example, the use of digital threads between a manufacturer and the installed equipment (product-service) increases the manufacturers' ability to "service" the equipment.

Efficiency: Successful ongoing process improvement programs. A proven capability to produce outputs to specification limits and within cost, quality, and time requirements consistently.

Convergence: The transfer of operating/production data between (and among) network partners monitors performance and speeds responses to correct specification problems.

Innovation (Product and Process): The application of "tacit knowledge" and the accumulation of knowledge through experience: "tacit knowledge" the kind of knowledge that only gets engaged and created in environments where there is longer-term, trust-based relationships. It is interesting to note how innovation in the automotive industry has developed in recent years. Many of the new product development (NPD) activities in the industry were based upon the commercialization of military solutions. For example, ABS braking systems evolved from nonskid devices from military aircraft systems designed to "read" runway surface conditions and to apply braking in stages, thereby ensuring deceleration in a "straight line" and the aircraft remaining on the runway rather skidding off and becoming damaged; GPS mapping systems and direction guidance was also initially a military ground navigation aid and is now a standard feature in medium-/high-range price vehicles. Initially as government-procured items and typically high-price/low-volume production items, they later became high-volume items (with a little less response accuracy) and are becoming standard across a wider price range. OEMs have distanced themselves from RD&D in these product-service sectors by relying upon SSME (specialist small-medium enterprise) suppliers for technological advancement that provides short-run competitive advantage.

Convergence: Collaboration is built upon trust. Trust is the accumulation of successful joint project outcomes.

Technological Integration to Create Network Competitiveness: The capability to conduct focused technological integration to create advanced solutions to market opportunities and competitive value advantage. Identifying and assembling relevant technological systems (scalable, modular, and flexible) combined with appropriate organizational structures, purpose-built processes, and work flows.

Convergence: The important inputs here are trust (the implicit understanding of partner capabilities) and the design of database capture that identifies partnership "qualifications," their contributions to projects, the success of the project, and the knowledge added to the partnership from the success.

Sharing Resources: A concept that is actually more complex than many organizations realize. Once understood, not just how important sharing is but also what "shared resources" encompasses, it can revolutionize how efficiently and successfully a value chain network can function. Resources are *hard* (time, people, capita, and equipment) and *soft* (innovative but focused ideas, commitment, determination to see successful results, and mutual trust).

Convergence: Managing resources requires a knowledge of what specific resources are required for a project, who owns them, and where they are located. Network data capture identifies resource bases as well as capabilities. Successful project management (in terms of performance, time, and cost) is beginning to apply this data-led approach.

Competitive Optimization: Relevant and optimal responses to stakeholder expectations, for example, customer serviceability time and content response, product range variety that satisfies customer choice requirements, and inventory availability levels. Competitive optimization requires the use of innovative thinking, for example, DIY retailers use paint colorants and mixing equipment to offer an extensive range of color options.

Convergence: The use of aggregate customer data enables the construction of product-service, inventory, and serviceability models capable of identifying (and managing) successful responses capability.

Approaches to Forecasting the Business Environment

Forecasting methods vary, and management should be aware of differences. They can be divided into two approaches, quantitative and qualitative methods. We identify each here but make the point that detailed knowledge and use experiences contribute to decisions concerning suitability. In general, it could be argued that quantitative methods are applicable to explore short-term situations in which product-services are established and there is tacit knowledge within the organization to guide judgment concerning forecasts. This contrasts with long-term forecasting in markets that, although they may concern existing product-services, what is not known, and should be established, is how they may develop. Value migration has two main characteristics: one is the changes occurring in consumer expectations and needs and the future influences that may affect them and, second, competitive business models that may offer new (alternative) benefits that have customer appeal. For these situations, qualitative methods using structured possibilities that can be considered by experts (having specific knowledge in an area being explored) and are capable of making judgement of "cause and effect" events that could have an influence on customers' emerging needs and behavioral responses are introduced.

Anticipating and Forecasting Change: Quantitative Methods

Two models are of interest for decision-making: time series analysis and causal modelling. Time series plots the behavior of a variable over time; by using experience and knowledge of the market, industry, and the relevant aspects of the economy, this identifies variations in the behavior of the variable monitored to adjust the likely future performance. The weakness of the approach is the implicit assumption that whatever has occurred in the past could likely occur again having a similar impact. Random variations that remain after adjustment for seasonal affects are without any known or probable cause. The assumption is made that future events will be influenced by the past events and should be dealt with. Moving average forecasting and exponential smoothing are used to improve the accuracy of the forecast. Causal modelling is more complex as it uses techniques to identify and understand the *strength of relationships between* network variables and the impact each has on the other using basic regression analysis. Past data, experience, and judgment find the "best fit" between two variables. The process is useful for management as it encourages exploration among all of the relationships that have occurred and to consider the implications. Slack et al. (2009) describe applications of both time series analysis and causal modelling and provide "need-to-know" explanations.

Using Delphi, "ExpertLens," Scenario Analysis and Planning

Building a business model suggests dealing with future events; it follows that decisions on its structure and the expectations of its output capability will essentially depend upon the ability to explore an uncertain future, and past data will have limited use. More useful will be considered opinion of relevant likely future events.

Scenario planning, or scenario analysis, is a strategic planning method that some organizations use to make *flexible long-term plans*. Scenario analysis asks "what if?" questions to explore alternative views of the future and to create realistic visions of the future around them. They can consider relevant long-term trends in economics, resources supply and demand, geopolitical shifts and social change, as well as specific important factors (to the organization) that drive change, such as developments in *industry dynamics characteristics*. They also should bring focus to the capabilities and positional requirements – *the value builders* – required for success and their implications for the *global value chain participants*. Shell has been using scenarios since the early 1970s to allow management to make better business decisions resulting in greater understanding of the business environment and how it might appear decades ahead.

Scenario analysis and planning offers a number of advantages over conventional strategic development processes by:

- Increasing the visibility of uncertainty by identifying opportunities and threats (risks) enabling the organization to prepare to take advantage of the opportunities and to avoid the threats.

- Identifying the *impact factor change trends* that are likely to have a strong influence on the outcome of strategic PRODUCT-service or product-SERVICE decisions (e.g., demographics, education, healthcare markets).
- Assisting an organization to manage and reduce business-related risk and uncertainties in a turbulent environment.
- Demonstrating how events can develop, interact with each other, and present an organization with unforeseen opportunity or threat. The application of statistical concepts (correlation coefficients, a statistical measure that calculates the strength of the relationship between the relative movements of the two variables). A useful check on the dependency of one event on another identifies market relationships that may influence marketing strategy decisions.
- Helping organizations understand how they should structure their response to either opportunities or threats.

Expert Panels

Aggregating expert opinion by the use of serial anonymous questionnaires that encourage discussion and debate has been successful in the creation of scenarios for strategic planning purposes.

The *Delphi method* is a forecasting process framework based on the results of several rounds of questionnaires sent to a panel of experts. Several rounds of questionnaires are sent out, and the anonymous responses are aggregated and shared with the group after each round. The experts are allowed to adjust their answers in subsequent rounds. Since multiple rounds of questions are asked and the panel is told what the group thinks as a whole, the Delphi method seeks to reach the correct response through consensus. Rand developed the Delphi method in the 1950s, originally to forecast the impact of technology on warfare. The method entails a group of experts who anonymously reply to questionnaires and subsequently receive feedback in the form of a statistical representation of the "group response" – after which the process is repeats itself. The goal is to reduce the range of responses and arrive at something close to an expert consensus. The Delphi method has been widely adopted and is still in use today (adapted from the Rand Corporation website).

The Delphi method seeks to aggregate opinions from a diverse set of experts, and it can be done without having to bring everyone together for a physical meeting. Since the responses of the participants are anonymous, individual panelists do not have to worry about repercussions for their opinions. Consensus can be reached over time as opinions are swayed.

However, there are disadvantages. While the Delphi method allows for commentary from a diverse group of participants, it does not result in the same sort of interactions as a live discussion. Response times can be long, which slows the rate of discussion. It is also possible that the information received back from the experts will provide no innate value.

ExpertLens is a similar approach to Delphi, incorporating elements of *the Delphi method,* and has been developed by RAND. ExpertLens can have wide application across such areas as public policy, healthcare, finance, and marketing, where expert panels are frequently used to help solve problems or predict an unknown future. ExpertLens allows expert panels to be done online in a robust way that resembles fact-to-face meetings but with lower costs and easier analysis of the information gathered. Existing options for gathering such opinions generally include convening meetings of experts where opinions are expressed face to face (the nominal group technique), organizing panels of experts who share their opinions without meeting in person (the Delphi method), and putting out an open call for input to a large community of people (the crowdsourcing method).

The first phase of an ExpertLens process requires participants answer a series of questions. In the second phase, they review the group's responses and discuss their answers using online discussion boards. In the third phase, participants re-answer phase one questions based on the additional information they received during the feedback and discussion in the second phase. The online nature of ExpertLens allows the results to be rapidly compiled and the findings to be analyzed quickly. The online process has been designed to reduce the costs associated with existing panels and to increase accuracy by reducing the potential for group process losses that can occur in large, diverse, dispersed groups. ExpertLens is iterative and does not require participant "consensus" as it uses statistical analysis of the data collected during the rounds of the process to determine what the group "thinks."

Given the expansive and rapidly changing nature of the business environment, it is likely that quantitative forecasting methods will become increasingly less informative and therefore of limited use at the early stages of building the business model. While the business model is a facilitating device for the identification, development, evaluation, and adoption of strategic direction, it is required to be able to use data inputs that will guide and determine implementable output. It is arguable that qualitative research input and artificial intelligence processing methodology will present the most useful outputs. An example of the output possible from expert opinion polling and scenario planning is shown in Fig. 10.5. The study was conducted by PwC and published in Strategy and Business (2015).

In the context of building a business model, the overall process should identify scenarios that provide positive guidance to the response capability the organization should expect to have to meet if its business model is to be successful, in relation to:

- Customer Responsive *Performance*-Based Value Engineering
- Sustainable *Profitability* Management
- Operating Effectiveness and Efficiency *Productivity*
- Management of *Network Partnerships*
- *Producibility*: Integrated and Coordinated Value Engineering and Delivery
- Meeting Sustainability Expectations: *Preservation*

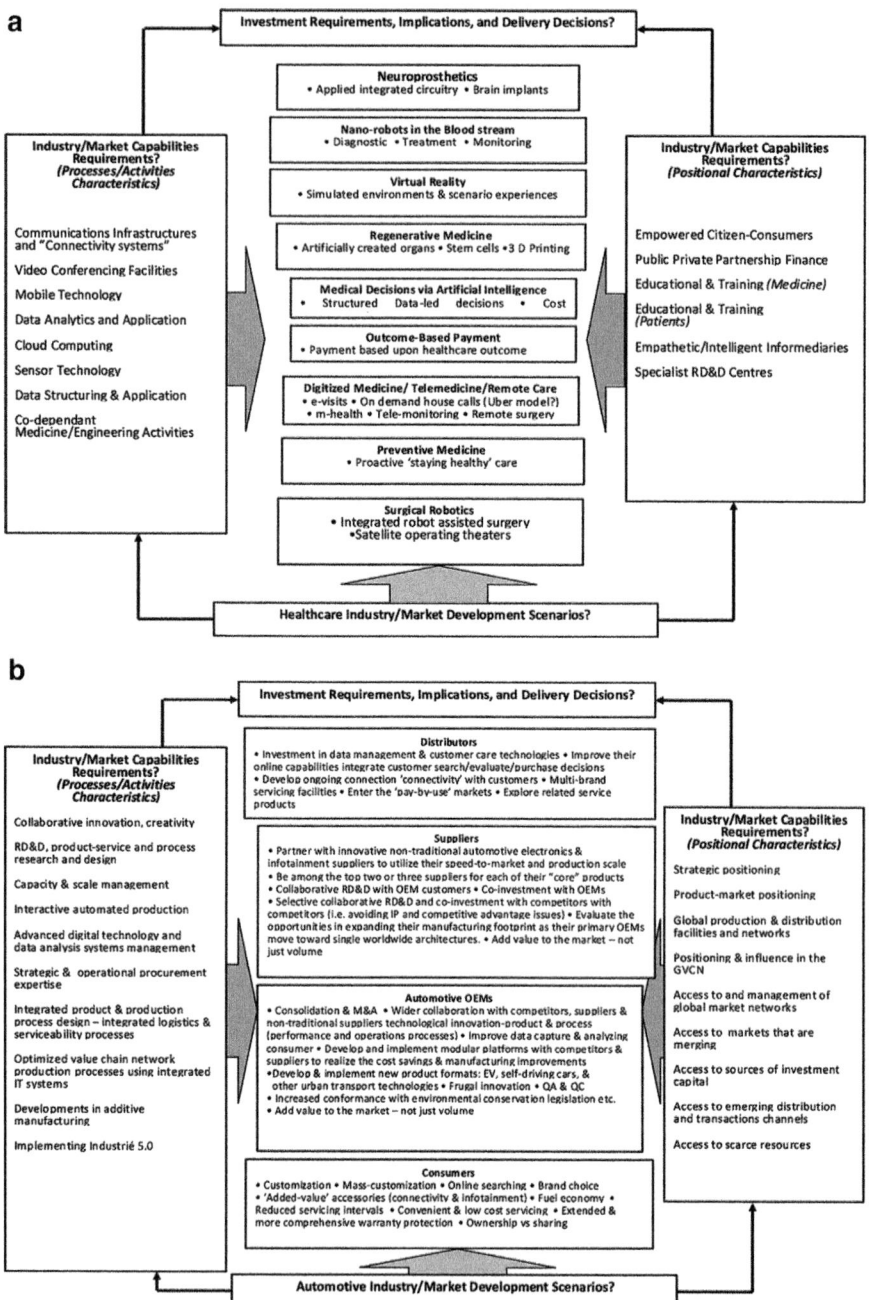

Fig. 10.5 (**a**) A forecast of developments in healthcare. (Adapted from: Strategy + Business PwC (2015)). (**b**) A forecast of developments in the automotive industry. (Adapted from: Strategy + Business PwC (2015))

Creating a Relevant Networked Business Model

Prior to building a networked business model, answers to some important questions are required:

- Norman (2001) suggested "'a new strategic logic'.... managers need to be good at mobilizing, managing, and using resources rather than at formally acquiring and necessarily owning resources. The ability to reconfigure, to use resources inside and particularly outside the boundaries of the traditional corporation more effectively, becomes a mandatory skill for managements." What is the relevance of this observation for future network business model design response capabilities?
- What are the market dynamics? Do end-users look for an expanding range of applications within the applications life cycle of a demand technology cycle? Aghina et al. (2018), found that recent changes in market dynamics such as: *rapidly evolving environments* (i.e., *all* stakeholders expectations are changing rapidly, investors are demanding rapid growth resulting in mergers and acquisitions and restructuring); *constant introduction of disruptive technology* (established businesses and industries are being commoditized or replaced through digitization, the innovative use of new models, and automation, such as machine learning, and robotics); *accelerating digitization and democratization of information* (an increase in volume, transparency, and distribution of information require organizations to rapidly engage in internal and external communication and complex collaboration with customers, partners, and colleagues). This is resulting in a move away from traditional organization structures which they label as "silo-based machines" towards "agile-led organisms" within which "boxes and lines" matter far less than a focus on direction, teams are built around end-to-end accountability, and rapid changes can occur enabled by flexible resources.
- What elements of the proposed business model design match competitive business models, are they matching the customers' most important priorities? How? What priorities are not well served? Why?
- How does/did the design compare to competitive networks? Do they reflect responses capability? What differentiates each of them? Are/were customers aware about that differentiation? Are they relevant to current and emerging customer needs?
- What is the commitment of value chain network partners? How successful is the value chain network? Are competitive networks outperforming the organization's network? If yes, how and why? Will the value chain network require change?
- Are competitors' business models based on the same assumptions?
- Can the proposed business model design capture value for all stakeholders? How sustainable and defensible is the proposed value capture design?
- How long will the business design be sustainable? How likely are changes in customer priorities that will require it be changed? Are there any changes in customer expectations or behavior patterns appearing? What alternative capability requirements are already being considered that meet the next cycle of customer priorities?

Process: Building the Business Model

Value Identification, Creation, Production, Delivery, and Capture: Response Capabilities

Response capability-based strategic and operational business models are described by components that identify the capabilities as essential for successful pursuit of specific identified opportunities. They are performance, profitability, productivity, partnerships, producibility, and preservation.

In the context of building a business model, the overall process should identify scenarios that provide positive guidance to the response capability the organization should expect to have to meet if its business model is to be successful:

Customer Responsive Performance-Based Value Engineering

Expected development in customer expectations for PRODUCT-service and product-SERVICE customer needs. Potential for value contribution: quantitative and qualitative added value evidence.

Performance Capability: What are the important customer value performance drivers? Can they all be met? What are the core value performance drivers? What is the impact on the existing value chain network capabilities in meeting them? Are additional capabilities required? If yes how can they be accessed and at what cost and how long would this be expected to take?

Sustainable "Profitability" Management

The evidence that suggests cost/service/price/value equivalent expectations considers, and satisfies, all stakeholders' financial objectives and its sustainability.

Profitability Capability: What are the profitability expectations of customers? What will be the impact on the margins of the value chain network?

Operating Effectiveness and Efficiency

What are the changing patterns of value migration and their implications for existing value production and delivery capabilities? Can they be identified? *Productivity Capability: Can an increase in the productivity response capability be achieved by restructuring the roles and tasks in the value chain network? Could this be achieved by considering a collaborative approach and expanding the partnership processes? Or does the value chain network require restructuring?*

Management of Network Partnership

A future view on business integration and collaboration; the implications on developing collaborative structures for the value chain, based upon the expected requirements on members' capabilities and capacities, and the need for change as customer needs change, *industry dynamics* change, and/or the means of delivering competitive value changes

Partnership (Management) Capability: Has (or will) the impact of the "convergence" (Industrié 4.0, Value Chain Network 2.0, and the acceptance of wider responsibilities, stakeholder value management). Is future success to be based upon new industries with new value creation, production, and delivery structures?

Integrated and Coordinated Value Engineering and Delivery

The likely impact of change on the "shape" of collaborative, seamless value production from product-service concept through to value delivery, serviceability, and value renewal. Changes in emphasis, structures, capabilities, and responsibilities

Producibility Capability: Does the existing producibility infrastructure model require minor modifications or is obsolescent becoming obsolete? As commented above, is future success to be based upon new industries with new value creation, production, and delivery structures?

Meeting Sustainability Expectations

The dimensions of stakeholder value at expected levels to ensure *organizational sustainability – longevity*. However, it will be necessary to ensure that the requirements of network organizations meet *environmental* sustainability expectations of government and communities: expectations of changes to UN's 17 Sustainable Development Goals (SDGs). Sustainability governance indicators and the GRI (Global Reporting Initiative), together with stronger legislation for environmental protection.

Protection Capability: Delivering stakeholder value at expected levels will ensure *organizational sustainability – longevity*. However, it will be necessary to ensure that the network organization meets *environmental* sustainability expectations of government and communities that may have implications for other capabilities, specifically performance, profitability, and productivity.

Figure 10.6 identifies the processes involved in developing an approach to strategic decision-making by relating the response capabilities required to achieve success from the networked organization to the strategic process; this is essentially a Capability Audit that is required before committing to building the business model. Knowing the responses capability is just the beginning of the business model building process.

Figure 10.6 is also useful as cross-check on interface issues and for each "capability" a metric should be set, for example, for *profitability* a range of EVA output can be set (in fact two will be set for the individual organization and an aggregate value representing an acceptable profitability range). Operating effectiveness can also use EVA as both individual and network performance measurement of *productivity*. In this context the EVA measures the use of input mix and the use of alternative production models. Measurement of the *management of network partnership effectiveness* requires a composite metric based upon the DuPont system.

$$\frac{EVA}{Revenues} \; X \; \frac{Revenues}{Assets} \; X \; \frac{Assets}{Equity} \; = \; \frac{EVA}{Equity}$$

| Economic Profit on sales *Efficiency* | Asset Turnover/Management *Effectiveness* | Financial Gearing/Leverage *"Risk" Management* | Return on Equity *"Stewardship"* |

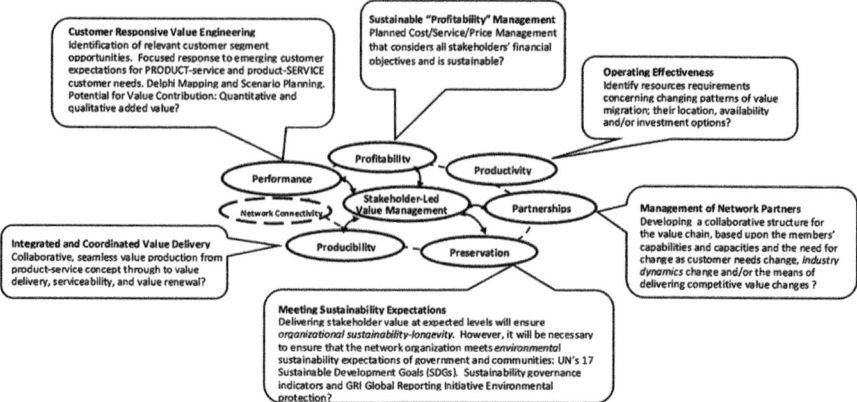

Fig. 10.6 Balancing expectations, capabilities, and costs

Each component equation identifies an important aspect of performance resulting in an overall measure reflecting "stewardship," reflecting the recognition of the overall responsibility of the management of the organization (both individual partner organizations and the value chain network, to the primary and secondary stakeholders).

Strategic and operational value-led choices (value identification), creating value, constructing and managing the value network, and capturing value each determine and contribute to the response capabilities. Essentially the business model is a "connected" sequence of activities involving value engineering and value delivery. Before undertaking the task of designing and building the business model, some consideration of the influences responsible for changing the value creation requirements, value delivery, and capture success processes should be considered, and they comprise:

Identifying Strategic and Operational Choices and the Implications for Value
- Identifying a realistic opportunity (reachable and remunerative)
- Identifying and "accessing" the relevant responses capability
- Creating a relevant and viable *value proposition* (for customers and network suppliers)

Creating Value: Value Engineering
- Relevant response capability
- Implementing the value proposition
- Optimal value chain positioning

Managing the Value Network
- Network strategy and structure: integration and viability
- Shared customer centricity commitment
- Synergistic, optimal financial, marketing structures

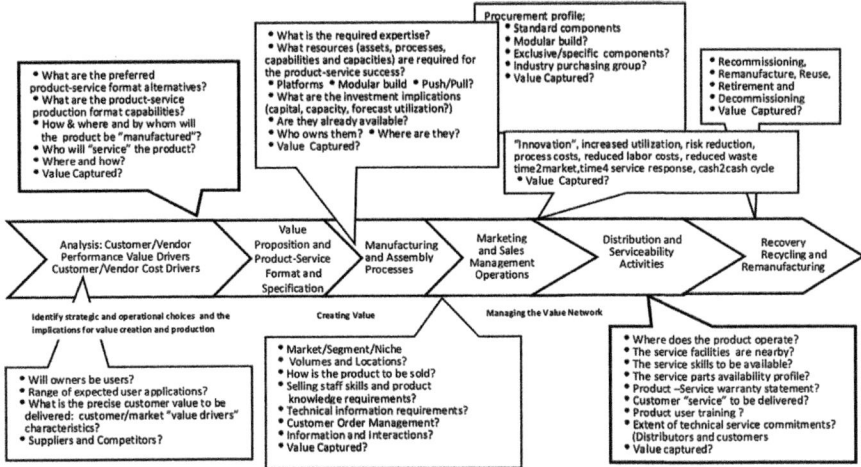

Fig. 10.7 The role of the value proposition in creating added value

Delivering and Capturing Value
- Optimal approach to stakeholder value expectations
- Equitable network reward structure
- Shared risk

The value proposition format has become the vehicle in which these tasks are addressed. Figure 10.7 identifies the sequence of tasks undertaken, and the data required, by the value proposition to identify the "offer" to customers and the implications for, and obligations of, the network partnership. Again, we are increasingly seeing the importance of an audit approach. By identifying the data requirements of all stakeholders (aggregate in the context of identifying response capabilities) and from the market-place (specifically customers' product-service expectations as they are the sources of revenue successes) the success/acceptance of the value proposition is reinforced. Recent events suggest this is not an approach taken by all large companies. General Electric attempted to build a "multi-focused" industrial conglomerate built upon their perceived strengths and perceptions of market opportunity; in the process they have appeared to incur huge amounts of debt and an organization requiring a very large response capability base (Sutherland 2018).

Convergence

The "convergence" of Industrié 4.0, Value Chain Network Management 2.0, and the wider concern of stakeholder management has produced a range of "management tools" (some might say management toys) that *facilitate* analysis and decision-making; it should not be assumed that it *produces* the business model. It offers new

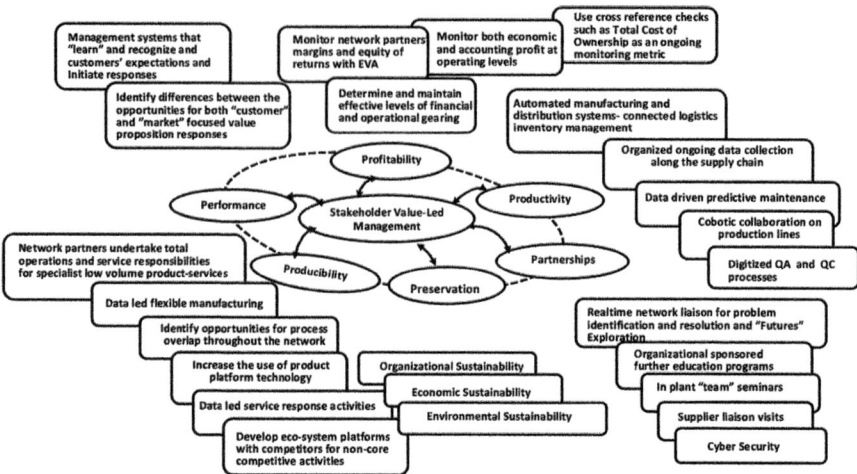

Fig. 10.8   Facilitating activities of a resource capability-led business model

insights into data relevance, collection, analysis, and decision-making. Figure 10.8 aligns activities and processes that are becoming digitized in the "connected" organization and links them to the planning response capabilities.

This chapter has demonstrated the logic underwriting the role of response capabilities in building the business model. It has avoided specific applications and has taken a broad perspective because companies and networks are all different; what they share is a need to respond to opportunities rapidly and for the response to be relevant. It emphasizes the importance of taking full advantages of facilities of digitization and the connectivity that digital thread and digital twin concepts offer; and furthermore, build upon the connectivity and collaboration that access to "big data" has made possible.

References

Aghina, W., DeSmit, A., Lackey, G., Lune, M., & Murarks, M. (2018). The five trademarks of agile organisations, *McKinsey Report,* January.

Ansoff, H. I. (1984). Implementing strategic management. Engelwood Cliffs, NJ: Prentice Hall.

Chandler, A.D. (1977) The visible hand: the managerial revolution in American business. Belknap Press, Cambridge, Mass.

Ellis, S. (2019, January 8). Top 10 predictions for worldwide supply chains in 2019. *IndustryWeek.*

Gyorey, T., Joachim, M. & Norton, S. (2010). The challenges ahead for supply chains. *McKinsey Quarterly,* November.

Hagel. J. Brown, J.S., & Davison, L. (2009, April 8). Introducing the collaboration curve. *Harvard Business Review.*

Housh, W. (2015, March 12). Choosing a business model that will grow your company. *Entrepreneur: Asia Pacific.*

Norman, R. (2001). *Reframing business*. Chichester: Wiley.

PCW. (2015). Strategy+Business. Retrieved from www.pwc.com/us/en/library.html

Slack, N., Chambers, S., Johnston, R., & Betts, A. (2009). *Operations and process management (2nd Ed.)*. Harlow, UK: FT-Prentice Hall.

Sutherland, B. (2018, January 1). General electric's new CEO Larry Culp turns the page on a horrible 2018. *Bloomberg/Australian Financial Review*.

Chapter 11
Managing the Business Model

Abstract Building the business model requires management to identify *the* relevant strategic opportunity, *and by identifying* the required capability responses and their availabilities and matching these to the opportunity. The "convergence" of Industrié 4.0 and Value Chain Network 2.0 and the "total" focus upon stakeholder value (not solely on shareholder satisfaction) offers new perspectives on what, how, when, and where this is undertaken. Managing the business model implements the strategic intention by selecting the relevant capability response characteristics required to achieve the objectives set by management. In this chapter we review the significant components of the "convergence" and how they impact on the capability responses. Capability characteristics play an important role in the operationalization of strategic capability responses. The availability of real-time data makes the concept of a "rolling value proposition" possible by identifying changes in operating requirements and applying them digitally. Real-time data analysis can be used to monitor margin management and "translate" the implications of change to short-term profitability throughout the network. Digital twinning provides data on operational efficiency; the data output can be used immediately to improve asset productivity and in the longer-term to make specification changes. Network partners are able to "connect" with the data to modify their inputs into product-service performance in both short- and long-term performance. The data outputs can also revise work practices and be added to training and upgrading of staff skills. The aggregation of performance data identifies the efficiency of the producibility infrastructure and identifies requirements for changes in partner contributions and, possibly, value chain positioning. Monitoring the use of resources, both production assets and consumables, is leading to changes in operational practices that can conserve these resources and lead to long-term design changes that build in remanufacturing requirements into product-service packages.

Keywords Managing business model · Value added and added value · Value contribution analysis · Activity-based management

Introduction

Managing the business model implements the strategic intention by selecting the relevant capability response characteristics required to achieve the objectives set by management. In this chapter we review the significant components of the convergence and how they impact on the capability responses. Figure 11.1 reminds us of the important drivers of capability responses: connectivity, collaborative networks, data collection and analysis, and flexible automation.

Capability characteristics play an important role in the operationalization of strategic capability responses. Figure 11.2 suggests examples of operational interpretations of strategic response capabilities. The availability of real-time data makes the concept of a "rolling value proposition" possible by identifying changes in operating requirements and applying them digitally. Real-time data analysis can be used to monitor margin management and "translate" the implications of change to short-term profitability throughout the network. Digital twinning provides data on operational efficiency; the data output can be used immediately to improve asset productivity and in the longer-term to make specification changes. Network partners are able to "connect" with the data to modify their inputs into product-service performance in both short- and long-term performance. The data outputs can also revise work practices and be added to training and upgrading of staff skills. The aggregation of performance data identifies the efficiency of the producibility infrastructure and identifies requirements for changes in partner contributions, and possibly, value chain positioning. Monitoring the use of resources, both production assets and consumables, is leading to changes in operational practices that can

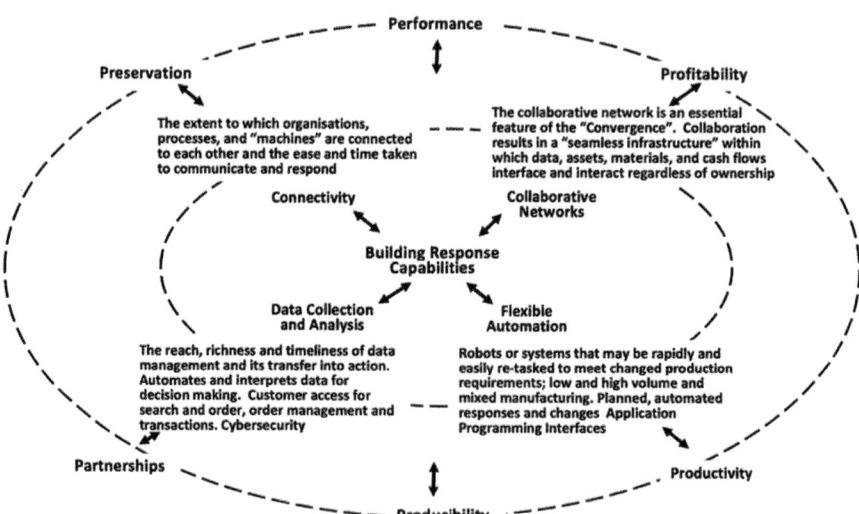

Fig. 11.1 Drivers of response capabilities

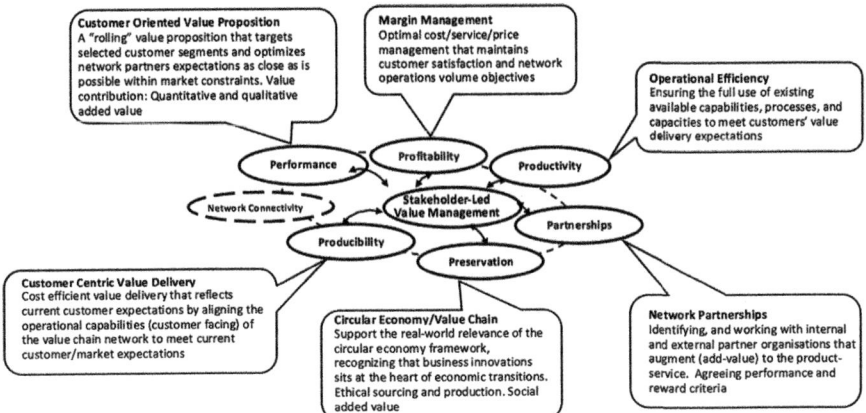

Fig. 11.2 "Ongoing operations" operating model

conserve these resources and lead to long-term design changes that build in remanufacturing requirements into product-service packages.

Working with the Convergence to Focus Capability Responses and to Create Stakeholder Value

Given the comments by Ellis (2019) (in the previous chapter, Chap. 10) and the number of and the focus of hardware and software producers, it is very clear that capability response-based strategy will rely very much upon Industrié 4.0, Value Chain Network 2.0, and stakeholder value-led management. Managing the business model and the operational implementation of the strategic direction of network-based organizations will require an understanding of the benefits they can deliver (and the constraints they may impose for some businesses). Bollard et al. (2017), at McKinsey, identified and addressed the implications of elements of the Industrié 4.0, Value Chain Network 2.0, and the refocused views of stakeholder value management. The McKinsey model identifies lean process design, business process outsourcing, advanced analytics, digitization, and intelligent process automation. These and other important developments are identified in Fig. 11.3 and discussed in the context of what they can deliver for the response capabilities developed and discussed in previous chapters: performance, profitability, productivity, producibility, partnerships, and preservation.

> *Digitization*; is the process of converting data or information into digital format: digitization (in the business context); is the use of digital technologies to change a business model and provide new revenue and value producing opportunities. *Performance* throughout the value chain network is enhanced; data content is increased and provides accuracy, faster time responses, and transparency. As a result, *profitability* increases because of improved revenue opportunities, as well as improved asset management and cost reductions.

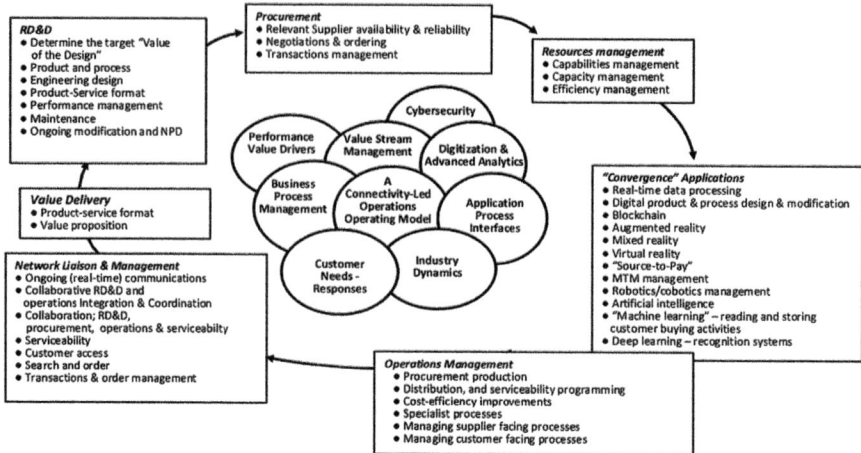

Fig. 11.3 A "connected" operational operations model. (Source: Adapted from McKinsey (Bollard et al. 2017))

Productivity improvements such as waste (labor time and materials usage – value destruction) are highlighted as data identifies time and inventory excesses throughout the network. *Process mining.* looks at data, which is recorded by information systems that support business processes, rather than actual logbooks, often process mining is the first attempt actually to use these data in a structured way. A systematic analysis of digital log traced through process mining tools offers enormous potential for all organizations that are struggling with complex processes. Digitalization provides opportunities for greater accuracy and detail in *partnerships* communications resulting in increased collaboration in RD&D and operations activities (e.g., digital threads providing network performance data and making real-time adjustments). A primary reason for the success of *producibility* operating infrastructure is due to the impact of digitalization on information and cash flows; this is enhanced by the use of *application programming interfaces (APIs)* software intermediaries that permit applications to communicate with each other which is essential for the successful transfer of data in partnership interfaces. *Preservation* programs benefit due to collaborative sourcing; this may be achieved by including "design-for-remanufacturing" during the design and development stages of product development. Alternatively, manufacturers working with organizations such as COVISINT may be an outcome; currently it is early days in the life of the circular economy.

High-speed mobile data: (specifically 5G) will provide rapid connectivity estimated likely to be 1000 times faster than 4G and is likely to enable industrial automation systems to operate at unprecedented speeds, increasing the time that real-time data analysis can be completed and decision-making realized. The implications for *performance* are for markedly improved response times for network order management and serviceability responses. There is an impact on *profitability;* shorter response times mean less inventory in supply response pipelines, faster and shorter cash2cycles, improved cash flow, and lower working capital. Faster data is having an impact on *productivity;* manufacturing capacity, labor, and materials utilization will be increased and with it the notional value of factor-based productivity values. *Producibility* is becoming more effective as high-speed mobile data flows identify shortages ahead of time rather than as they occur, providing opportunity to use alternative supply sources. *Partnership* communication and collaboration is becoming more effective; track and trace software provides accurate time and location details providing time for rescheduling work and materials flows, if necessary, based on accurate data. *Preservation* (organizational and environmental sustainability) is supported by fast and accurate information, for example, data exchange between industry partners is identifying global sources of "rare earth" materials measuring quality digitally and "broadcasting" problem awareness and solutions rapidly and is improving effectiveness and efficiency within organizations and materials owners.

Artificial intelligence (AI): is based upon the theory and development of computer systems capable of performing tasks normally requiring human intelligence, such as visual perception, speech recognition, decision-making, and translation between languages. The impact on *performance* is in quality control of production (quality control and management) and administrative (order and transactions management) processes. *Profitability* improvement occurs as consistency and reliability of product-service quality reduces/eliminates returns and reworking problems; administrative applications improve cycle time management. Cashflow improvements have positive impacts on asset management. By reducing and supporting human involvement, *productivity* improvements become significant as wastage is reduced. *Partnership* collaboration can increase as the consistency and reliability of materials and information is standardized throughout the network. Much the same can be said concerning *producibility;* the notion of an effective and efficient operation infrastructure is dependent upon the "standardization" and application of AI technology. Effective *preservation* policies should include a contribution from AI that ensures standardization of production processes, and wherever possible, targets for ethical production. An example of AI working successfully is image processing used in the production of electronic circuit boards; every circuit board is photographed as it passes down the production line, and images are passed to an "AI agent," which has been trained to identify any defect on the circuit board ("Harness AI to accelerate innovation" 2019). *Big data*: can be considered here as it is an important input into AI and algorithm models providing extremely large data sets that may be analyzed computationally to reveal patterns, trends, and associations, especially relating to human behavior and interactions. Also considered is *advanced analytics,* comprising *descriptive analytics* (this provides many of the what, where, who, how, as well as the when answers particularly for the question of what happened). In the *Diagnostic analytics* stage, historical data can be measured against other data to answer the questions of *why* and *how* something happened. *Predictive analytics* tells *what* is likely to happen; *prescriptive analytics* explores optimal outcomes.

Machine learning: is an application of artificial intelligence (AI) providing systems with the ability to learn and improve from experience automatically without being explicitly programmed. Machine learning focuses on the development of computer programs that can access data and use it to learn for themselves. Machine learning can help by using historical data to make better business decisions by building algorithms that discover patterns in data and construct mathematical models using the output. The models are used to make predictions on future data. For example, a current example is the application of a machine to predict how likely a customer is to purchase a particular product based on their past behavior. *Performance* benefits include a decrease in the demand/supply response time in customer satisfaction with additional benefits of reduction in relevant industry holding and the (almost) elimination of irrelevant SKU items, presenting opportunities for building a customer "first-choice" positioning based upon reliability. *Profitability* benefits follow as costs can be profiled around the algorithm models, eliminating irrelevant marketing costs as well as the inventory and its associated logistics costs. *Productivity* improvements accompany the profitability benefits; "knowledge" of the *what* leads to *where* and *when* of consumer behavior from the models and efficiencies in organizational supply chain responses to customers. *Partnerships* are strengthened; suppliers and distributors favor effective and efficient network partners, and transparency increases collaboration. *Producibility* effectiveness is enhanced. The increasing accuracy of algorithm-based models offers stronger data with which to plan the producibility infrastructure and its managing its operating efficiencies. *Preservation* objectives of ethical sourcing and production impact upon the environment can be set with more confidence than hitherto.

Machine2Machine (M2M): direct communication between devices using physical or wireless connection. Machine to machine communication includes sensors or meters to communicate the data it records (such as temperature, inventory level, etc.) to application software to monitor and adjust industrial processes based on managing manufacturing processes or ordering and transaction management. Interconnected wireless networks improve *performance* in various areas in a number of industries (automotive and consumer durables, healthcare equipment) and production "machinery" that require just-in-time supply chain responses and increasingly predicted maintenance. M2M applications have a positive impact on *profitability* as networked organizations undertake transformation connecting geographically dispersed people, devices, sensors, and machines to corporate networks. Industries such as oil and gas, agriculture (arable, dairy, etc.), defense, public city, and urban developments are in this category, which attempt to meet customers' value-adding expectations. *Productivity* improvement opportunities are improved throughout the value chain network; manufacturers costs are reduced as customer centricity becomes more focused on specific outcome needs; customers' productivity increases as process outcomes and costs respond to supplier customer centricity. *Partnerships* are strengthened as the ability to satisfy increasingly industrial end-user financial and marketing objectives indicates that partnership resources meet their targets for fixed asset and working capital utilization. Strategic and operational *producibility* infrastructures increase their cost-effectiveness and efficiency by focused investment in specific M2M networked systems that are designed to optimize response times and network materials flows. *Preservation* benefits from more accurate material input specifications throughout the value chain network.

Cloud technology: or *cloud computing* is the delivery of *computing* services – servers, storage, databases, networking, software, analytics, intelligence, and more – over the Internet, "the *cloud*," to offer faster innovation, flexible shared resources, and economies of scale. Cloud computing is the use of various services, such as software development platforms, servers, storage, and software, in a business context. The ability to pay on demand and to scale resources quickly is largely a result of cloud computing vendors being able to pool resources that may be divided among multiple clients. The impact on capability responses is significant. *Performance;* (maintaining focus on the business); businesses are realizing that running an IT department is not their core competency buying cloud services, is often more cost effective, more reliable and enables organizations to reallocate their limited resources to growing their business. Furthermore, agility can be impaired for businesses with significant technology investments that can find themselves unable to take advantage of shifts in the market or respond to competitive pressures because the capital, people, and time are not available in the measure needed to react. Cloud services remove these barriers, allowing businesses to continually adapt their technology needs to their business without the costs that would normally have to be considered with an on-site data center. D2C (device-to-cloud) IoT solutions will soon be available for distributed data orchestration to simplify IoT infrastructure in the same way that cloud services simplified IT infrastructure. Enterprises will have a single, tightly integrated D2C platform for deploying the IoT devices, wireless connectivity, cloud-based network management, security, and other infrastructure needed to build and deploy IoT solutions, so they no longer need to be wireless or embedded experts to gain access to their IoT data. *Profitability,* large capital investments can be minimized or eliminated altogether in favor of period payments. Capital can be protected by keeping capital and operational expenses to a minimum and can be directed into intangible assets that increase business effectiveness such as brand building and customer relationships management. It follows that unlikely *productivity* improvements are available; businesses that have peak seasons or different seasonal staffing demands can benefit from cloud services by temporarily accessing capacity for the seasonal business peaks, avoiding the purchase of hardware and/or software that would otherwise go unused during the slower times of the year. *Partnerships*: collaboration is improved with the "access from anywhere" capability of cloud computing, and this broadens the capabilities of the network. *Producibility*: being able to extend the boundaries of the business is one of the major benefits of cloud services and clearly benefits the building of industry/network production infrastructures with access to shared portals and applications, and data is available to authorized users anywhere there is Internet access. The impact on *preservation* (environmental sustainability) is linked to productivity in the use of the cloud; portals can be established to encourage ethical sourcing and production activities throughout the partnership.

Immersive technology enables mixed reality: *Augmented, mixed, and virtual reality* is technology that attempts to emulate the physical world through the means of a digitally simulated environment thereby creating a sense of immersion. Immersive technology enables mixed reality; in some uses, the term "immersive computing" is effectively synonymous with mixed reality as a user. Augmented reality (AR) is a type of interactive, reality-based display environment that takes the capabilities of computer-generated display, sound, text, and effects to enhance the user's real-world experience. The word "augment" means to increase, extend, or make better; AR can be understood as a form of virtual reality (VR) *where the real world is expanded or enhanced through the use of virtual elements*, usually overlaying those elements on the view of the real world through the use of a visual device. VR comprises *a three-dimensional, computer-generated environment* which can be explored and interacted with by a person. That person becomes part of this virtual world or is immersed within this environment and while there is able to manipulate objects or perform a series of actions. *Performance*: AR applications offer both customer and network opportunities to create added value. In collaboration with a supplier, customers are able to access the IKEA website. IKEA launched their augmented reality catalogue in 2013 to enable shoppers to visualize how certain pieces of furniture could look inside their home. The "app" measures the size of the products against the surrounding room and fixtures to offer a true-to-life size where possible. Very, a UK retailer does a weather check at a user's location to present climate-specific offers on its website's home page. In industrial product-service markets, AR applications are used to plan the engineering programs for new products and to manage quality control of ongoing production using scanning to inspect production line products. In networked value chains, this application proves to be invaluable in identifying rectification problems and where best to perform the rectification. *Profitability* can be enhanced in two ways: customers can use AR applications to match their needs against alternative value propositions and make appropriate decisions, and the second follows the observation of Leszinski and Marn (1997) describing the notion of price/value equivalence; customers evaluate market offers on the basis of the price being asked by a vendor with the value being offered. Customer research can identify the value performance drivers that are used in this evaluation and demonstrate how the customers' expectations are to be delivered. The positive impact on *productivity* can be increased. AR identifies the issues likely to be problems during manufacturing by eliminating them in the planning stage; quality control is managed using remote digital equipment such as cameras. *Partnership* collaboration is extended by AR presentations of product and production development ideas in which suppliers and distributors make constructive contributions to a project. *Producibility* infrastructure design benefits from the partnership approach by using immersive technology to "build" an infrastructure that can be used to explore interaction and interactivity processes. Agricultural applications of immersion technology explore production methods that are linked to the environmental sustainability aspects of *preservation*.

Shared knowledge, knowledge transfer networks: organized approaches that share knowledge to perform a task, solve a problem, and deliver a service; the approach is both intraorganizational and inter-organizational. Organizations have increasingly recognized that knowledge constitutes valuable intangible assets (aka. intellectual property) for creating and sustaining competitive value advantage. Knowledge sharing activities are generally supported by knowledge management systems. Technology constitutes only one of the many factors that affect the sharing of knowledge in organizations, such as organizational and network culture, trust, and incentives. *Performance* capabilities are enhanced by sharing knowledge this constitutes a major challenge in the field of knowledge management because internally, some employees tend to resist sharing their knowledge (intellectual property) with the rest of the organization, and some totally resist sharing with external network partners as it is considered to be a source of competitive value advantage. In the digital network-based structures, websites and applications enable knowledge or talent sharing between individuals and/or within teams. The individuals can easily reach the people who want to learn, share their talent, and get rewarded. For example, KTN is a UK government established organization with the purpose of enabling organizations to build better links between science, creativity business, and business partnerships, which brings together businesses, entrepreneurs, academics, and funders to develop new products, processes, and services (https://www.ktn-uk.co.uk/). *Profitability* improvements accrue by channeling knowledge into innovative RD&D, thereby improving the ability to create competitive value advantage. *Productivity* increases follow; regular exchanges of experience (on intra- and inter-organizational levels of the business) benefit operations management; these also increase *partnership* collaboration. The "seamless" production infrastructure, essential if *producibility* is to be optimized to meet stakeholder expectations, consistently requires an ongoing approach to knowledge transfer networks. Successful implementation of *preservation* policies and strategy is another capability benefiting from shared knowledge by network partners.

Blockchain: is a *database* that is shared across a network of computers. Once a record has been added to the chain, it is very difficult to change (an aspect of cybersecurity); an entry made will exist in perpetuity. To ensure all the copies are the same, the network makes content checks. Blockchains are being used to underpin cyber currencies such as bitcoin, but many other possible uses are emerging. Transactions are recorded into a record. The record is checked by the network, known as *nodes* to ensure validity. Once accepted the record is added to a block. Each block has a unique code, a hash. It also contains the hash of the previous block in the chain. The block is added to the blockchain; the hash codes connect the blocks together in a specific order. Hash codes keep records secure. A hash code is created by a mathematical function that takes digital information and generates a string of alphanumerics. Regardless of the size of the file, the hash function always generates a code of the same length. Any change to the original input will generate a new hash; a changed hash breaks the chain. Therefore, attempts to hack into the chain would have to recalculate the block content; a tedious and very, very long process and changes would be rejected by the many computers used in the verification process. Entry into blockchain requires permission; blockchain establishes tests for computers seeking to join and add records; the tests are known as consensus models. Current uses for blockchain include cryptocurrency, banking, supply chains, healthcare, voting, and property records. (See Reuters Graphics for an explanation of blockchain protocols and processes). Blockchain offers networks *performance* characteristics capabilities of transactions, convenience, and secure data storage. *Profitability* and *productivity* capabilities are both enhanced by the benefits of a secure transactions log and data storage "document." *Partnerships* can use blockchain with confidence and improve collaborative activities. The benefits offered by blockchain fit the needs of a production infrastructure such as the *producibility* model in which secure transactions and data management play an important role. The application of blockchain to a *preservation* capability is likely to be increasingly important as the blocks will contain information on uses of scarce/rare materials. An example is given by provenance (https://www.provenance.org/). The organization argues every product has a story and helps brands and retailers build customer trust through transparency, using blockchain technology that empowers shoppers. Provenance is a digital platform that enables businesses to gather and present information about products and supply chains supported by verified data – this information can be connected to a store, packaging, and online so that consumers can see the origin, journey, and the impact of products.

Cybersecurity: relies on cryptographic protocols to encrypt emails, files, and other critical data. This not only protects information in transit but also guards against loss or theft. In addition, end-user security software scans computers for pieces of malicious code, quarantines this code, and then removes it from the machine. Electronic security protocols also focus on real-time detection. Many use heuristic and behavioral analysis to monitor the behavior of a program and its code to defend against viruses or "Trojans" (from Trojan horse), a type of malware that is often disguised as legitimate software. T*rojans* can be employed by *cyber*-thieves and hackers trying to gain access, which change their shape with each execution (polymorphic and metamorphic malware). Security programs can confine potentially malicious programs to a virtual bubble separate from a user's network to analyze their behavior and learn how to better detect new infections. Cybersecurity concerns the defense of computers, servers, mobile devices, electronic systems, networks, and data from malicious attacks. It's also known as information technology security or electronic information security. Cyber-security threats affect all industries, regardless of size. The industries that reported the most cyberattacks in recent years are healthcare, manufacturing, finance, and government. Some of these sectors are more appealing to cybercriminals because they collect financial and medical data, but all businesses that use networks can be targeted for customer data, corporate espionage, or customer attacks. Cybersecurity can be applied in a variety of computing contexts, from business to mobile computing, and can be divided into a few common categories. *Performance*: security – is the practice of securing a computer network from intruders, targeted attackers, or opportunistic attacks. It is equally applicable to networked organizations such as value chain networks. It provides both customers and suppliers with safe communications channels and protects the integrity and privacy of data, both in storage and in transit, increasingly a primary requirement. *Profitability* benefits from a secure communications environment; intellectual property can be protected and utilized to ensure optimal competitive value advantage. *Operational security is important for productivity* including the processes and decisions for handling and protecting data assets and production outputs. The extent of "permissions" users have when accessing a network and the procedures that determine how and where data may be stored or shared are all included. *Partnerships* are dependent upon strategic security: information on future product-service innovation and potential target markets and customers requires to be transmitted using protected communications facilities. Cybersecurity is particularly applicable for ensuring *producibility* activities and processes proceed without interference or interruption; without this assurance the benefits offered are unlikely to be delivered. *Preservation*: confidential knowledge concerning access to, and use of, scarce materials and processes requires protection if competitive value advantage is to be maximized.

Organizations should consider their digital transformation journey carefully; the following approaches help maximize the value of both current and future technology investments in the value chain network. They revolve around thinking about the future of the business, the likelihood of industry disruption, and the specific things technology would be able to do for your value chain network:

- Invest in technologies that provide current efficiency/effectiveness for the network and will enable future capabilities that may be as yet undefined.

- Not "technology for technology's sake," or because of fashion and/or fad" but for identifying and resolving business problems or seizing on opportunities.
- Work with all relevant partners to accelerate your connectivity capabilities and who share the objectives of stakeholder-led value management.
- Review the value chain network partnership to ensure it is ready for increasing levels of digitally enabled products and processes.
- Evaluate the "organizations'" relative maturity in the adoption of new technologies and, more importantly, its' ability to translate those technologies into digital transformation.

Value Contribution as an Approach to Operations Planning and Control

Value Engineering, Value Analysis

Value analysis is a systematic interdisciplinary examination of factors affecting the cost of a product or service in order to devise means of achieving the specified purpose most economically at the required standard of quality and reliability (British Standards Institution 1992). Value analysis is critical assessment by an organization of every feature of a product to ensure that its cost is no greater than is necessary to carry out its functions. It is an approach to improving the value of an item or process by understanding its constituent components and activities and their associated costs. It then seeks to find improvements to the components by either reducing their cost or increasing the value of the functions.

Value engineering is concerned with new products. It is applied during product development. The focus is on reducing costs, improving function, or both, by way of teamwork-based product evaluation and analysis. This takes place before any capital is invested in tooling, plant, or equipment.

This is particularly significant because up to 80 percent of a product's costs occur during the rest of its life cycle and are only "variable" at the design development stage; after this they become "fixed." This is understandable the design of any product determines many factors, such as tooling, plant and equipment, labor and skills, training costs, materials, shipping, installation, operations, maintenance, as well as decommissioning and recycling and remanufacturing costs. There are questions that are applicable to both value analysis (existing product-services) and value engineering (potential/planned product services):

- Can nonstandard parts be redesigned so that they become standard parts? If so, what will be the impact on product-service performance as well as on cost?
- Can the manufacturing process be shared across a wide range of product-service outputs?
- Can two or more parts be replaced by one part?

- Can component materials be substituted for lower-cost materials (or materials that will increase efficiencies by lowering production process costs), lower end-product-service operating, and/or maintenance costs?
- Can the organization work with network partners to manufacture components at lower costs by improving production methods or by increasing volumes?
- What impact does economies of scale, the experience effect, and collaborative costs have on value analysis?
- What cost savings are available when product groups are built on platforms?
- What investment opportunity costs are involved? Could the organization benefit by outsourcing the total manufacturing task to a "low-volume" capability partner?
- Has the "Total Cost of Ownership" been thoroughly identified?

Added Value and Value Added

A potential source of error is confusion of the terms *added value* and *value added;* this was raised in Chap. 5 (profitability). Value added is the principal basis of expenditure taxation in a number of countries. Value added is therefore not the same thing as added value; the former is the difference between the value of the material and semi-produced inputs which a firm buys in and the value of the output which it sells. It is therefore equal to the firm's net output. Often the terms are used interchangeably, and this is incorrect and confusing. In this book our interest is in both value added (the increase effect of economic efficiency) and added value (the impact of a change in product or process that can have a positive impact on a customer that increases the customer's value advantage in the market).

In the context of a value chain network, both are important. The underlying principle of value chain networks is to optimize the overall cost of value engineering and value delivery. It follows that if a task that is currently undertaken later in the production process can be completed earlier at lower cost (possibly due to economies of scale or more likely because of platform-based manufacturing), then this occurs and adds value to the network output at that point in the process. Clearly this has cost implications, and the reduction in costs benefits the network partnership and the end-user customers. Thus, added value has both qualitative and quantitative aspects. Furthermore, the added value produced and delivered has the potential to benefit both purchaser and vendor; clearly given the underlying principle that networks optimize the overall cost of value engineering and value delivery, the added value feature would only be added to the value proposition if it can be produced cost-efficiently, i.e., the incremental revenues generated will cover the incremental costs involved, providing mutual benefits. Product-service features that reduce costs have positive impact on value added as they increase productivity. Therefore, we have:

Cost-Led Attributes
    ↘
        +Added Value = + $Value Added = $Value Contribution = Competitive Value
    ↗   Advantage
Revenue-Led Attributes

Figure 11.4 illustrates a range of activities that add value to value chain members by introducing differentiation attributes (exclusive to the network) that increase the attractiveness to end-users of the product-service; here the assumption is that in real terms, the end-user will gain more utility from the differentiation characteristic, and consequently members of the network will increase revenues and profit. Figure 11.4 also suggests that increasing efficiencies (reducing materials, labor, and service-ability operating costs, therefore increasing productivity) will also add $ value added.

Identifying opportunities to work with network partners and end users is suggested by Fig. 11.5 that identifies five activity areas where opportunity may be found; these are revenue generation opportunities, operating efficiencies, asset productivity, operating effectiveness and risk management. Examples of *revenue generation opportunities* often occur when government and/or consumers become more aware of healthcare requirements. Governments introduce "medicare" systems, and insurance organizations provide healthcare insurance plans. These are not new. However, what is very recent is pet healthcare, veterinary services, and pet healthcare insurance. In this rapidly growing market, healthcare diagnostics have been applied to animal healthcare diagnostics, and animal healthcare insurance is offered by animal interest organizations (the Royal Society for the Prevention of Cruelty Animals), as well as retail FMCG chains, and the PetSmart, Banfield veterinary service partnership in the USA (Banfield.com). *Operating efficiencies* have been increased by digitization and connectivity; demand and supply response times are reducing rapidly. *Asset productivity* is being addressed by "serviceability-based" products where the PRODUCT-service is being replaced by product-SERVICE portfolios, discussed earlier. *Risk management* is being reduced by the "serviceability-based" products, releasing capital for investment in intangible products, such as customer-service initiatives (performance management programs) and by collaborative product-service and process RD&D.

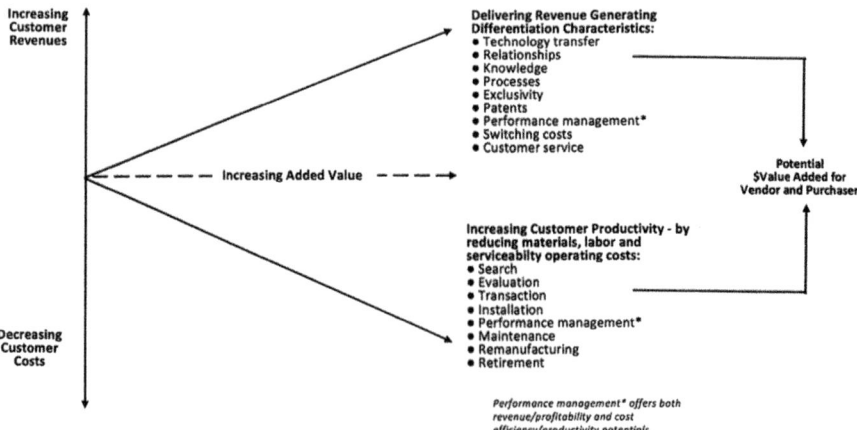

Fig. 11.4 Value added can be delivered by revenue enhancement and/or increasing customer productivity resulting in added value

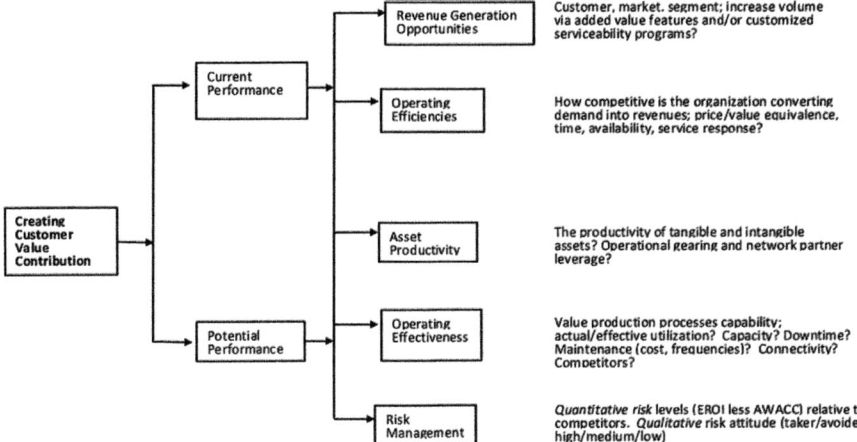

Fig. 11.5 Identifying mutual opportunities for adding value and value contribution

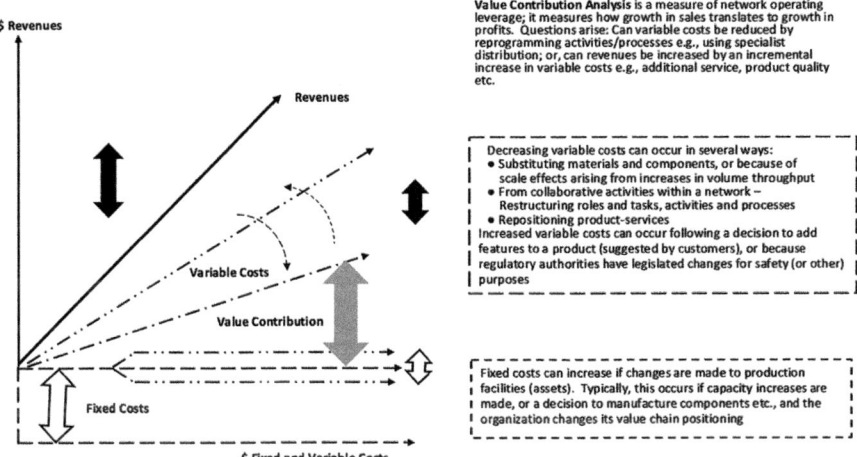

Fig. 11.6 Changes to operational gearing occurring due to organizational-based product-service sourcing decisions

Value Contribution Analysis

Figure 11.6 represents graphically the process. Value contribution must offer both vendor and customer measurable positive gain in value added. In essence it is a process of measuring the risk involved in changing operational gearing and its impact on risk. Chapter 5 introduced the concept of gearing: an organization working in a market in which revenues are very responsive to changes in consumer expenditure (fast fashion apparel) will be concerned to ensure that it operates with low gearing (low fixed costs/high variable costs) and tends to outsource much of its product-service production. Companies in less sensitive markets (FMCG food) tend to have higher fixed costs. Value contribution aims to increase the economic profit of both. Value contribution analysis is a measure of network operating leverage; it measures how growth in sales translates to growth in profits. Questions arise. Can variable costs be reduced by reprogramming activities/processes, e.g., using specialist distribution, or can revenues be increased by an incremental increase in variable costs, e.g., additional service, product quality, etc. The decisions are led by the anticipated increase in economic profit for both customer and vendor organizations. A fruitful area is to consider processes where "value overlap" occurs; for example, the construction industry has developed pre-formed inputs ranging from sealed double-glazed window units to complete kitchen and bathroom installations.

It follows that when a firm is generating *economic profit* (considered here to be operating profit less a charge for the use of capital involved in the production of the operating profit), it is creating and appropriating added value. Provided the value of the sales generated is more than sufficient to cover the cost of all the resources used by the collective, the added value becomes $ value added. This surplus represents an addition to the value of all the resources tied up in the networked coalition.

Value contribution is calculated based on the operating result after tax and the opportunity cost of invested capital. The operating result shows the economic performance of the organization and is initially a pre-tax figure. Volkswagen uses the various international income tax rates levied on the relevant companies but assumes an overall average tax rate of 30% when calculating the operating result after tax. The Volkswagen Group's financial target system centers on continuously and sustainably increasing the value of the company. To ensure the efficient use of resources in the automotive division and to measure the success of this, they have been using a value-based management system for several years, with return on investment (ROI) as a relative indicator and value contribution, as the key performance indicator linked to the cost of capital, and as an absolute performance measure: Volkswagen explained their approach in their 2013 Annual Report.

For Volkswagen the economic return on investment serves as a consistent target in strategic and operational management. If the return on investment exceeds the market cost of capital, there is an increase in the value of the invested capital and a positive value contribution. The concept of value-based management (economic return on the annualized weighted average cost of capital) allows the success of the organizations and individual business units to be evaluated. It also enables the earnings power of products, product lines, and projects – such as new manufacturing plants – to be measured. Value contribution may appear not to be a "pure" traditional accounting method, but we develop it here to use it as method of evaluating and measuring the economic performance of operations.

Contribution analysis is an approach for assessing causal questions and inferring causality in real-life program evaluations. It offers a step-by-step approach designed to help managers arrive at conclusions about the contribution their program has made (or is currently making) to particular outcomes. The essential value of contribution analysis is that it offers an approach designed to reduce uncertainty about the contribution the intervention is making to the observed results through an increased understanding of why the observed results have occurred (or not!) and the impact of structural and process changes and other internal and external factors.

Contribution analysis is particularly useful in situations where the application is "experimental", e.g., new product-services, a change in product-service formats, as well as ongoing product-services, product-service range effectiveness. Contribution analysis helps to confirm or revise a theory for changing processes, process sequences, or restructuring the value chain.

Five steps are required to produce a credible contribution methodology:

Identify what the analysis is attempting to determine; is the analysis looking to evaluate an existing or developing opportunity? The difference will be the data available. Ongoing activities now have large amounts of relevant data; some of it in real time and therefore timely and accurate. "Futures"-based data will, at best, be as good as attribute-based costing (aka. ABC2) can develop.

Develop a producibility model; map the processes of the existing infrastructure and of available alternatives, or for a new venture, identify the component partners of the theoretical infrastructure and collaborate with relevant personnel and identify potential alternative infrastructure models. Identify the activities involved (i.e., the required process technology applications), the investment required and its costs, the role and tasks of the labor input skills and experience (and costs), and materials required (characteristics, handling problems, and costs). Compare the results of this stage with known competitive producibility infrastructure models.

For existing activities a comparison of the existing model's results with the results of competitive models (and of alternative options) is informative; as are comparisons of market results and stakeholder expectations achievements; customer satisfaction performance; fit4purpose, time responses (product-service and serviceabilty responses), comparative price/value equivalence; distributor responses, PDM (physical distribution management) services, serviceability (supplier support), margins and cash flow, role in distributor product-service range. Identify infrastructure cost profiles and activity levels.

For future prospective activities, compare the existing model results with competitive models and the alternatives; identify best practice competitive producibility infrastructures and any existing similar infrastructures (value, volume, rate of sale, cube, and weight characteristics). Compare with cost profile estimates of the alternative infrastructure models that have been developed.

Identify qualitative data; identify the qualitative benefits that exist throughout the network, for example, the influence of brands, of experience and reputation, of views of primary stakeholders, and of secondary stakeholders concerning the social value added by the networks ongoing activities.

Aggregate the findings; conduct a SWOT analysis – undertake a simple strengths, weaknesses, opportunities, and threats analysis to reach a decision concerning investment in a new venture or for an ongoing product-service or whether or not to consider possible operational changes.

Value Contribution Analysis: Managing the Business

The Volkswagen model offers an approach to *managing the business*. Value contribution is calculated using operating profit after tax and the opportunity cost of invested capital. EVA (operating profit less tax and the annualized weighted average cost of capital) shows the economic performance of the business by deducting tax and an annualized weighted average cost of capital. For global operations local taxation rates would be used. Invested capital is calculated as total operating assets (property, plant and equipment, intangible assets, inventories, and receivables) less noninterest-bearing liabilities (trading payables and payments and receivables). The cost of capital is calculated by multiplying the invested capital by an annualized weighted average cost of capital to give the opportunity cost of capital. As Volkswagen suggests, the concept of value-based management only covers operating activities; assets relating to investments in subsidiaries and associates and the investment of cash funds are not included when calculating invested capital. Interest charged on these assets is reported in the financial results. This approach makes the EVA model ideal for measuring and managing the economic profitability of a business and of a value chain network.

A modified operations model is shown in Fig. 11.7. It differs only by adding activities that can be considered as direct costs, for example, RD&D, transactions management, physical distribution, data management, promotions, and advertising. While at first collecting this data may appear daunting, the availability and detail of activity are becoming much easier with the "connectivity" facilitated by development of digital activities. Surrogate data can also be obtained by asking for quotations

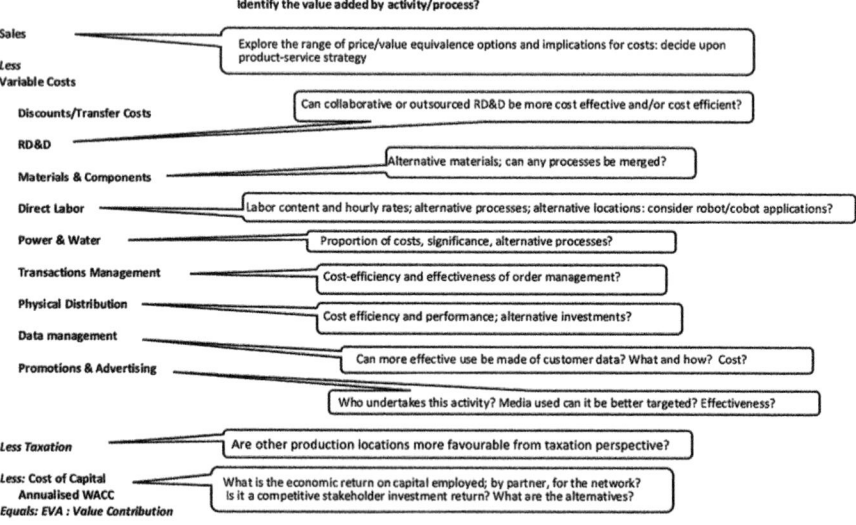

Fig. 11.7 Adapting the operating statement to produce value capture and contribution

for services (transactions management and physical distribution management from specialist service organizations).

Figure 11.7 can also be used for planning purposes. It offers a "what if?" approach, and when used on a spreadsheet program, the "what if?" options can be explored for their impact on value contribution/EVA. Competitors' activities may be analyzed and compared as the data used to calculate EVA performance is published in annual financial statements, and furthermore, Stern Stewart & Co. publish the EVA performance of listed companies as well as offering a consultancy activity (Note: The value contribution model corresponds to the economic value added (EVA®) approach of Stern Stewart & Co. EVA® is a registered trademark of Stern Stewart & Co – see www.sternstewart.com).

Applying an "Activity-Based Management" Approach

An "activity-based management" approach using ABC1 or ABC2 (attribute costing) is also an available approach. The integration of strategic marketing, management accounting, and operations management activities, with all working within an overall set of parameters determined by corporate management (and includes reference to the organization's own performance value drivers), is proving to be a promising development given the availability of real-time data. Many of the responses to the customer's "ideal" service requirements are (initially) proving to be a problem in this respect. For example, the requirements for inventory management and delivery services are beyond the current capabilities of the organization, and it is in this area that a service partnership involving the possibility of an alternative network structure would appear to be required.

The evaluation process views customer performance value drivers as the basic input into product-service planning and management using *attribute-based costing*. Figure 11.8 is an intraorganizational model used to identify the sequence of processes involved. Identifying *customer expectations* and using these to create a perspective of their *value drivers* is an essential start to the overall process. These are interpreted as the *attributes* (or the customer-/market-based essentials to order winning criteria) around which the conventional activity-based analysis then takes place. In order to develop a *"feasible offer,"* a *value proposition* has to meet the *"price/value equivalence"* requirements of the customer organization. This requires a study of the relative importance of each performance value driver to the customer. The authors' research suggests a "Pareto" relationship in which typically "some twenty per cent of the value driver's account for eighty percent of the customer performance." The next stage of the evaluation involves exploring the resource issues requirements of customers' performance value driver performance expectations. This involves identifying the *required capabilities*, assets, processes, and capacities, by investigating both internal and external sources. At this stage of the project, the *"economic viability"* (economic profitability) of a number of alternative options for delivering the value proposition are identified and explored. Activity

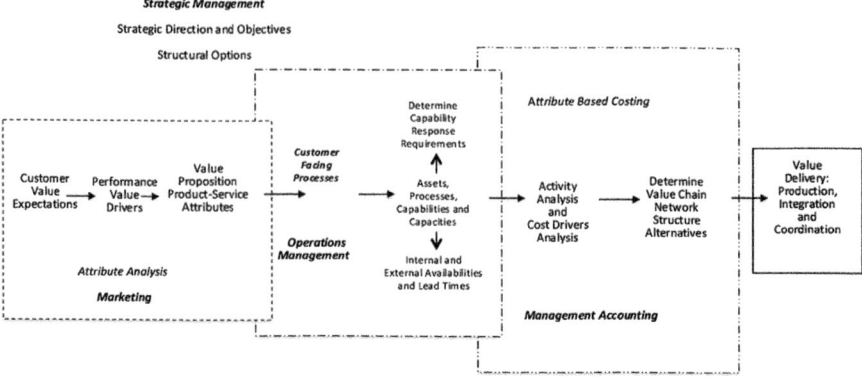

Fig. 11.8 Adapting attribute-based costing (ABC2) for evaluating business operations model management options in a value chain network

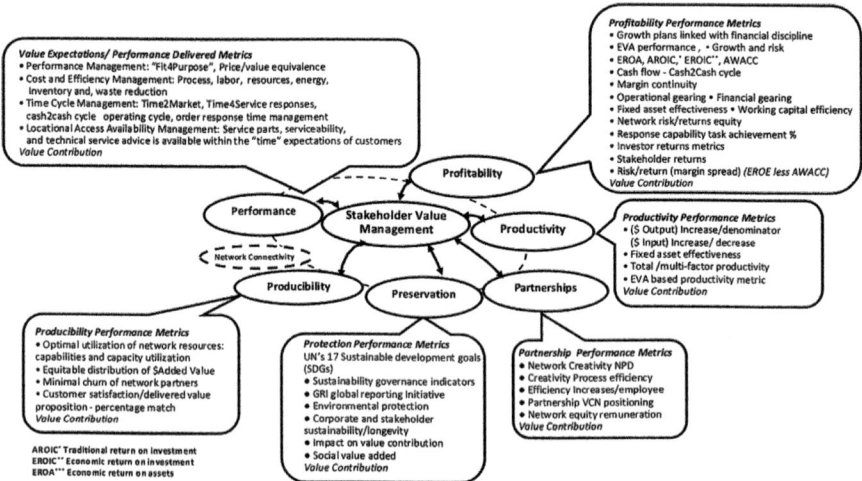

Fig. 11.9 Setting organizational performance metrics for capability responses to meet customer expectations

Based Management comprises *activity analysis, activity-based costing and attribute-based costing involved in managing* the existing business model and exploring the viability of potential strategic directions. The activities should be considered at three network levels, exploring the use of internal resources, collaboration with specialist suppliers, and the use of third- and fourth-party service partners.

Convergence...

Managing the business model requires an approach with formal structure and some formal measures: method, roles and tasks, and capability response performance metrics. These are suggested by Fig. 11.9. Capability response activities will, of necessity, require both quantitative and qualitative performance metrics, and, as we will demonstrate in the following chapter, they will interface and interact; correlation analysis will be used to demonstrate synergy between two or more capability responses.

References

Bollard, A., Larrea, E., Singla, A., & Sood, R. (2017, March). *A next-generation operating model for the digital world*. New York: McKinsey.

British Standards Institution. (1992). www.bsigroup.com

Ellis, S. (2019, January 8). Top 10 predictions for worldwide supply chains in 2019. *IndustryWeek*.

Harness AI to accelerate innovation: A 2019 perspective for manufacturers. (2019, January 14). *UK Manufacturer*.

Leszinski, R., & Marn, M. (1997, February). Setting value, not price. *The McKinsey Quarterly, 1*, 98–115.

Chapter 12
Working with the Convergence to Achieve Competitive Value Advantage

Abstract *Capability response analysi*s is very much about what and why and whenand how and where and who. Capability response analysis is becoming an essential feature of strategic analysis in a digitally connected time of rapidly changing stakeholder expectations. Given the recent problems of General Electric, it is arguable that there were very few, or no, questions asked concerning *what* was needed to be successful as a significant player as a "themed conglomerate." The approach being taken by IKEA to explore its strategic direction suggests a pensive *what* and *why* and *when* and how and *where* and who. *Industry dynamics differs;* it is searching for changes that will impact upon *all* organizations; it is for each of them to analyze and explore the implications of changes and shifts of emphasis in their definition of their business environment; it is for the organization to define their future business environment and the opportunities it offers. IKEA are clearly taking a very broad view of the opportunity space they are considering. In retrospect General Electric appears not to have taken such an approach. Observation of large automotive manufacturers suggests they have identified changes in ownership and use patterns of behavior, vehicle power sources, and whether or not there will be the need for a vehicle driver.

Keywords Convergence · Value advantage · Fulfilment execution system (FES) · Business ecosystems

Introduction

While the convergence of Industrié 4.0, Value Chain Network Management 2.0, and stakeholder value management is relatively new, and considering them as a "package" that can be applied as a means by which effective strategy decisions and their efficient operational implementation may be effected, is also new, the means by which the benefits of their convergence may be explored is not new, as the prose of Kipling (1865–1936) suggests:

© Springer Nature Switzerland AG 2020
D. Walters, D. Helman, *Strategic Capability Response Analysis*,
https://doi.org/10.1007/978-3-030-22944-3_12

I KEEP six honest serving-men
(They taught me all I knew);
Their names are What and Why and When
And How and Where and Who.
I send them over land and sea,
I send them east and west;

I let them rest from nine till five,
For I am busy then,
As well as breakfast, lunch, and tea,
For they are hungry men.
But different folk have different views;
I know a person small
She keeps ten million serving-men,
Who get no rest at all!

She sends'em abroad on her own affairs,
From the second she opens her eyes—
One million Hows, two million Wheres,
And seven million Whys!
But after they have worked for me,
I give them all a rest.

Rudyard Kipling (*The Elephant's Child, The Jungle Book*)

Capability response analysis is very much about what and why and when and how and where and who. Capability response analysis is becoming an essential feature of strategic analysis in a digitally connected time of rapidly changing stakeholder expectations. Given the recent problems of General Electric, it is arguable that there were very few, or no, questions concerning *what* was needed to be successful as a significant player as a "themed conglomerate." The approach being taken by IKEA to explore its strategic direction suggests a pensive *what* and *why* and *when* and *how* and *where* and *who* (Milne 2019). The IKEA CEO, Torbjorn Loof, suggested an interest in an approach similar to the German company Zalando, an online fashion retailer selling clothes from multiple brand sources, none of whom are owners of the German website. Loof suggested a comprehensive activity: "It is (also) about how you connect. If you take home furnishings, for instance – how you connect with communities, how you connect knowledge, how you connect the home, It's not only furniture, it's paintings, it's the do-it-yourself part. There are many different constellations that can and will evolve over the years to come" (Milne 2019).

In a recent republished interview with Richard Rumelt (McKinsey Classics 2011, first published November 2007, Strategy's strategist: An interview with Richard Rumelt), the author was asked during a discussion concerning strategy as a process, how identified opportunities were managed: What capabilities do companies need in order to take advantage of these ideas? And how do we know which changes are important and which resources to combine? Rumelt replied: "It is a key issue – the next frontier. And it is under researched, underwritten about, and under understood. I call it "strategy dynamics," qualifying his statement with:

Strategy dynamics studies how those changes would shift each dimension of an industry. Would the industry become more concentrated or less? More integrated or less? Would there be more product differentiation or less? More segmentation or less? Given consumer desires and available technologies, how should the industry or business look in, say, ten years? Where are the economic forces trying to take you? Should your strategy ride those forces or fight them? (McKinsey 2011).

Strategy dynamics considers the impact of changes in the business environment on the structure of industries that will impact upon *all* organizations within an industry grouping. *Industry dynamics* considers the changes occurring within each of the components (i.e., knowledge, technology, processes, relationships, regulation, and geopolitics). It is for each organization (individual and network structures) to analyze and explore the implications of changes and shifts of emphasis in the *industry dynamics* components and to define their future business environment and the opportunities it offers. IKEA (above) are clearly taking a very broad view of the opportunity space they are considering. In retrospect General Electric appears not to have taken such an approach. Observation of large automotive manufacturers suggests they have identified changes in ownership and use patterns of behavior, of vehicle power sources, and whether or not there will be the need for a vehicle driver.

Clearly there are some interesting initial questions to be asked, for example:

How important is the influence of industry dynamics? How different is the impact across industries? The answers are likely to be very different between stable food processing and fast fashion.

Do changes in Industry Dynamics change the characteristics of Value Builders? For example, a shift of emphasis in government policy on say healthcare may have significant issues for the industry infrastructure. Danish healthcare has undergone major changes with the application of "e-communications" and repurposing of hospitals, thereby changing the required "use" behavior of users, (Margo 2019).

How does "capability response analysis" guide strategic direction? Capability response analysis appears to be becoming an essential component of strategic direction decisions as trends in demographics, socioeconomics, employment structures, geopolitics (populism), and globalization are indicating change.

How can the combined capabilities offered by the convergence of Industrié 4.0, Value Chain 2.0, and stakeholder value-led management be focused on industry dynamics to identify opportunities, to analyze their likely format, and to suggest possible response capability portfolios? "Big data," real-time supplier and customer relationship management, and "rolling value propositions" are currently available and being used; costs are lowering, and speed and accuracy are increasing suggesting that improvements in "topline" (revenues) and "bottom line" (costs) activities will increasingly become more effective and efficient.

How will the future volatile global value chain impact market demand? What capability responses will be required to compete successfully? Emerging markets are active in manufacturing a number of industrial and consumer products. China has significant lead in industrial robotics as well as consumer durables and large online retailing organizations; India will soon be adding industrial hardware to its strengths in software suggesting a restructuring of the global value chain.

Major Global Industries Are Implementing Convergence Successfully

Convergence: Current and Future Developments in Healthcare

The future of healthcare is changing mainly through digital technologies, such as artificial intelligence, VR/AR, 3D printing, robotics, or nanotechnology. In medicine and healthcare, digital technology could help transform unsustainable healthcare systems into sustainable ones, equalize the relationship between medical professionals and patients, and provide cheaper, faster, and more effective solutions for diseases – technologies could win the battle for us against cancer, AIDS, or Ebola – and could simply lead to healthier individuals living in healthier communities.

Artificial intelligence: has the potential to redesign healthcare. AI algorithms use medical records to design treatment plans or create new drugs. *Atomwise,* a research partnership, has been deployed to help invent new potential medicines for more than 50 disease targets working with leading research groups and several large pharmaceutical companies and uses advanced software and hardware to explore therapies from a database of molecular structures. Recently, the start-up launched a virtual search for safe, existing medicines that could be redesigned to treat the Ebola virus. They found two drugs predicted by the company's AI technology which may significantly reduce Ebola infectivity. Imagine what horizons open for humanity if early utilization of AI results in such amazing discoveries.

Virtual reality: is impacting both patients and physicians. *Embodied Labs* have created *"We Are Alfred,"* a VR technology-based video to show young medical students what aging means and experience how it feels to live as a 74-year-old man with audio-visual impairments. The developers' ultimate goal is to solve the disconnection between young doctors and elderly patients due to their huge age difference.

Augmented reality: differs from VR in two respects – it reinforces reality, and it puts information into visual context. These distinctive features enable AR to become a driving force in the future of medicine, both on the healthcare providers' and the receivers' side. In case of medical professionals, it might help medical students prepare better for real-life operations, as well as enables surgeons to enhance their capabilities. *Medsights Tech* have developed software to test the feasibility of using augmented reality to create accurate three-dimensional reconstructions of tumors. It presents surgeons with X-ray views – without any radiation exposure, in real time. In the case of patients, AR might help them describe their symptoms more accurately, or pharma companies might offer more innovative drug information to patients.

Healthcare trackers, wearables, and sensors: the future of medicine and healthcare is closely connected to the empowerment of patients as well as health carers to undertake care of their own health through technologies and retake control over our own lives managing weight, stress levels, and cognitive capabilities or to reach an overall fit and energetic state.

Disruption in drug development: currently, the process of developing new drugs is long and expensive. There are ways to improve it ranging from artificial intelligence to better organizational procedures. The most revolutionary is the concept of in silico trials, individualized computer simulations used in the development, or regulatory evaluation of a medicinal product, device, or intervention.

Nanotechnology: nanoparticles and nanodevices will soon operate as precise drug delivery systems, cancer treatment tools, or "minute" surgeons. Researchers from the Max Planck Institute have been experimenting with exceptionally microsized – smaller than a millimeter – robots that quite literally swim through bodily fluids and could be used to deliver drugs or other medical relief in a highly targeted way. These "microbots" are designed to swim through the bloodstream, around the lymphatic system, or across the surface of the eye.

Robotics: developments range from robot companions through surgical robots, pharmabotics, disinfectant robots, and exoskeletons (the application of robotics and biomechatronics towards the augmentation of humans in the performance of a variety of tasks). With the help of these devices, paralyzed people can walk and enable rehabilitation of stroke or spinal cord injury patients. They can enhance strength so that it allows a nurse to lift an elderly patient.

Additive manufacturing (3D printing): the long list of successfully 3D-printed objects demonstrates the potential this technology holds for the near future of medicine, ranging from patient-specific limbs, printing tissues with blood vessels, bones, and synthetic skin. Surgeons use data taken from MRI and CT scans to build 3D models of the patient's anatomy they will operate on. The dental industry is also one of the biggest users of 3D printing. Using scans taken from the actual patient's mouth, dentists and dental laboratories are able to build accurate and tailored solutions to fix dental problems, for example, 3D-printed aligners which slowly move the teeth into a desired position as well as 3D-printed crowns and bridges. Patient-specific implants (PSIs) can be very expensive and time-consuming to make using traditional manufacturing techniques, but with additive manufacturing, this can be done precisely in massively reduced time lines. 3D printing really holds its own for complex organic shapes which traditional manufacturing techniques struggle with. Some of the applications of 3D printing outside of traditional manufacturing include the possibility to 3D print organs. This could be one of the most revolutionary uses of 3D printing and one with real scope to change the world as we know it. If it was possible to print new organs and body parts using your own cells as the base, then millions of people desperate for transplant donors would no longer have to wait for a suitable match. In relation to 3D-printed designer drugs – current drug and medicine production are on a mass scale, and the medicines are not targeted but more catchall solutions which work with varying effectiveness and intensity for different people. Additive manufacturing opens up the possibility of printing designer medicines which are specific to you, your body, and your needs. Additive manufacturing enables not just personalization but also, importantly, cost reduction and gains in success. Many more potential applications are likely to emerge – not just of existing technologies and materials, but as new medical-specific technologies are designed and developed, 3D printing will continue to change the face of the medical ("10 Ways Technology is Changing Healthcare", 2017).

Convergence: Current and Future Developments in Agriculture

There have been a number of applications of "convergence" characteristics technology/Industrié 4.0 in agriculture. Primarily they have been directed towards improving efficiency and, consequently, productivity. They have included:

Remote equipment monitoring: Aided by mobile application, this helps farmers survey several acres of fields from rote locations.

Smart irrigation: Sensor- and sprinkler-based technology for optimal water usage.

Smart, autonomous tractors and remote equipment monitoring: GPS enabled driverless tractors that can run on optimized field routes, thereby reducing soil erosion and saving fuel costs.

Agbot (agricultural robots aka. agribots): A fleet of agricultural robots used to automate agricultural processes, such as harvesting, fruit picking, plowing, soil maintenance, weeding, planting, and irrigation.

Soil sensors: Monitoring moisture and nutrient levels in the soil.

Smart logistics and warehousing: "Farm-to-fork" connectivity for agricultural products.

Livestock biometrics: Collars with GPS, RFID, and biometrics can automatically identify and relay vital information about the livestock in real time.

Crop sensors: Instead of prescribing field fertilization before application, high-resolution crop sensors inform application equipment of correct amounts needed. Optical sensors or drones are able to identify crop health across the field.

Building on existing geolocation technologies: Future swath control could save on seed, minerals, fertilizer, and herbicides by reducing overlapping inputs. By pre-computing the shape of the field where the inputs are to be used and by understanding the relative productivity of different areas of the field, tractors or agbots can procedurally apply inputs at variable rates throughout the field.

Rapid iteration selective breeding: The next generation of selective breeding where the end result is analyzed quantitatively and improvements are suggested algorithmically.

Precision agriculture: Farming management based on observing (and responding to) intra-field variations. With satellite imagery and advanced sensors, farmers can optimize returns on inputs while preserving resources at ever larger scales. Further understanding of crop variability, geolocated weather data and precise sensors should allow improved automated decision-making and complementary planting techniques.

Robotic farm swarms: The hypothetical combination of dozens or hundreds of agricultural robots with thousands of microscopic sensors, which together would monitor, predict, cultivate, and extract crops from the land with practically no human intervention. Small-scale implementations are already on the horizon (Zappa 2014).

Convergence: Current and Future Developments in Automotive Manufacturing

Reducing margins and fierce competition have focused automotive manufacturers on optimizing manufacturing processes in order to reduce costs and improve flexibility and time to market. From the current processes used in manufacturing, the industry is expected to move towards a digital planning/manufacturing approach *with the aim of increasing efficiency without having to add personnel on the floor.* Currently, *autonomous final assembly uses* robots that can achieve higher levels of volumes and precision than their human counterparts. German (notably Volkswagen) OEMs, for instance, have achieved almost 95 percent automation in the "body in white" (unpainted vehicle bodies) assembly. Efforts focus on making body in white lighter and more modular in design, and major OEMs have set clear targets on weight reduction for the future. Important to consider though is how this is expected to affect the manufacturing process itself, as suppliers develop smarter and more lighter machines that can be easily relocated throughout the assembly line and can be adapted to work with different platforms.

With *powertrain manufacturing the focus is not just on reducing material, however, but also on the number of operations* required to manufacture components; for example, the average number of operations to manufacture a crankshaft in Europe is 15 as against 25 in China. Rising labor costs are an issue that is pushing manufacturers to a higher degree of automation, even in these emerging markets. There has also been an increased use in machines with parallel kinematics for higher flexibility on smart assembly lines.

Advanced manufacturing methods such as micro manufacturing, machine vision, and smarter robots as OEMs move towards digital factories. The future of factories will revolve around (Peters et al. 2014):

Smart clouds – the next trend in cloud computing where flexible customized clouds can address a particular business need depending on requirements.

Industrial cybersecurity – cyber threats have the potential to disrupt safety, impact productivity, and cause loss of intellectual property.

The enterprise ecosystem – the convergence of ERP (enterprise resource planning), PLM (product lifecycle management), and MES (manufacturing execution systems) will enable considerable optimization throughout the product life cycle and is critical for automotive manufacturing.

Remanufacturing – current economic benefits suggested by existing salvaging technology are available, but the long-term benefit is to recover end-of-life products to "as-good-as-new" – with "cradle-to-cradle" product design for manufacturing.

Barkai and Manenti (2011) suggested current market trends would require the future production environment to be highly adaptable and reconfigurable to respond to rapid changes in market demand, technology innovation, and changing regulations. Flexible manufacturing technologies are employed by most automakers and are a critical ability in this process and the foundation for profitable growth, but these alone will not suffice in a long-term strategy to fend off the competition. The authors suggest a practical "design anywhere, make anywhere, sell anywhere" strategy is needed and propose:

> Factories of the future will be a global network of production facilities managed as single virtual factory. This type of manufacturing network consolidates multiple resources and capabilities to form an end-to-end fulfilment network that we call fulfilment execution system (FES) (Barkai and Manenti 2011)

A FES (fulfilment execution system) is an approach to a coordinated management of demand, capacity and resources, and outbound order fulfilment across the entire network of manufacturing plants and along the supply chain. Data gathered is connected to corporate-level intelligent decision support tools, creating visibility and intelligence on operational data. It enables manufacturers to identify problems, isolate root causes, understand the state of execution processes, and adopt corrective actions quickly across multiple plants. The authors' proposal takes us beyond the *marketspace/marketplace* work by Rayport and Sviokla (1994) in which they suggested the traditional marketplace interaction between physical seller and physical buyer are being eliminated. See Fig. 12.1.

The digitization and connectivity characteristics that have been applied to each of the industries identified earlier have improved product-service performance and profitability by also having a large impact on productivity. Structured data flows

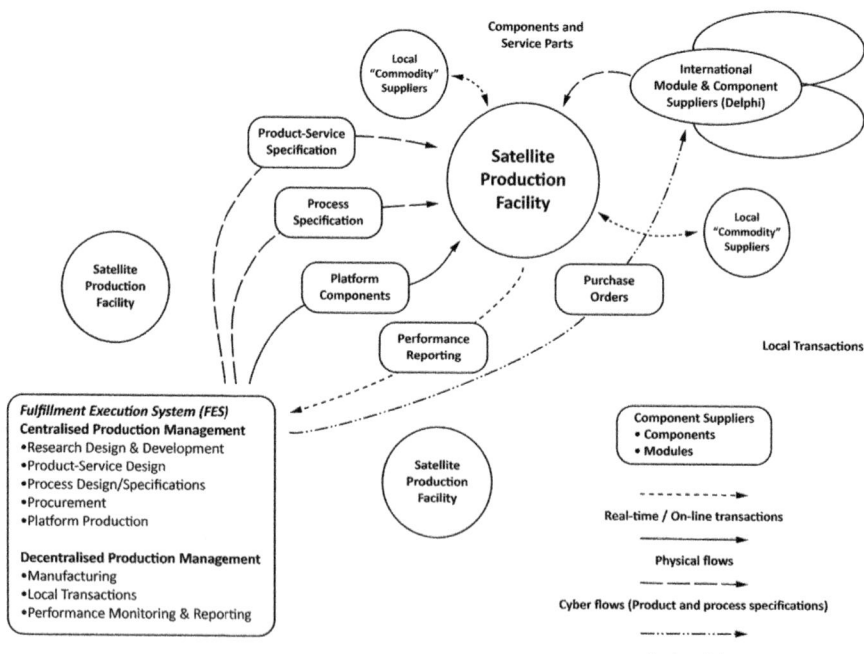

Fig. 12.1 Industrié 4.0/cyber-physical networks

have resulted in improvements in arable and dairy farming by not only identifying problems, which hitherto were never easy to identify, but activating solutions, for example, water and fertilizer applications. The increasing agility of robotics has improved productivity in automotive production, almost 95 percent automation in the "body in white" assembly. Efforts are being focused on platform technology, making "body in white" lighter and more modular in design. Artificial Intelligence is being applied successfully across a range of activities in healthcare, diagnostics, imaging, and pharmaceutical drug research and development, saving cost and time and improving end results.

Developing a Response Capabilities Profile: Response Capability Scoping

Customers are becoming more discerning and demanding. The business environment is becoming increasingly dynamic, and therefore a very clear view of the characteristics of essential response capabilities requirements becomes critical. Before a credible strategy and structure value chain network can be constructed, a "scoping exercise" is essential. The attraction and purpose of network organizations is the effective use of the leverage of network members' assets and efficient operations; it is unlikely these will occur unless opportunities are met by appropriate resources. Response capability comprises six interrelated components; these we have explored throughout this text; they are performance, profitability, productivity, partnerships, producibility, and preservation. Response capability scoping has seven separate but coordinated steps:

First, what is the customers' expectation of value? A number of questions are raised at this stage each aiming at determining the demand characteristics of the potential market and its segments. Another important question concerns the complexity of the market and the influence this has upon customer expectations, their performance value drivers, and current competitive responses.

Second is the role of *demand and development* activities. In a number of product-service-markets (e.g., pharmaceuticals and aerospace), RD&D expertise is becoming specialist and is often the role of a partner organization, often a specialist small-medium enterprise (S-SME). The questions that typically arise are to consider whether this activity should be in-house or outsourced; here, concerns over time to market as well as investment in RD&D become paramount. A decision to outsource the activity leads to further questions concerning who and how the task will be managed.

The *third* consideration concerns *procurement and manufacturing*. Again, a number of questions are raised; customer expectations may require complex *solutions to problems* as well as a product-service (hardware) solution, and this may require a choice between these options (the problem IBM confronted). Alternatively, the requirement for choice may suggest that all (or part) of the procurement and manufacturing be outsourced. An expectation for differentiation may also raise the outsourcing question concerning the use of specialist internationally branded component suppliers (Intel for computers).

A *fourth* question concerns the role of *marketing and sales operations.* We now have "big data" detailing not only market volumes and locations, but also patterns of flow are identifying optimal customer availability and order cycle time requirements that are to be met. Identifying and responding to these is becoming a critical component of competitive advantage, and this data is now seen as a major role for marketing and sales operations.

The *fifth* decision concerns *service operations management – "serviceability"* and the importance of service to the success of the value offer throughout its life cycle should be ascertained in considerable detail. Service support may be required during the "specification" process and extend throughout the life of the product – even to its "value renewal" (its traditional replacement, or its repurposing and/or remanufacturing) and retirement. The PRODUCT-service/product-SERVICE characteristics of many B2B markets should be considered at this stage rather than left to chance.

The *sixth* consideration concerns *logistics and supply chain response management* (LSCRM): inbound (expectations of OEM assemblers of their suppliers) and outbound (customer expectations of vendors) are important considerations, and whether or not procurement and manufacturing are internal or external to the organization have an impact on end-user service and their relationship to the network organization. The connectivity (of Industrié 4.0) and the increasing scope and intensity of network collaboration (Value Chain 2.0) are facilitating real-time data flows. All of the LSCRM service characteristics identified will be important to a greater or lesser degree depending upon the product category being serviced. Reverse logistics, the "management of mistakes," is often overlooked and should be addressed, and a "policy" is established and made known to customers; the impact of automated order management should be reducing order management errors.

The *seventh* activity is to develop a *network resource capability profile* identifying the resources required (and their ownership) necessary to meet *the customers' expectations of value.* A scoping exercise will have identified the capability responses required and relate these to assets, processes, and capacities required for success. It should also detail the extent of the integration (strategic effectiveness) and coordination and collaboration (operating efficiencies) that will be required together with the performance expectations (contributions to profitability and productivity) of each activity. The exercise should include identification of the most likely *constraints* and to put in place countermeasures. Increasingly these issues are becoming resolved in real time; Fig. 12.2 illustrates the principle of digital data flows for industrial product-service products. By monitoring ongoing activity using digital feedback, customer users can be advised of performance efficiency levels (and how underperformance may be addressed), predictive maintenance, digitally managed product modification, and value renewal requirements (predicted useful life cycles).

A Structured Approach to Capability Response Analysis

A structured approach to capability response analysis is required; Fig. 12.3 outlines a model for this purpose. The model comprises a sequential approach commencing with identifying *significant impact factor change trends,* trends such as demographic changes, educational standards and content, healthcare quality, costs and availability, legislative and regulatory developments, political developments, employment structures and availability, impacts of globalization in areas of interest, and economic performance (GDP, balance of payments, forex, etc.). Trends that are exhibiting recurring change or unusual results should be identified and analyzed.

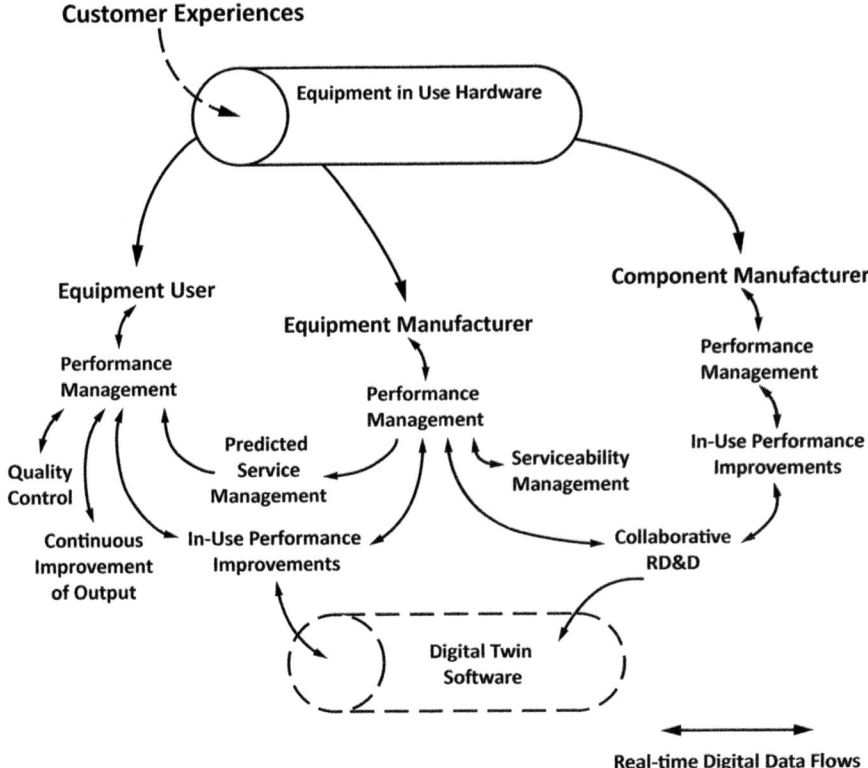

Fig. 12.2 Real-time manufacturing incorporating digital thread inputs and digital twin models for evaluation

Important aspects of *industry dynamics,* specifically those that are known to have an impact on the business and its network partners, and other factors that demonstrate changes in influences should be identified. Changes in industry dynamics may well require a close look at *value drivers* (suggesting new approaches to customer expectations for the existing business activities) and *value builders* (the means by which a competitive value advantage may be established for the future). *Value migration* is likely to influence the approach to future growth, either because of changing expectations and/or new competitive business models. This review will provide input for *future directions.*

Monitoring Industry Dynamics

The "New Economy" (1995–2005), the period during which the holistic independent aspects of business models changed managements' attitudes and behavior, was rapidly modified by the cyber-physical approach facilitated by Industrié 4.0. Markets that were becoming globalized become all embracing, and relationships

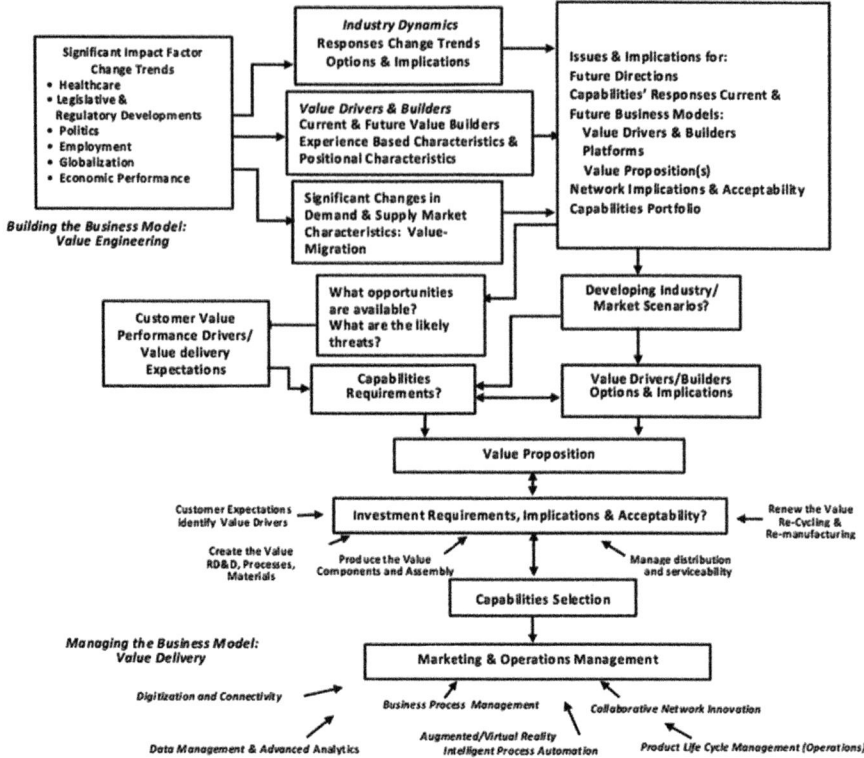

Fig. 12.3 Identifying and managing capability response

with suppliers, customers, and competitors are undergoing constant change. New business models emerged in which competitive advantage is based upon *managing processes* that facilitate rapid and flexible responses to "market" change, in which new *capabilities* were based upon developing unique relationships with partners (suppliers, customers, employees, shareholders, government, and, often, competitors). Normann (2001) discussed "a new strategic logic" suggesting that: "...managers need to be good at *mobilizing, managing,* and *using* resources rather than at formally *acquiring* and necessarily *owning* resources. The ability to reconfigure, to use resources inside and particularly outside the boundaries of the traditional corporation more effectively becomes a mandatory skill for managements." Iansiti and Levien (2004) also made a significant comment: "Strategy is becoming, to an increasing extent, the art of managing assets that one does not own." Arguing that business networks are ubiquitous in our economy, they suggest: "...the pervasive networked nature of our business environment has triggered a significant evolution in the design of business operations and in the role of managers." Industrié 4.0 offers an opportunity to evaluate alternative value propositions and select the most suitable optimal response.

Future Direction: Scenario Analysis and Planning

Future competition will be between value chain networks, and this has, without doubt, been a major consideration in network developments. A question does need to be asked (and answered); how does the organization (individual or part of a network) identify and manage risk? Lindgren and Bandhold (2003) addressed the problem by exploring the role of scenario planning by linking a future objective to a potential scenario. Scenarios are vivid descriptions of plausible futures. A scenario *is not* a *forecast* (a description of a relatively unsurprising projection of the future); *nor is it a vision* (a desired future); it is a well-worked answer to the question: What can *conceivably* (or can be realistically expected) happen? Or what would happen if? As such it differs from both a forecast and a vision both of which (the authors suggest) can conceal risks, thus making risk management possible.

Lindgren and Bandhold (2003) suggest scenario techniques are applicable for a number of purposes. When used for planning with an explicit purpose of developing implementable results, scenarios that probe industrial processes, technological developments and applications. Scenarios probing changing consumer attitudes, behavior and expectations can provide very useful input into strategy and structure models for R&D, business model, product and process development. Scenario techniques can be used to generate filters for ideas and project evaluation by exploring the robustness of existing business concepts, strategies, or product-service proposals. They may also be used to drive change; the process of developing scenarios is also a learning process for the management team that are involved because they encourage rigorous questioning of their logic and validity.

Scenario planning, also called scenario thinking or scenario analysis, is a strategic planning method that some organizations use to make *flexible long-term plans*. Scenario analysis asks "what if?" questions to explore alternative views of the future and to create plausible visions of the future around them. They can consider long-term trends in economics, resources supply and demand, geopolitical shifts and social change, as well as important factors that drive change, such as developments in *industry dynamics* such as applied; knowledge management, technology management, process management, relationship management, regulatory compliance and, geopolitics. They also can bring focus to the capabilities and positional requirements – *the value builders* – required for success and their implications for the *global value chain participants*, resulting in greater understanding of the business environment and how it might appear decades ahead.

We introduced scenario planning and analysis in Chap. 10 and explored its advantages and discussed the Delphi technique used for developing scenarios. Figure 12.4 identifies the data required for creating a scenario for evaluation concerning its attraction – opportunities – and the response capabilities required for entry.

Lindgren and Bandhold (2003) suggest we need to understand and learn from past success and failure and to learn from what has happened (feedback, evaluation of historical results); we also need information concerning the future (scenarios,

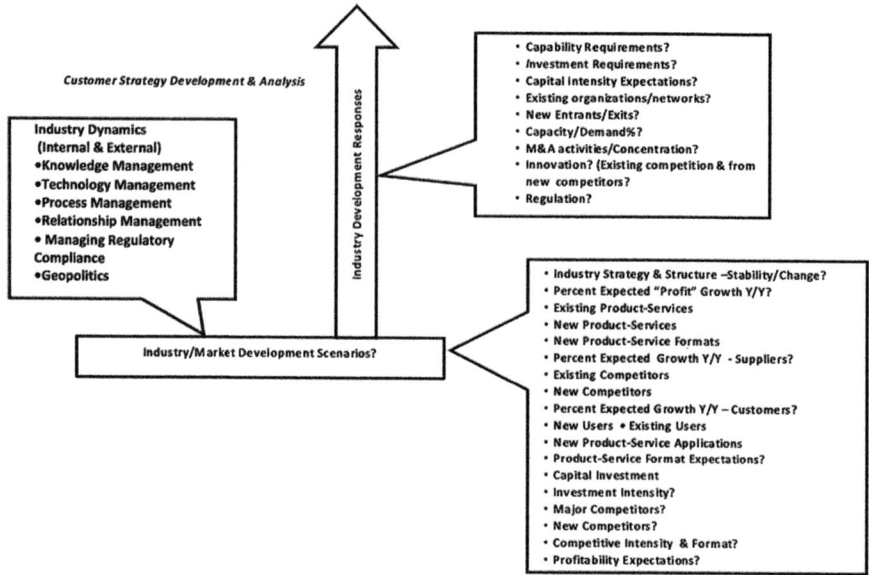

Fig. 12.4 Developing strategic direction using scenario analysis

trends, and forecasts) if we are to make decisions concerning future strategy. They offer a model, TAIDA™ for this purpose (tracking, analyzing, imaging, deciding, and acting):

Tracking: traces changes and signs of threats and opportunities. Recruit expert input that can provide perspectives (opportunities and constraints) that may be overlooked by planners within the organization. For example, healthcare engineering is benefiting from research being conducted by neurologists and electronic communications engineering with a view to developing prosthetics applications.

Analyzing: the organization should explore its responses and how it's perceived strengths and weaknesses facilitated success or created failures for the organization as an individual unit as well as its network partners and the competition. What has been learnt? The organization should then generate potential scenarios that offer alternative future outcomes that lead to answers to questions such as: What are the implications for the future of current problems (travel, health conditions, manufacturing, etc.)? Are there connections across customer groups, user types, product-service applications, and delivery "technologies"? What are the implications?

Imaging: identify possibilities and consider possible influences using experience and intuition. Match these with the *desired futures*. Explore feasibilities and viabilities.

Deciding: selecting and rejecting – consider the information; using potential opportunities and threats, identify choices and strategies, and consider competitor responses.

Acting: set objectives and performance metrics, identify resource requirements and their availabilities, and implement the selected strategies.

The advantages of scenario planning are the inputs from expert knowledge, some of which may not been considered during the process of strategic planning. As technological capabilities advance, a question concerning the intellectual, physical, and psychological capacities of users or future users needs to be considered as part of the development process.

Responding to Performance Value Drivers

Performance value drivers were discussed in Chap. 4 (Performance); here we build on the material in Chap. 4 by considering topics that are concerned with identifying and developing response capabilities. Essentially it is the responsibility of *marketing* to provide a profile of the customer "value-in-use" characteristics. Of particular interest is the rate of flow of product developments, their use, and service expectations. The demand data that is generated will provide input to *operations* to plan the logistics management (materials and service flows, information content, and service support), which will ensure the timely management of transactions and cash flows of the required response. For the response to be successful, it is essential that a collaborative working arrangement be established between marketing and operations. This is particularly necessary for efficient materials and information flows. The proposal that a *value proposition* is used as the vehicle for the liaison role is based upon the premise that it (the value proposition) not only reflects customer expectations but *also* identifies the resource deployment required for success. Operations are tasked with evaluating response alternatives against the organization's overall strategic and operating plans.

Creating value incurs cost, and for many organizations, there is a decision to be made concerning the precise relationship between the value delivered to the customer, the value generated for the organization and its partner stakeholders, and the *cost of creating, producing, communicating, delivering, and servicing the value*. It follows that the relationship between value drivers and cost drivers is important. Scott (1998) commented:

> Since time immemorial there have been two sorts of activities in companies; those that drive value creation and those that drive unproductive cost...

Scott suggested that the harsh reality of globalization and the accompanying increase in competition has forced most companies into making efficiency gains. However, given that the persistence of competitive pressures makes the speed of efficiency gains in production and the speed of market responsiveness necessary to compete. And:

> Cost structures are shifting dramatically year by year as new producers come on line and new technologies propel shifts in business processes. Everything is moving faster and will continue to accelerate. Today's competitive "paradigms" will be tomorrow's old hat. (Scott 1998)

Scott's comments were made some 20 years ago, and what they describe has not changed very much, if at all, there is now the means by which they may be better

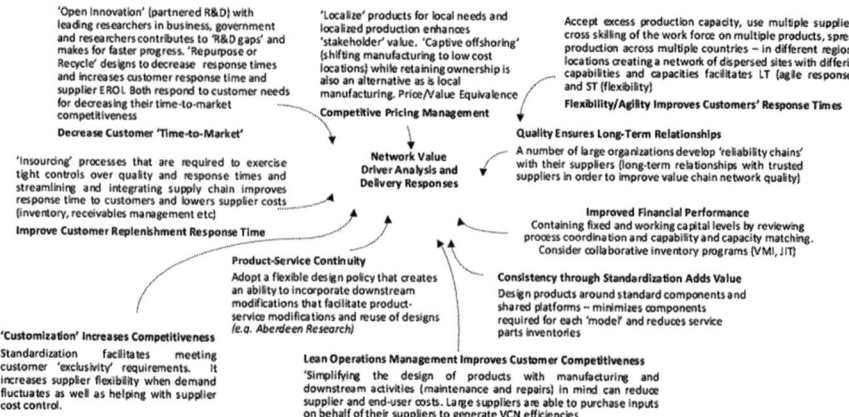

Fig. 12.5 Value driver analysis: Capability responses

managed. The demand/supply response cycle has been reduced; real-time order replenishment is reducing response time and therefore inventory holding costs. AI is being applied to customer purchasing and product usage behavior to produce predictive forecasts that reflect the dynamics and costs involved in order cycle management and identifying cost-efficient alternatives. Network collaboration and communications are resulting in transparency and access to partner activities, useful when unforeseen problems occur and often preventing their occurrence.

The availability of real-time information flows and data thread links makes the identification of the causes of, and evaluation of, alternative response delivery options more effective strategic options (i.e., value engineering) and more efficient operations (i.e., value delivery) to respond to. Figure 12.5 illustrates this point. In the diagram the bold titles for each of the items are a reported *customer value driver;* the supporting copy identifies an optimal, cost-efficient solution, reached by using a software-based analysis of solutions, and that created competitive advantage for customers and network members. For example, "*Lean Operations Management Improves Customer Competitiveness*" was addressed by simplifying design of products with manufacturing and downstream activities (maintenance and repairs) in mind to reduce supplier and end-user costs.

Responding to Value Builders

Performance value builders were discussed in Chap. 4 (Performance); here we add to that discussion by considering topics that are concerned with identifying and developing *response capabilities*.

Baghai et al. (1999) discussed the role of capabilities from a growth perspective. They argued that simply to focus on operational skills or competencies is limiting

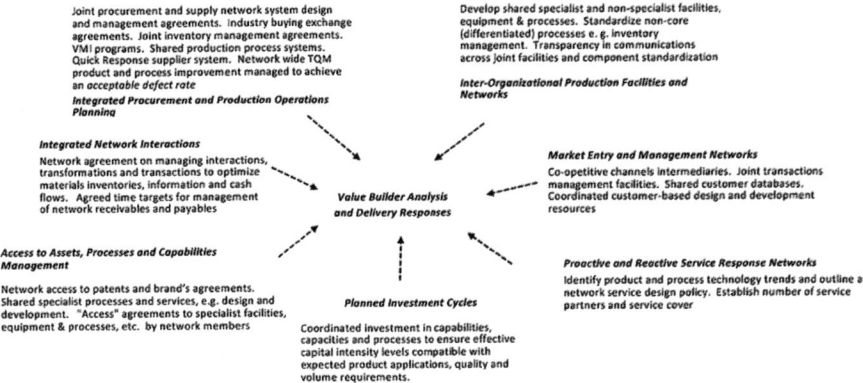

Fig. 12.6 Value builder analysis: Capability responses

and that a broader definition that includes all the resources useful in developing competitive advantage is required. They suggest that in addition to operational skills, privileged assets, growth-enabling skills, and special relationships are equally important. *Privileged assets* are physical or intangible assets that are hard to replicate and confer competitive advantage on their owner. These are infrastructure, intellectual property, distribution networks, brands and reputations, and customer information. *Growth-enabling skills* are considered to be acquisition management, deal structuring and negotiating, financing, risk management, and capital management. *Special relationships* include the linkages with customers and suppliers, government, and "regulators." We added to these in Chap. 4 by introducing the "capability balance sheet" (Olvé et al. 1997).

Understanding your customer's value builders is an essential input into identifying the costs of customer relationship management (see Fig. 12.6). Typically, these will be based on the scenario analysis and as a result of discussions with the customer organizations. The output from the discussions will be of major significance to both customer and supplier organizations, as, based upon the outcome, there will be a need to consider strategy, structure and partnerships, and future investment programs.

Tracking Value Migration

The Slywotzky and Morrison (1997) comment, which during the 1980s and the 1990s many fundamental business characteristics changed resulting in very different ways business is done and what determines business success, remains a sound suggestion. They argued that during this period, considerable amounts of market value have migrated from old business designs to new. The high stock market value businesses are not necessarily the largest organizations but are the innovative organizations whose creative business models include:

- High customer relevance
- An internally consistent set of decisions about scope (market positioning, value propositions, and value chain positioning activities they have chosen to perform)
- An outstanding value capture methodology or profit model
- An exclusive means of differentiation and strategic effectiveness that create investor confidence in future cash flows
- An organizational structure that is designed to support the company's business model design

An additional and significant aspect is:

- Close attention to customer and market shifts that identify changing customer behavior and changes in their expectations, changes that are likely to become part of their choice criteria for a significant future and therefore should be "read" and understood and used to modify the value proposition and the business model design

Slywotzky and Morrison (1997) provided examples of organizations that identified changing patterns of consumer demand satisfaction and created new business designs to capture the potential value. They cited General Electric (GE), and IBM, who moved away from the traditional product-led value proposition to a product-service offer that included finance, insurance, consulting, and management. *Hilti* (construction industry equipment and tools) became aware of the imitation and eventual commoditization of their product and repositioned their value proposition to offer a product-SERVICE. In computing the authors identified the "deintegration" of the value chain and the emergence of specialist component manufacturers and acceptance of standardization of components. Coca-Cola has "reintegrated" its value chain by acquiring bottlers and distributors. GE subsequently undertook extensive merger, acquisition, and alliance arrangements to build a credible and powerful base in the renewable and clean energy industry. However, shareholder dissatisfaction with financial results (2016–2017) resulted in two changes in CEO and considerable rationalization of the product-service portfolio.

An example of a customer-led shift was identified by Passariello (2011) who reported on the decline of the hypermarket, a business model introduced by Carrefour some 50 years ago. The hypermarket model included fast-moving consumer goods (FMCG) and consumer durables. Carrefour CEO, Lars Olofsson, commented that customers were "…either coming to us less often or shopping at supermarkets." Passariello (2011) noted that this was commonplace among Western European hypermarkets as customers are now purchasing fewer durable products, and when they do, they prefer specialist stores. Furthermore, customers can often find lower food prices closer to home. Olofsson repositioned the hypermarkets to reflect changing consumer expectations; non-food ranges were reduced, and food categories expanded. In-store merchandising was redesigned with wider aisles, eye-level displays, attractive cosmetics counters, and color-coded shopping

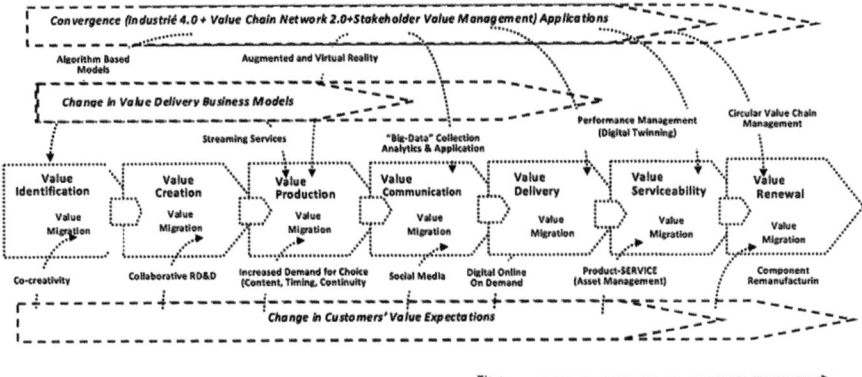

Fig. 12.7 Effects of value migration: changing consumer expectations, suppliers using efficiency to increase their value capture, and the application of technology

category areas. Passariello adds that post the financial problems of 2008–2009, the increase in fuel (petrol) price, the slow but continuous demographic shift (aging populations), increased female employment, and smaller families have led consumers to favor the convenience of local supermarkets rather than the price-led offers of the hypermarkets. Specialist retailers such as Darty (electronics) and H&M (clothing) responded to the lower spending power of consumers with reduced prices; Carrefour made a loss on the non-food categories that occupied some 50 percent of selling area. Subsequently other superstore/hypermarket operators responded more positively by developing catchment and "metro"-based stores offering edited product ranges to meet immediate needs.

Technology applications also have a major impact on value migration. Spector and Tractenberg (2011) reported on the problems of Borders, the book retailer. They commented: "The bookseller suffered a series of management gaffes, piled up unsustainable debts and failed to cultivate a meaningful presence on the internet or in increasingly popular digital e-readers." Customers became accustomed to receiving book purchases mailed to their homes or alternatively downloaded to electronic readers. They suggest one of Borders' biggest mistakes was to outsource its Internet operations to Amazon some years ago, a share buyback program and an overseas expansion program that increased the company's debt. Figure 12.7 illustrates the influence technology is having in developing business models, changing consumer purchasing habits and expectations, and the application of technology on value migration. It is difficult to isolate the separate impact of each as it is likely that they are related to each other. For example, online shopping (a technology application) has emerged as a response to diminishing time budgets (changing consumer expectations), and these have been identified by manufacturers and retailers and have resulted in transformed business models.

The Value Proposition

While not using that terminology, one of the earliest attempts we are aware of to articulate what the *value proposition* means in practice is a McKinsey staff paper from 1988 by Lanning and Michaels (1988) which outlined the formulation of a *value proposition*. This suggested that "… the heart of running a business is choosing a winning value proposition, then gearing all functions in the business to profitably deliver it." A well-articulated value proposition replaces the traditional "Mission Statement" – it describes how organizations will achieve their objective(s) by excelling at customer satisfaction and be "World Class" in their markets.

A value proposition describes the *bundle of product-services* that create value for a specific market, segment, or customer. It is the reason *why customers select one company over another – and the reason why they return.* The value proposition is a response, a solution, to an identified customer need. It consists of a package of products and/or services – or possibly a reimagined PRODUCT-service that has become a product-SERVICE because of changes in customers' business models or product-service applications. *Typically, the value proposition is an offer to a specified target.* The acceptance of the notion of "customer centricity" implies that the value proposition is a purpose-led bundle of benefits to targeted customers.

Some value propositions may be innovative solutions to problems; others may be market entrants offering alternative performance management, cost management, time management, or locational benefits – in other words an alternative business model application. They all should include reference to acquisition costs.

A *value proposition* should:

- Reflect an understanding of the customers' business(s).
- Substantiate the "value claims."
- Document the value delivered.
- Make the customer value proposition and supplier implications an important input into market planning.

A checklist for developing a company's value proposition should include:

- A market that is clearly identified and stakeholder expectations clearly understood.
- Evidence of adequate and ongoing demand.
- *Response capabilities* requirements for satisfying customer experiences and journeys are understood and acquired.
- The value proposition is explicit, specific, and clearly "stated."
- "Total price" to the immediate customer is clear.
- A clear statement of how this value proposition is superior to competing solutions in the market.
- Solution is producible, profitably and productively for all stakeholders.
- Evidence of acceptable stakeholder returns in terms of investment made and the risk taken.
- Consistent with social value objectives and sustainability.

- The most effective of several value propositions considered.
- Clear, well-articulated, simple, and understood by the stakeholders.

Haier: An Example of the Use of a Value Proposition Indicating Strategic Direction

The importance of the value proposition is that it identifies the strategic marketing positioning of the organization as well as its position in the value chain network. Haier's market value proposition clearly demonstrates this role. Haier has gradually broadened since Zhang Ruimin became CEO in the mid-1980s. The company first took the role of a category leader, maintaining top market share because of its reputation for quality in China. Then it became a customizer (adapting its products to customer demands) and a solutions provider (helping consumers manage issues like water quality and home design). Haier now sells not just home appliances but related services, adapted to consumer demand in China, and, increasingly, other markets. Haier excels by developing differentiating capabilities.

Convergence: Managing the Capability Response Portfolio

Capabilities Mix: An Exercise in Optimization and Leverage

Stakeholder value management is about meeting the expectations of a holistic group with differing but interconnected interests. In the opening chapters, we identified stakeholders as occupying two groups: *primary stakeholders* who have an active role in creating, producing, delivering, and capturing value (either as a positive input into their own business as a product service moved towards the end-user or as the end-user who satisfied a particular need) and *secondary stakeholders'* whose interests are less direct; their concerns are about broader issues, contribution to "community" needs, to creating employment, contributing to the economic health of a nation (its GDP), and generally improving living standards of a nation.

Our focus is on the primary stakeholders, typically members of value-producing networks, in which decisions are made concerning the allocation of resources to create product-services that meet the expectations of the primary stakeholders within the (often the well-articulated) expectations of the secondary group of stakeholders.

Clearly it is not possible to maximize each of the capability response requirements, but the *convergence* of Industrié 4.0, Value Chain Network 2.0, and stakeholder value-based management is increasing operating efficiencies as the examples (above) of applications in healthcare, in agriculture, and in automotive production demonstrate.

Selecting Capability Responses: Interfaces and Interactions

Figure 12.8 identifies possible objectives, leverage potential, and possible constraints of the capability responses of the primary stakeholder group who, typically, are partners of a market-focused value chain. Business is confronted by an intensively dynamic environment; change and disruption are a way of life, and their principle objective is survival, or, as expressed in earlier chapters, organizational survival. Within a competitive and disruptive business environment, individual and network business models require an approach to planning both current and future activities that assesses end-user customer expectations with a high degree of reliability. Capability response analysis provides this facility. It is not an alternative strategy model; rather it is a "prequel" that identifies the requirements of organizations that will meet the needs of all members of the value chain network (suppliers, distributors, and serviceability providers) that combine their resources, to meet the satisfaction of end-user purchasers

Capability response analysis is concerned with value creators, producers, and consumers.

There are core objectives for each; observation suggests priority is given to profitability and performance; this is understandable because without profits and acceptable product-service "fit," there is no sustainable business. However, when the facilitating characteristics of the convergence components are explored, opportunities for leverage are identified.

More detail of capability characteristics is required if the approach is to offer a prequel to either strategic or operational planning; Figs. 12.9, 12.10, 12.11, 12.12, 12.13, and 12.14 provide a generic approach. It is essential that the interfaces and interactions between each of them is explored, the purpose being to identify problems that might occur (for example, we might conclude from a hypothetical situation) that within a specific value chain network value delivery system is a robust and reliable activity only to find that a competitor has gained a competitive value advantage by digitizing the product-service. Newspapers are an excellent example.

Performance (superior product-service performance) is the dominant response capability that if made, the "lead response capability" would provide the current differentiation to which the network responds. Figure 12.9 raises a number of considerations. For example, if we consider the other capability response areas, one at a time, some questions will arise. Organizational longevity is reliant upon a revenue, profit, and cash flow stream; typically this has been built around a value proposition that has been successful in creating satisfaction to the network's customers and stakeholders through a level of value contribution delivered by the network.

The product-service format may be under threat from a new format. Large customers are finding they need to strengthen brand reputation and do so through increased customer service activities. A competitor enters the market with the view that *performance* should be considered in a broader context, the organization, and proposes an alternative business model. The value proposition then becomes about

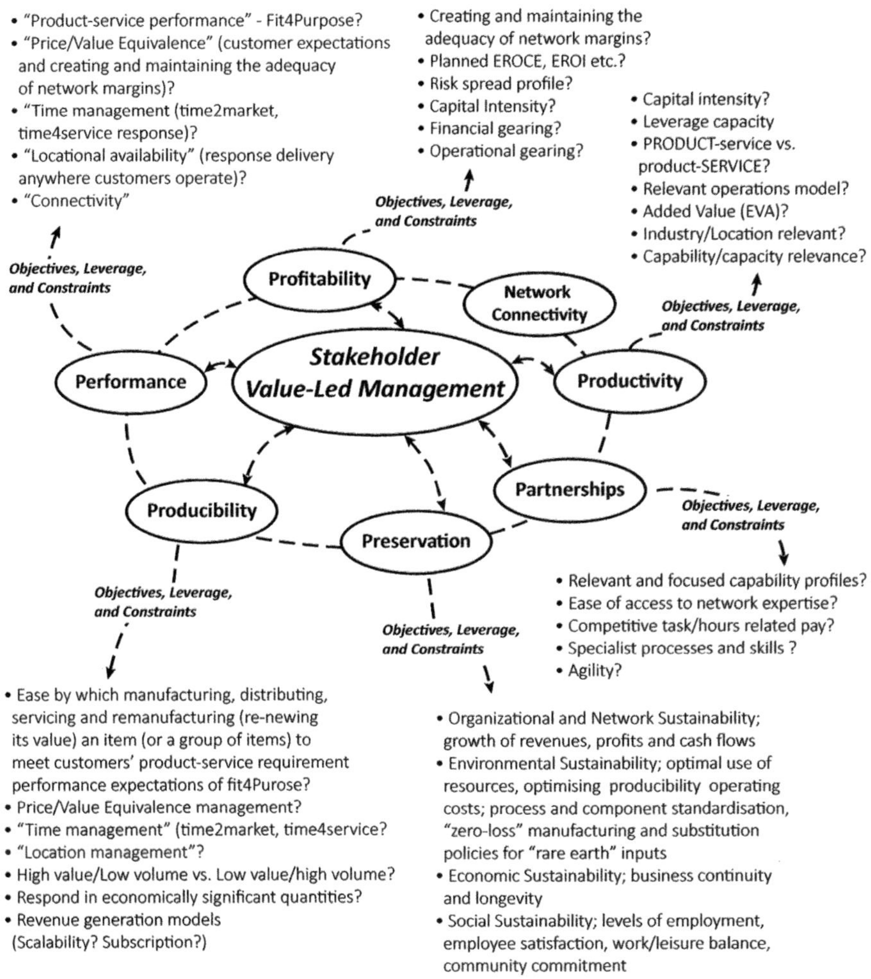

- "Product-service performance" - Fit4Purpose?
- "Price/Value Equivalence" (customer expectations and creating and maintaining the adequacy of network margins)?
- "Time management (time2market, time4service response)?
- "Locational availability" (response delivery anywhere customers operate)?
- "Connectivity"

- Creating and maintaining the adequacy of network margins?
- Planned EROCE, EROI etc.?
- Risk spread profile?
- Capital Intensity?
- Financial gearing?
- Operational gearing?

- Capital intensity?
- Leverage capacity
- PRODUCT-service vs. product-SERVICE?
- Relevant operations model?
- Added Value (EVA)?
- Industry/Location relevant?
- Capability/capacity relevance?

Objectives, Leverage, and Constraints

Objectives, Leverage, and Constraints

Objectives, Leverage, and Constraints

Profitability

Network Connectivity

Performance

Stakeholder Value-Led Management

Productivity

Producibility

Partnerships

Preservation

Objectives, Leverage, and Constraints

Objectives, Leverage, and Constraints

Objectives, Leverage, and Constraints

Objectives, Leverage, and Constraints

- Relevant and focused capability profiles?
- Ease of access to network expertise?
- Competitive task/hours related pay?
- Specialist processes and skills ?
- Agility?

- Ease by which manufacturing, distributing, servicing and remanufacturing (re-newing its value) an item (or a group of items) to meet customers' product-service requirement performance expectations of fit4Purose?
- Price/Value Equivalence management?
- "Time management" (time2market, time4service?
- "Location management"?
- High value/Low volume vs. Low value/high volume?
- Respond in economically significant quantities?
- Revenue generation models (Scalability? Subscription?)

- Organizational and Network Sustainability; growth of revenues, profits and cash flows
- Environmental Sustainability; optimal use of resources, optimising producibility operating costs; process and component standardisation, "zero-loss" manufacturing and substitution policies for "rare earth" inputs
- Economic Sustainability; business continuity and longevity
- Social Sustainability; levels of employment, employee satisfaction, work/leisure balance, community commitment

Fig. 12.8 Identifying and leveraging organizational and network response capabilities

serviceability, and the supplier's PRODUCT-service becomes a product-SER-VICE, the output of the PRODUCT. In serviceability (output reliability supported by predictive maintenance that may be made available globally, by the provider or by a service organization), the existing service structure becomes redundant. *Productivity* metrics change; basic output/input, typically based upon labor efficiency measures, is replaced by service-led metrics (and this is one reason why we are suggesting EVA as surrogate productivity measure). *Partnerships* change as new supplier and customer relations become important – and different. *Preservation* also has a different orientation and sets of rules. Intraorganizational sustainability that may have been a critical feature of previous success for some considerable time may be replaced by inter-organizational sustainability as another member of

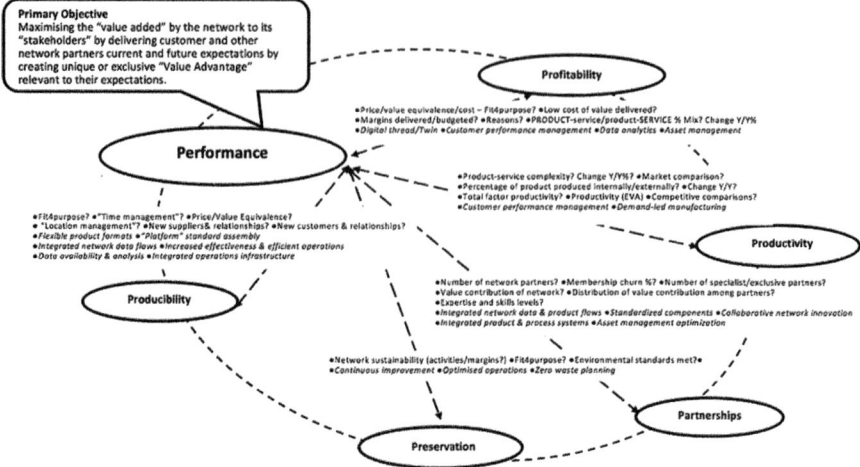

Fig. 12.9 Interactions and interfaces: Performance

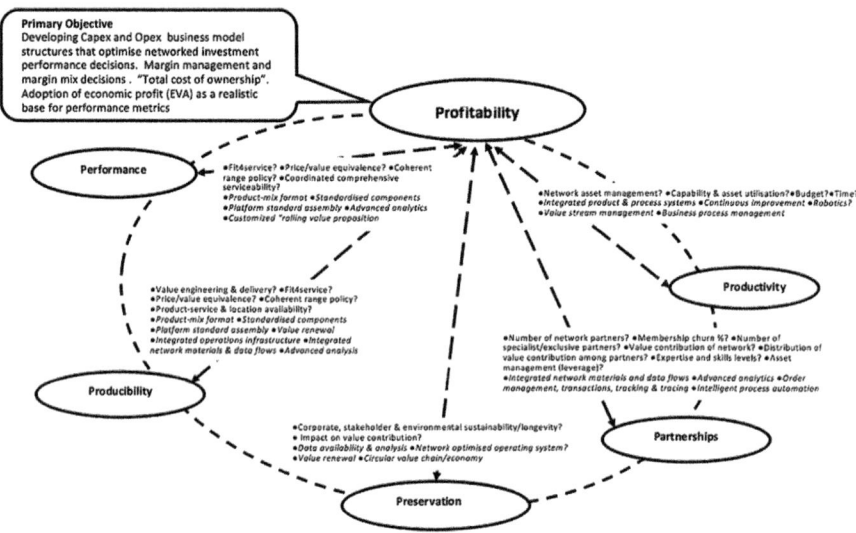

Fig. 12.10 Interactions and interfaces: Profitability

the value chain network assumes importance in delivering *performance*. The production value infrastructure, *producibility,* almost certainly may require a shift of emphasis as it is a process-optimized activity; the issue being that while the end-user receives much the same utility from transactions aspects of the delivery, processes may change.

Profitability (Fig. 12.10) is a major capability response; it is responsible for the success and viability of each member of the network; this has major implications for

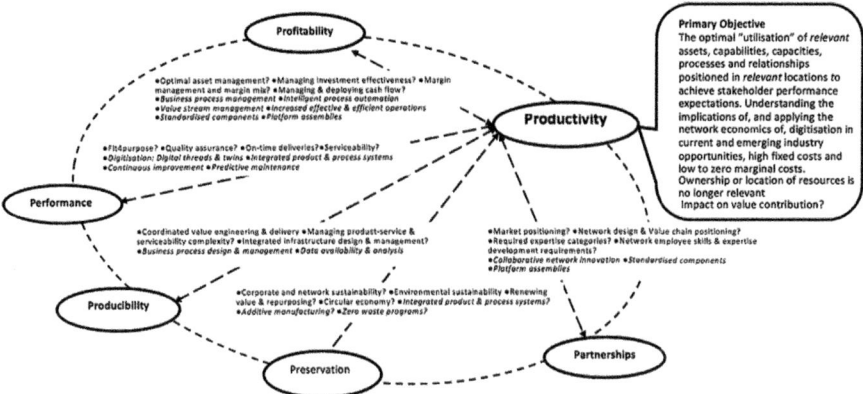

Fig. 12.11 Interactions and interfaces: Productivity

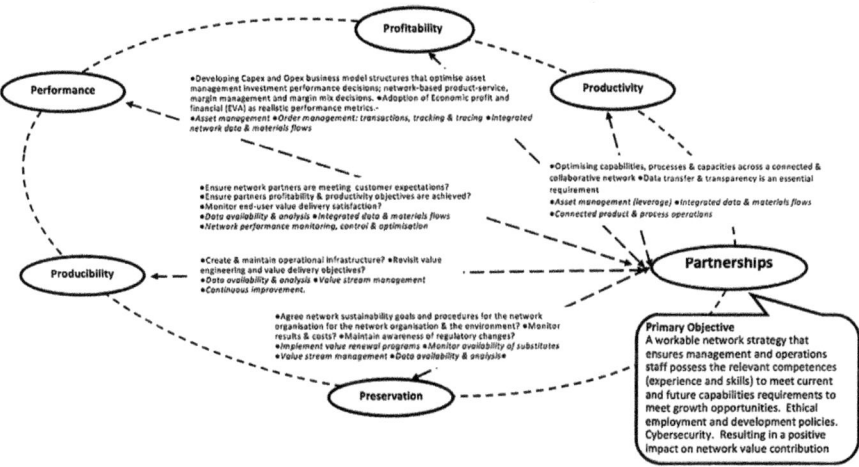

Fig. 12.12 Interactions and interfaces: Partnerships

primary stakeholders and not inconsequential concerns for secondary stakeholders. Without strong revenue, profitability, and cash flow streams, primary stakeholders face difficulties in both short and long term. Clearly *performance* decisions are paramount, and the fit4purpose and price/value equivalence offers made by the value proposition must be delivered and be affordable. *Productivity* and profitability are closely linked. Value can be added by increasing revenues, but it can also be increased by improved supply chain responses. A review of *total cost of ownership* can result in cost savings (often the longer order and production life cycles of off-shore suppliers have significant impact on "total" productivity); the in-transit inventory has a cost that is borne by the customer and is totally nonproductive. Often a review of inventory holding indicates significant wastage. A review of *partnership*

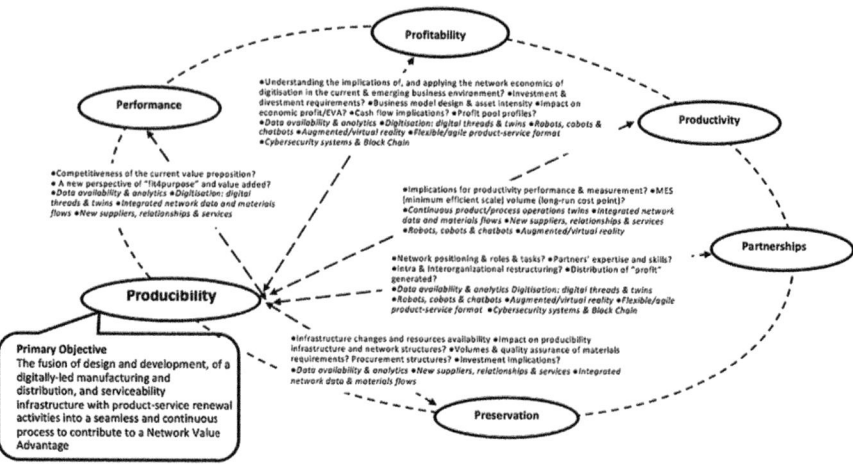

Fig. 12.13 Interactions and interfaces: Producibility

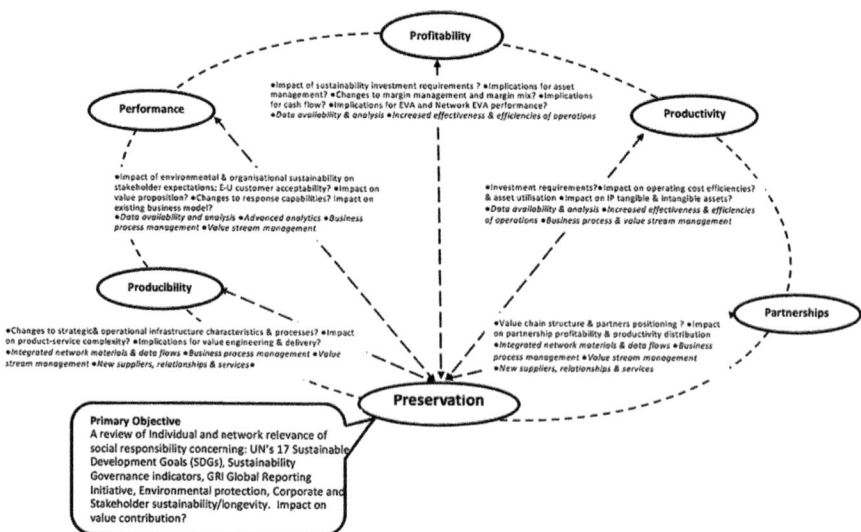

Fig. 12.14 Interactions and interfaces: Preservation

total cycle times is rewarding. Given the advances in data transfer from connectivity and the increased collaboration in Value Chain 2.0, materials flows should be optimized, and these costs contained. Sustainability strategies *(preservation)* addressing labor and materials wastages have an impact, but often excessive use of energy and water adds to costs. An optimization review of physical distribution operations also results in more efficient methods as well as cost savings. The *raison-d'etre* behind *producibility* is effective (strategic operations) and efficient (operational implementation) of the value production infrastructure. It is a planned activity that ensures an optimal solution to achieving customer (and subsequently stakeholder) satisfaction, and as a consequence, profitability *and productivity* objectives.

Productivity (Fig. 12.11) is concerned with the utilization of relevant assets of both the individual and the network organization. Decisions are based upon relevance to the planned outcome: stakeholder satisfaction. Productivity decisions are based upon meeting the value proposition offer: fit4purpose specifications, delivery time promises, and customer-specified location. *Partnership* (network design) through leverage facilitates and reflects the productivity decision; the qualifications requirement for membership is based upon relevance to satisfactory delivering value; ownership of the assets and processes is not relevant; relevant experience and employee abilities are essential, as is the overall ability to work within a network. *Profitability* decisions interact with productivity, and both are influenced by the profile of the market, customers, and the product-service characteristics; the revenue/profit/cash flow model will reflect product-service characteristics, complexity, life-cycle duration, and replacement rate. *Performance* has two considerations; it must address the customers' expectations concerning price/value equivalence *and* the organizations' expectations of cost/value equivalence. Often these requirements are similar; component and module build suppliers in automotive manufacturing expectations overlap concerning value delivery in this respect. *Producibility* features strongly; end-user product-service characteristics are determined by the all-embracing value proposition, its activities reflect the fit4purpose expectations of the network, and it also follows that quality control, time, location, and cost/price value equivalence are critical metrics. The interface with *preservation* identifies both organizational and environmental sustainabilities; low tolerance wastage rates are expected as are successful production activities.

Integrated, coordinated, collaborative, and interactive *partnerships,* Fig. 12.12, are an essential capability response. Because of their role in delivering the value proposition within time and cost "budgets," they require ongoing monitoring and adjustment. The application of digital connectivity has made (and continues to make) partnership performance reliable. The development of the visual abilities of sensors has increased the accuracy and speed of quality assurance and has decreased costs; this feature has had a considerable impact on productivity and profitability. *Productivity* increases have occurred as network connectivity has improved coordination of material flows; other improvements, such as collaborative procurement and platform technology, have increased the cohesion of the value chain. The cohesion effect can also improve *profitability;* improved asset base utilization implies increased returns on system investment; the cohesion throughout the value chain has benefited from collaborative RD&D. *Performance* (planned versus actual) is increasingly benefiting from partnership collaboration (enhanced by data transfer accuracy, distribution, and analysis); "promises" made by the value proposition are being delivered. Furthermore, where practical, the application of digital thread and twinning technology are resulting in "rolling" value propositions; product-services in use are being modified digitally. The optimized operating infrastructure of *producibility* manages relationships with and among partners because of the ability to synchronize materials, cash, and data flows; producibility also optimizes operations that create and manage wastage objectives, increasingly required by organizations to meet voluntary *preservation* objectives as well as efficiency targets to maintain competitive positioning.

Producibility makes significant contributions to the success of the value chain. It adopts an overarching role; its responsibilities are for managing both product-service development (value engineering) and delivering the value proposition (value delivery). See Fig. 12.13. In effect it ensures the efficacy of the capability response approach. Producibility coordinates *performance* activities by developing an optimal infrastructure that delivers a valid value proposition that meets and maintains stakeholder satisfaction. The production infrastructure creates and manages performance criteria that conform with objectives the network has undertaken to meet. Management groups in each capability response area have made decisions on structures and activities that will contribute to the network success. For example, *profitability* decisions concern the structure of the business model and its capital intensity based upon market structures, competitive intensity, and activities. *Productivity* plan materials flow around availability and delivery locations, together with a data management specification that helps manage the processes involved. Clearly, there is a link to the *partnership* network, much of which may be external but has a commitment to success, where there may be a lack of expertise and/or skills changes to the network structure, these will be identified and rectified. Producibility management assumes responsibility for identifying constraints that may be imposed (or suggested) by a *preservation* capability.

An increasing interest in sustainability has led to a range of definitions that can be collected under the label of corporate citizenship and cover: environmental, organizational, economic, and social sustainability. Some organizations have responded to corporate citizenship and have constructed statements outlining their responses; Chap. 9 provides more detail on the companies and their responses. As can be expected, the responses have been influenced by academic inputs and by government; an extensive contribution has been made by the European Union. *Preservation* has inputs to each of the other capabilities. For *performance* there are concerns that product-service and processes conform with "best practice" as offered as examples by organizations identified as being notable corporate citizens. Often organizations are pressured by stakeholder activists to comply with UN recommended practice; this tends to lead to product-service designs being influenced by environmental sustainability recommendations. This clearly has implications for both *profitability* and *productivity* guidelines and suggested best practice on waste controls can impact on cost and possibly profitability due to margin dilution. *Partnerships* operate on a basis of meeting an organizational value proposition requiring target cost and cycle time performance, to add a further (voluntary) sustainability constraint may create a value disadvantage unless all competitors have been persuaded to participate. Networks program balanced workflows around agreed production cycles, and participation in any sustainability activity may prove dysfunctional. *Producibility* is a production infrastructure that works with the value chain partners and experiences problems when adding non-value-producing activities. Work on going at the Ellen MacArthur Foundation suggests that "sustainability" should be part of the design task, not an afterthought. However, it is interesting to note that the Ellen MacArthur foundation (www.ellenmacarthurfoundations.org) has been successfully pioneering the concept of the circular economy. The extent of support from large corporations gives an idea of the powerful impact the Foundation has. The topic has been discussed in Chap. 9. Figure 12.14 illustrates some of the major interface areas.

Working with Response Capabilities and "the Convergence"

We suggest an approach to identifying appropriate capability combinations requires structure:

Determine network objectives (strategic and operational), and for each, set a range of essential and desirable end-user customer outcomes. Essential outcomes are those critical to the long-term success of the network organization's value delivery performance; desirable outcomes are more flexible and relate to product-service characteristics, the price/value positioning, and contribute to the value added and are included provided the incremental cost is covered either in margin terms or due to an increase in sales volume.

Identify alternative capability responses (or combinations of capabilities) that can meet the objectives and identify any constraints they may have (e.g., capacity and cycle time) and how these may be overcome. Solutions may be resolved by leverage decisions that involve changes in network process responsibilities (experiential or positional capability characteristics), or they may be resolved by seeking more effective/efficient matching of network roles with facilities offered by the Industrié 4.0, Value Chain Network 2.0, and stakeholder value management *convergence.*

Manage the stakeholders' expectations. Clearly, many organizations will be confronted by a considerable number of alternative capability responses. To reduce these to a manageable number, two stages are suggested: one is for the capabilities to be divided into primary stakeholder and secondary stakeholder groups (this a reasonable approach because it will be remembered that secondary stakeholders' interests comprise a monitoring of the organizations concerns: community awareness (contributions to education, etc.), environmental sustainability (pollution control, recycling/remanufacturing), governments' (taxation commitments and relevant legislation), regulatory compliance (construction limits, etc.), and geopolitics (adhering to the laws, regulations, and business practices requirements). The capabilities addressing primary stakeholder interests may also be divided into *customer facing capabilities,* the activities that take place during value production, delivery, and serviceability; these will require close analysis and evaluation of alternatives and *customer supporting capabilities.*

Allocating the Capability Response Budget

Given that the customer value drivers have been identified and a value proposition is being considered, some decisions need to be made concerning the allocation of costs to fund the response to the value offer. Marketing and operations management

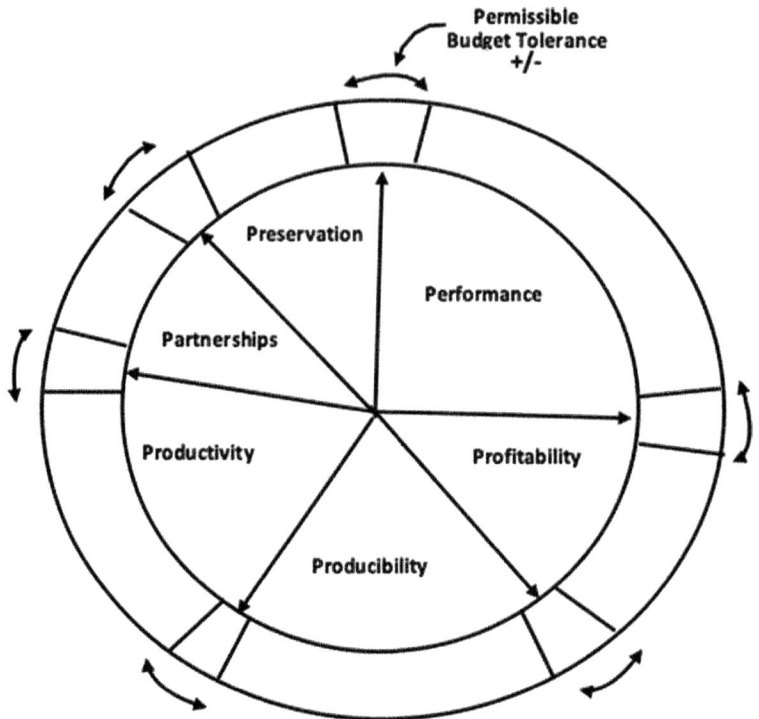

Fig. 12.15 Allocating the capability response budget

Table 12.1 Allocating the capability response budget – an example

	Percent	Percent of permissible budget tolerance
Performance product-service fit4purpose	25	±5.0
Profitability margin management	20	±2.5
Producibility coordinating network production	20	±7.5
Productivity facilities operations	15	±10.0
Partnership network development	10	±10.0
Preservation remanufacturing operations	10	±15.0

together should consider the capability response budget requirements to fund the value proposition. The value proposition has a twofold purpose; it communicates to the customer how their expectations will be met, and it communicates to the network organization the activities required to deliver the value proposition and the costs involved. See Fig. 12.15.

In the hypothetical situation described by the diagram, liaison with the customer has resulted in estimates of the types of characteristics of the response capabilities and the budget emphasis being allocated for the value delivery. The example (Table 12.1) suggests:

The table is indicative of the process rather than amounts to be allocated. Much depends upon the location of the product-service in its marketing life cycle (new PRODUCT-services or perhaps the transformation of a customer need to a product-SERVICE format may require an emphasis on fit4purpose and serviceability organization costs); the structure and form of competition may be important. Productivity/network costs may increase as and when product complexity and product range format changes are introduced. The availability of real-time data and its analysis using digital threads and digital twins may increase operational effectiveness but will require budget increases to do so. Using Fig. 12.15 together with the analysis from Figs. 12.9, 12.10, 12.11, 12.12, 12.13, and 12.14, alternative capability responses may be explored in the context of their efficacy and their cost. Figures 12.9, 12.10, 12.11, 12.12, 12.13, and 12.14 provide data on interrelationships between capability responses and the convergence factors available for their implementation. The percent of permissible budget tolerance suggests the expected impact of the respective capability response; a low tolerance factor suggests the activity is important, and the activity budget should not be reduced beyond the suggested level; it also signifies the view that larger increases than those suggested are unlikely to have significant impact.

Convergence …

To end this last chapter, we are concluding with an edited version of an article from the Dassault Systèmes website (Wilson 2018):

A new era in manufacturing tech is dawning – one that is deserving the name "renaissance." The Industrial Internet of Things (IIoT), superfast Internet, mobile communications, big data, virtual and augmented reality, digital twins, and 3D printing are among them that are pushing our mass production methods into the future. This Industry Renaissance is global, and it's bringing new ways, both real and using virtual technology of inventing, learning, producing, and trading (Bruno Latchague, Senior EVP, Global Brands, Dassault Systèmes). It is shaking all the existing sectors of the economy and society. We see new categories of industrial firms creating new categories of solutions for new categories of customers.

New technologies and methods are creating disruptions in industrial manufacturing. "Digital twin" technology, for example, involves creating a virtual copy of a product, a component, a production line, a factory, or even an entire city. This simulation allows manufacturers to virtually implement and test equipment and processes before implementing them in reality.

Advances in virtualization technology now mean that digital twin test beds are more accurate than they were before. Manufacturers can tweak their ideas in virtual form without needing to build physical versions first – saving money and time. Virtual testing can take place as part of a digital experience platform that links the diverse players in a production cycle – from the resource extraction stage to production and then onto the safe disposal of by-products. Such a platform functions as a central information-sharing hub, letting participants better collaborate with each other. The result is synergy that improves manufacturing. A supplier may notice that a factory is using a suboptimal material for a component and suggest a better alternative; or a consultant, exposed to data that gives a picture of the entire system, may discover management flaws and suggest solutions. Each stakeholder benefits from this whole-system view.

One especially compelling example of virtualization tech is Dassault Systèmes' digital twin of the city of Singapore itself. Using an array of images and data – including geometric, topological, environmental, road traffic, and other information – the company's technicians created a detailed model of the city-state. The digital model allows users to explore the impact that proposed projects will have, optimizing logistics, infrastructure, environmental and disaster management and security – and beyond.

The manufacturing sector has in recent years "been heavily challenged" by a deep transformation of the economy (Guillaume Vendroux, CEO of DELMIA, Dassault Systèmes). DELMIA delivers solutions to model, execute, and optimize global operations. That transformation has to do with a shift towards the "economy of the experience." Theorized as the successor to the service economy, the experience economy involves consumers' placing increasing value on the "experiences" that goods or services can provide them, in distinction to placing it on the ownership of material objects themselves. This shift has left the manufacturing industry, which by definition specializes in physical products, at a crossroads; it has triggered a very deep movement to totally transform our way of doing things.

"Knowledge and know-how will become the value of each company and the differentiating factor between them. Industry needs to find ways to capture knowledge store it and provide it to their teams in order to continuously improve their work", (Wilson 2018). At present, the CEO said, industry is seeing something of a gap at the managerial level when it comes to awareness of how to use these new technologies to create value. Organizations need to overcome that gap. Meanwhile, traditional boundaries between different industrial and business sectors and disciplines are breaking down. That means both managers and workers need to diversify their skill sets and function as specialists in multiple areas. Boosting employers' understanding of technological advances and the potential they offer remains a challenge.

The original "Renaissance" encompassed a range of developments: cultural achievements, philosophical transformations, and bold scientific discoveries, in addition to technological breakthroughs. Though legendary individuals played their part in that era, it really was the product of various phenomena. Today's Industry Renaissance is much the same. It's the product of a number of technologies that are coming together to create powerful multiplier effects. The common denominator is data – our era's powerful new data collection and analytics capabilities and their ability to create connections between people and assets. (Based on Industry of Tomorrow, BRANDVOICE, Dassault Systèmes. An edited version of this blog appeared in Forbes, Wilson 2018).

References

10 Ways technology is changing healthcare. (2017, March 21). *The Medical Futurist*.

Barkai, J. & Manenti, P. (2011, January 12). The assembly plant of the future. *Industry Week*.

Baghai, M., Coley, S., & White, D. (1999). Turning capabilities into advantages. *McKinsey Quarterly Number, 1*, 101–109.

Iansiti, M., & Levien, R. (2004). *The keystone advantage: What the new dynamics of business ecosystems mean for strategy, innovation and sustainability*. Boston: Harvard Business School Press.

Lanning. M.J. and Michaels, E.G. (1988). A business is a value delivery system. *McKinsey*.

Lindgren, M., & Bandhold, H. (2003). *Scenario planning: The link between the future and strategy*. Basingstoke, UK: Palgrave Macmillan.

Margo, J. (2019, February 19). Why Denmark is reducing hospitals while we are building more? *Australian Financial Review*.

McKinsey, (Quarterly, June 2011). The perils of bad strategy, Richard Rumelt Interview.

Milne, R. (2019, February 13). Ikea looks to launch sales platform that would include rival products. *Financial Times*.

Normann, R. (2001). *Reframing business*. Chichester: Wiley.

Olvé, N., Roy, J., & Wetter, M. (1997). *Performance drivers*. Chichester: Wiley.

Passariello, C. (2011, February 12). Carrefour thinks again as tastes shift from hypermarket to hometown shopping. *The Australian*.

Peters, S., Lanza, G., Ni, J., Pei-Yun, Y., & Colledani, M. (2014). Automotive manufacturing technologies – an international review. *Manufacturing Review, 1*, 10.

Rayport, J. & Sviokla, I. (1994, November/December). Managing the marketspace. *Harvard Business Review*.

Scott, M. (1998). *Value drivers*. Chichester: Wiley.

Slywotzky, A., & Morrison, D. (1997). The profit zone: How strategic design will lead you to tomorrow's profits. New York: Wiley.

Strategy's strategist: An interview with Richard Rumelt. (2007, November), *McKinsey Quarterly*.

Spector, M. & Tractenberg, J.A. (2011, February 12). Chap. 11 for Borders, new chapter for books. *The Wall Street Journal*.

Wilson, C. (2018, November 13). Industry of tomorrow. *Forbes*.

Zappa, M. (2014, May 6). 15 Emerging agriculture technologies that will change the world. Retrieved from www.businessinsider.com.au/15-emerging-agriculture-technologies-2014-4#YMhopyH SaZyQTEwK.99

Chapter 13
Concepts and Cases

Abstract This text has made the case for undertaking *response capability analysis* before deciding upon strategic direction and structure. We argue that in the current business environment not only are strategic (and for that matter shorter term operational) opportunities increasingly available but so too are the ways and means of competing for them. The "convergence" of Industrié 4.0, Value Chain Management 2.0, and meeting stakeholder perspectives of value has created an increasing number of alternative solutions for competing in the marketplace. The case studies in this chapter have been constructed such that they identify activities that are decision points concerning response capabilities. The chapters dealing with the response capabilities, both existing and new concepts, have been introduced and explored; some of the concepts may not have been encountered in previous business school programs, or were, but not in the approach taken here. We have presented a wide range of topics and would argue that all are relevant and should be used when thinking thorough an overall response to an opportunity. It can be expected that future market opportunities are likely to present themselves in a manner that suggests numerous solutions. The benefit of the response capability approach is that it opens up these alternatives. This suggests that response capabilities are not only useful for creating new innovative solutions but can also be used to search for alternatives around a specific capability. Case studies do not represent superior ways of resolving issues, nor do they represent inferior practice; they reflect ongoing or recent situations that have occurred. But they do present the reader with an opportunity of exploring the situation by asking: if they knew then what they know now, what could the end result be? The chapter begins with a list of key concepts and definitions – this may assist the reader in thinking through the case studies and is a useful summary of the key ideas that are examined in this publication.

Keywords Performance · Profitability · Productivity · Producibility · Partnerships and Preservation

Concepts

The chapter begins with a list of key concepts and definitions – this may assist the reader in thinking through the case studies and is a useful summary of the key ideas that are examined in this publication.

Enterprise Architecture: is a conceptual blueprint that defines the structure and operation of an organization. The intent of an enterprise architecture is to determine how an organization can most effectively achieve its current and future objectives.

Capability Responses: the approach views the firm as a portfolio of capabilities that evolve in response to the (perceived) demands of the business environment. An "enterprise" comprises one organization or collaborative network of firms; response capabilities are characteristics that reflect an understanding of the marketplace and of the opportunities it offers and the characteristics of capabilities essential for successful engagement. The core business response capability model components of *performance* (value engineering, value delivery, and the value proposition; stakeholder-performance expectations, fit4purpose, etc.), *profitability* (financial viability, sustainable economic profit, positive cash flow), *productivity* (total factor productivity, optimal utilization of capital, labor, materials, and service inputs, and EVA (economic value-added)), *partnerships* (expertise and skills of value generating employees, workplace cultures, and management styles, among all network partners), *producibility* (the seamless intra- and inter-organizational infrastructures of sequential processes and activities of value engineering and value delivery that creates, produces, delivers, and captures value), and *preservation* (a socioeconomic, responsible, use of environmental resources in creating stakeholder value, and the concerns of environmental and corporate sustainability and responsibility). Similarly, the response capability approach can set prerequisites; for example, organizational policy, not strategy, can establish ground rules concerning levels of returns, such as a margin of the amount by which an EVA return should exceed the annualized weighted cost of capital or perhaps the amount of increase in capital intensity the organization is prepared to undertake as further investment.

Capability Response Analysis: is a management process that identifies the relevant capabilities that will be required to undertake a successful response to a market opportunity. The response capability approach can reinforce a market (and value chain) positioning, as well opening up alternative market opportunities. The analysis is conducted and based on the expectations of potential customers in order that these are fully understood and can be used as input for planning a response. However, the vendor organization also has expectations based upon the same capability criteria (not the same characteristics), but they are determined by the expectations of the organizations' stakeholders.

Capability Response Management: is the ability to structure, combine, and leverage internal and external resources for the purpose of creating a capability-led response to a market opportunity to create value for stakeholders by maximizing competitive value advantage. Organizational policy, not strategy, can establish prerequisites or ground rules concerning response capability levels of response, such as a margin of the amount by which an EVA return should exceed the annualized weighted cost of capital *(profitability)* or perhaps the amount of incremental increase in capital intensity the organization is prepared to undertake as further investment in pursuit of an opportunity *(producibility and productivity)*.

Serviceability: is a "total service" package comprising collaborative RD&D (exclusive product-service format specification, development, and selected manufacturing activities), installation and operator training, data-driven performance management, predictive maintenance, product-service renewal/retirement, and replacement.

Time-to-Market/Time-2-Market: the amount of time required to deliver a product-service from concept to reliable, operational fit4purpose.

Industrié 4.0: is the name widely accepted as the current trend of automation, connectivity and data exchange in manufacturing technologies. It includes cyber-physical systems, the Internet of things, cloud computing and cognitive computing. It has been commonly referred to as the fourth *industrial* revolution.

Stakeholder Value-Led Management: whereas shareholder value management characteristics had a narrow focus, with objectives driven by quantifiable metrics reflecting share market valuation; stakeholder value-led management thinking tends to be deeper and broader and may include such characteristics as, sustainable, visionary, and competitive-thinking approach, involving a multi-dimensional view of the organization and its responsibilities.

Convergence: the management of a relevant combination of the characteristics of Industrié 4.0, Value Chain Network Management 2.0, and stakeholder value-led management that deliver a superior value proposition in competing in existing and new product-service-markets.

Chapter 1

Value Proposition: is a positioning statement responding to customer expectations developed from an understanding of their product-service purchasing and their applications activities: a "value proposition" spells out a response to these. A rolling value proposition describes the impact of real-time digital data flows now available from operating product services that identify product modifications that can be effected digitally.

Value Migration: occurs when customer needs change or a new business design begins to encroach on an existing configuration of providers.

Value in Networks: is generated by collaboration on shared objectives and coordinated operations. By leveraging assets and allocating roles and tasks, a network approach optimizes its resource capabilities in terms of investment in PRODUCT-service/product-SERVICE offers, resulting in a value proposition that meets customer expectations, optimizes capabilities, and is economically viable.

Shared Value: is creating economic value in a way that also creates value for society by addressing its needs and challenges and involves understanding and building stakeholder expectations into the network organizations' value proposition.

The Sharing Economy: is an economic model often defined as a peer-to peer economy (P2P) based activity of acquiring, providing, or sharing access to goods and services that are facilitated by a community-based online platform.

Customer Journey: the complete sum of experiences that *customers* go through when interacting with your company and brand. Instead of looking at just a part of a transaction or experience, the customer journey documents the full experience of being a customer.

Customer Experience: the product of interactions between an organization and a customer over the duration of their relationship at all points of contact.

Customer Cohorts: usually share common characteristics or expectations within a defined time-span or an experience. Its members share a significant experience at a time, a group of people who share a defining characteristic, typically who experienced a common event in a selected period, such as birth or graduation, representing a cross section of the population at intervals through time.

Chapter 2

Strategy Dynamics: studies how changes in strategy dynamics would shift each dimension of an industry. Would the industry become more concentrated or less? More integrated or less? Would there be more product differentiation or less? More segmentation or less? Given consumer desires and available technologies, how should the industry or business look in, say, 10 years? Where are the economic forces trying to take you? Should your strategy ride those forces or fight them? Strategy dynamics is focused on current and past observations.

Industry Dynamics: is a framework comprising *six* generic *influencers*: knowledge management, technology management, process management, relationship management, regulatory compliance, and geopolitics. It is future oriented.

Chapter 3

Response Capabilities: an approach that views the firm as a portfolio of capabilities that evolve in response to the (perceived) demands of the business environment – *performance management* (the value engineering and value delivery proposition), *profitability, productivity, partnerships, producibility* (the strategic and operational infrastructure – the processes and activities of value engineering and value delivery), and *preservation* (sustainability issues: organizational, economic, social, and environmental). An "enterprise" comprises one organization or collaborative network of firms.

Support Response Capabilities: facilitate the application of resource capabilities solutions that are an essential component to the model. They will be structured around *leadership and professional capabilities,* identifying direction and applying managerial abilities that understand the nature of current market demand requirements and identifying the relevant response capabilities (*digital capabilities, human capabilities,* and *financial engineering capabilities*).

Capability Planning: involves a functional analysis of operational requirements and capabilities that are identified based on the tasks required. Once the required *capability* "inventory" is identified, the most cost-effective and efficient options to satisfy the requirements are sought.

Value Contribution: if the economic return on investment exceeds the market cost of capital, there is an increase in the value of the invested capital and a positive value contribution.

Value advantage: support capabilities make a significant contribution in developing and delivering the value proposition, by assuring value contribution continuity.

Chapter 4

Performance Management: requires some explanation. It is mutually bilateral in as much as it requires both customer and vendor agreement on expectations and satisfaction for it to work. Short-term performance management comprises modifications to a product-service and possibly to the activities and processes that are responsible for its "production." The long-term perspectives will be based upon ongoing developments of the digitization of product-services and of processes and activities offering a framework for developing business futures.

Operational Performance Value Drivers: are short- to medium-response actions that respond to the market based on existing capabilities. These might build competitive value necessities, correct competitive value disadvantage, and create competitive value advantage largely using the existing business model.

Strategic Performance Value Builders: are long term and may have no relationship with the current business model. They comprise *experience capability characteristics* (skill sets, expertise) and *positional capability characteristics* (features/characteristics the organization has (or is) investing in, which have a long-term payoff).

Chapter 5

Profitability: has many interpretations, often competing with other metrics as a means of measuring corporate performance in meaningful interpretations across a range of interests, strategic effectiveness, investment strategy, operating efficiency, and shareholder returns.

Economic Profit: is considered to be operating profit less a charge for the use of capital involved in the production of the operating profit. Depreciation is not considered to be an accurate charge for the use of capital.

Economic Value-Added (EVA): EVA is net operating profit after tax (NOPAT) less a capital charge for the invested capital employed in the business. EVA is arguably a short-term measure of performance as it is based upon short-term performance.

Value Contribution: is calculated based on the operating result after tax and the opportunity cost of invested capital. The operating result shows the economic performance of the organization and is initially a pre-tax figure. *Value contribution margin analysis* is a measure of operating leverage; it measures how growth in sales translates to growth in profits.

Operational Gearing: is the extent to which an organization commits itself to high levels of fixed operating costs (such as dedicated locations, plant and equipment, RD&D, and management and labor expertise) as compared with variable costs (materials, proprietary components, and semiskilled labor).

Operating leverage: measures the degree to which a firm or project can increase operating income by increasing revenue. A business that generates sales with a high gross margin and low variable costs has high operating leverage.

Financial Gearing: is the measure of the comparative levels of equity and debt that finance the revenue generating activities of the organization. Financial gearing is measured by the ratio of total debt to total assets.

Financial Engineering: has a number of interpretations. Financial engineers design, create, and implement new *financial* instruments, models, and processes to solve problems, taking advantage of new *financial* opportunities including repurposing existing manufacturing facilities.

Profit Pools: profit pools can be defined as the total profits earned in an industry, identified at defined points along the industry's value chain (i.e., value chain activities, organizational activities (RD&D, manufacturing, marketing, etc. or specific processes)).

Risk: comprises business response model risk, business response model risk, alliance risk, and capability risk.

Returns Spread: returns spread (return on assets managed *less* the current cost of capital).

Chapter 6

Productivity: the effectiveness of productive effort, especially in industry, as measured in terms of the rate of output per unit of input.

Multifactor Productivity: is a model that seeks the most likely combination of resources that will successfully meet the expectations of both customers and suppliers – an optimal stakeholder solution.

Productivity Pools: can be defined as the total productivity of an industry or a value chain at defined points (i.e., value chain activities, organizational activities (RD&D, manufacturing marketing, etc. or specific processes)).

Platforms: are a business model format that creates value by facilitating exchanges between two or more interdependent groups, component suppliers and original equipment manufacturers (OEMs) assemblers, and consumers and producers. In order to make these exchanges happen, platforms harness and create large, scalable networks of users and resources that can be accessed on demand. Platforms create communities and markets with network effects that allow users to interact and transact. Platforms don't own the means of production; they create the means of connection.

Chapter 7

Producibility: is the process by which value is created and captured by the network and industry infrastructure. It is a total design activity that includes all relevant activities within the value chain network and creates intra- and inter-organizational partnerships to achieve stakeholder satisfaction. It is an operating infrastructure network.

Design to Value (DTV): is a portfolio-optimization approach that helps management understand precisely which product features are important to consumers and whether consumers are willing to pay for those features.

Design for Value and Growth (D4VG): extends DTV to provide "exceptional customer experiences" for value generation and growth by producing products with characteristics that create customer loyalty (and therefore repeat purchases).

Fulfilment Execution System (FES): is an approach to a coordinated management of demand, capacity and resources, and outbound order fulfilment across the entire network of manufacturing plants and along the supply chain.

Chapter 8

Operational Partnering: is typically cost-led and robust for as long as the value advantage prevails.

Strategic Partnering: individual organizations create networks of organizations, each of which contributes essential core capabilities, in order to create a value chain network that can compete with other networks successfully.

Value Engineering: is concerned with new product-services. It is becoming an integral activity in new product development. The focus is on reducing costs, improving functional characteristics, or both, by way of teamwork-based product evaluation and analysis.

Value Delivery: is concerned with *existing* products, it includes the overall operational activity of distribution, support service (and serviceability), and remanufacturing/repurposing activities.

Total Cost of Ownership (TCO): is a quantitative, evaluative analysis of existing and potential partnerships. While most of these costs often are itemized separately on a company's financial statements, many are either ignored or unknown. Comprehensive analysis of the cost of ownership is a common practice for business dealings.

The Global Value Chain (GVC): offers a capability to coordinate sequential production activities on an inter-organizational basis, across international borders. One view of this activity is the outsourcing approach whereby offshoring reduced variable costs, typically labor and perhaps materials costs. A strategic perspective expands this view to include situations (e.g., Dyson the UK consumer durables manufacturer) in which the entire operations activity is located to achieve an optimal cost structure.

The "Smile Curve": identifies the distribution of value adding activities by plotting labor compensation ($ per hour) in the participating countries (reflecting the value-added by each country), indicating high versus low value-added activities, against the aggregate production cycle length from RD&D activities through component manufacturing and assembly to end-user consumers and after sales support activities.

Chapter 9

Sustainability (Organizational and Environmental): the ability to be maintained at a certain rate or level (an organization individual and/or network profitability growth) and environmental sustainability (the preservation of "stocks" of scare resources by recycling and/or by identifying substitutes).

Governance: refers to oversight and decision-making related to strategic direction, financial planning, and bylaws – the set of core policies that outline the organization's purpose, values, and structure. Governance decisions should provide guidelines for management.

Social Responsibility: corporate social responsibility is a self-regulating business model that helps a company be socially accountable to itself, its stakeholders, and the public. By practicing corporate social responsibility, they can be conscious of the kind of impact they are having called corporate citizenship.

Value Renewal: is an aspect of the design process that identifies components of a product that can be recovered after the product has undergone planned use that can be recovered and remanufactured and included in a "new" product.

Sustainability Equilibrium: stewardship and the management of resources such that assets will produce value in the short and long term. Maintaining and enhancing the quality and productivity of assets in the midst of changing economic and ecological conditions.

Circular Economy: is an alternative to a traditional linear economy (make, use, dispose) in which we keep resources in use for as long as possible, extract the maximum value from them while in use, and then recover and regenerate products and materials at the end of each service life (make, use, renew). Also known as the circular value chain.

B Corporations: make a commitment to corporate social responsibility by agreeing a charter of social and environmental development that is purpose-driven and creates benefit for all stakeholders, not just shareholders.

Corporate Citizenship: corporate citizens are organizations practicing stakeholder value-led management.

Chapter 10

Capability Audit (1): a review of the fit4purpose of current capabilities *performance management* (the value engineering and value delivery proposition), *profitability, productivity, producibility* (the strategic and operational infrastructure – the processes and activities of value engineering and value delivery), *partnerships*, and *preservation* (sustainability issues: organizational, economic, social, and environmental) that deliver the value proposition in competing in existing product-service-markets.

Capability Audit (2): analysis and assessment of fit4purpose capabilities (performance, profitability, productivity, partnerships, producibility, and preservation) *required* to present a competitive value proposition in emerging product-service-markets. The audit includes an exploration of "costs" (fixed costs (investment in tangible and intangible assets) and "availability" of the resources (do they exist? where? and can they be leveraged by forming some form of an alliance?).

Convergence Management: the management of a relevant combination of the characteristics of Industrié 4.0, Value Chain Network Management 2.0, and stakeholder value management that deliver a superior value proposition in existing and new (or alternative) product-service-markets.

Convergence of Technology Management: applied technology management. The design of a technology mix that creates competitive value advantage (product components, service components and activities, and delivery).

Product Life Cycle Management (PLM) Operations: is the process of managing the entire life cycle of a product from product concept, RD&D, engineering design, and manufacturing to serviceability design and implementation and renewal (remanufacturing/repurposing) and the ultimate disposal of manufactured products. PLM an integrative activity of people, data, processes, and business systems provides a real-time information for production and serviceability organizations and customers.

Product Life Cycle Management (PLM) Marketing: is an important concept in marketing. It identifies and describes the stages a product-service goes through from its concept development, introduction, and market activity stages until it finally is removed from the market.

Scenario Planning: sometimes called "scenario and contingency planning" is a structured way for organizations to think about the future. A group of experts with specific knowledge in topics that will impact on future markets/industries sets out to develop a small number of scenarios – "futures" – about how the dynamics of change might unfold and how this might affect a market/industry development issue that may impact their decisions concerning an organization's future developments.

Chapter 11

Value-Added and Added Value: value-added is the principle basis of expenditure taxation in a number of countries. Value-added is therefore not the same thing as added value; the former is the difference between the value of the material and semi-produced inputs which a firm buys in and the value of the output which it sells. It is therefore equal to the firm's net output. Often the terms are used interchangeably, and this is incorrect and confusing. In this book, our interest is in both in value added (the increase effect of economic efficiency) and added value (the impact of a change in product or process that can have a positive impact on a customer that increases the customer's value advantage in the market within the context of market attractiveness).

Value Contribution Analysis: value contribution (calculated as dollar EVA) must offer both vendor and customer measurable positive gain in value-added (EVA). In essence it is a process of measuring the risk involved in changing operational gearing and its impact on risk.

Activity-Based Management: the integration of strategic marketing, management accounting, and operations management activities, working within an overall set of parameters determined by corporate management, for evaluating the costs of the possible responses to the customer's "ideal" product-service requirements.

Chapter 12

FES (Fulfilment Execution System): is an approach to implementing a coordinated management of response capabilities, capacity, and resources, in meeting outbound customer order fulfilment requirements across the entire network of manufacturing plants and along the supply chain.

Business ecosystems: is a network of organizations and activities – including designers, manufactures, suppliers, distributors, customers, competitors, government agencies, and so on involved in the delivery of a specific product or service through competition and cooperation.

Cases

This text has made the case for undertaking *response capability analysis* before deciding upon strategic direction and structure. We argue that in the current business environment, not only are strategic (and for that matter shorter-term operational) opportunities increasingly available but so too are the ways and means of competing for them. The "convergence" of Industrie 4.0, Value Chain Management 2.0, and meeting stakeholder perspectives of value has created an increasing number of alternative solutions for competing in the marketplace. The case studies in this chapter have been constructed such that they identify activities that are decision points concerning response capabilities.

In the chapters dealing with the response capabilities, both existing and new concepts have been introduced and explored; some of the concepts, existing or those developed, may not have been encountered in previous business school programs, or were, but not in the approach taken here. We have presented a wide range of topics and would argue that all are relevant and should be used when thinking thorough an overall response to an opportunity. It can be expected that future market opportunities are likely to present themselves in a manner that suggests numerous solutions. The benefit of the response capability approach is that it opens up these alternatives. This suggests that response capabilities are not only useful for creating new innovative solutions but can also be used to search for alternatives around a specific capability. For example, preservation may be one response for which there are constraints such as the reuse of resources, requiring product-services to be designed such that a fixed amount of the retired product will be remanufactured and form the basis of a second generation of the item. Similarly, the response capability can set prerequisites; for example, organizational policy, not strategy, can establish ground rules concerning levels of returns, such as a margin of the amount by which an EVA return should exceed the annualized weighted cost of capital or perhaps the amount of increase in capital intensity the organization is prepared to undertake as further investment. Nonfinancial constraints may also be policy issues, such as environmental sustainability and/or environmental damage the organization is prepared to accept from any new processes or activities it undertakes.

Response capabilities are *performance* (stakeholder-wide, performance, fit4purpose, etc.), *profitability* (financial viability, economic profit), *productivity* (total factor productivity, optimal utilization of capital, labor, materials, and service inputs), *partnerships* (expertise and skills of value generating network members), *producibility* (the cost-effective and efficient network operating infrastructures that create, produce, and deliver value), and *preservation* (a socially responsible use of environmental resources used in creating stakeholder value).

Capability Response and Support Capability Response Activities comprise:

Performance
- Product-service fit4purpose specification
- Serviceability support
- Quality specifications
- Delivery frequency and reliability
- Availability (product-service and location)
- Customer-Supplier-Performance-Management

Profitability
- Impact on "total-stakeholder-investment" effectiveness
- Economic profit and financial performance (EVA)
- Impact on optimal capital intensity and asset management
- Margin management and margin mix
- Cash flow management
- Impact on industry/market "profit pool" optimization (a shift in the value chain roles)
- Network value contribution (EVA) and the distribution equity of economic profit, i.e., share of $ value of value advantage

Productivity
- Availability of relevant capabilities, capacities, and processes
- Network relationships (value chain network positioning)
- Network digitization and connectivity capability
- Availability ownership and location of resources
- Productivity metric, $\sum$Network, and individual organizational EVA

Partners
- Managerial capabilities requirements
- Labor force operations expertise and skills requirements
- Availability of relevant management if required
- Availability of relevant operations labor if required
- Innovation and creativity; product-services and processes

Producibility
- The creation, production, delivery, and capture of value
- The impact of changes (and potential changes) to the value proposition
- Changes required to current and planned infrastructure model
- Implications for market and value chain network positioning
- Changes to product complexity and serviceability complexities
- Infrastructure connectivity and collaboration requirements
- Impact on the value contribution from an alternative infrastructure

Preservation
- Current implications on the market/industry of environmental sustainability requirements
- Impact of forecast changes on corporate and network processes and activities
- Implications for current and future costs

Returning to the case material – throughout each case the response capabilities appear in bold type with question marks, providing the opportunity for the reader to explore the situation recorded and questioning a "what if?" alternative. Prior to undertaking the case tasks, we offer a reminder of the nature of response capabilities with examples in each response capability category.

Small Robot Company

The Small Robot Company, a British agritech start-up, commenced operations in November 2017. The company combines the power and precision of robots and artificial intelligence (AI) to improve the way that food is produced and minimize chemical usage. Agritech is a start-up commercializing a deceptively simple idea: small robots not big tractors. The company has focused on changing the way that technology is used to manage agricultural production, small robots not big tractors, because big tractors are neither efficient nor environmentally friendly. Small Robot has worked with farmers to understand how it can answer their needs and make this into reality. Small Robot comprises a group of farmers, engineers, scientists, and service designers with a deep knowledge on farming, robotics, AI, and service design. The technology builds on 15 years' research by Professor Simon Blackmore, the world's leading expert in precision farming. Small Robot is dedicated to building technology that will make farming profitable, more efficient, and more environmentally friendly. A farming service designed by farmers for farmers that uses robotics and AI to deliver this dream.

Personnel

Ben Scott-Robinson, an experienced entrepreneur working at the cutting edge of user centered design and Sam Watson Jones, a fourth-generation farmer, were inspired by the work of Simon Blackmore in the National Centre for Precision Farming at Harper Adams and his vision for replacing much of the work done by tractors in fields with a series of highly accurate, smart, lightweight robots. Sam's first step was to interview a large number of farmers around the country, firstly on the pain points around their current, tractor-centered systems and secondly on what specific outcomes they would look for from a new model of farming. His discovery was that farmers felt poorly served by existing machinery manufacturers and new technology delivered as either hardware or software products.

Small Robot Company Objectives

The company's primary objective is to make farms more profitable and increase yield and efficiency, through using small robots instead of tractors. This is also kinder to the soil, is kinder to the environment, protects biodiversity, and enables permaculture at scale. The robots will be offered through a "farming as a service" (FaaS) model, which uses robotics and AI to "digitize the field" and deliver precision farming. Robust per-plant data creates a "profit map" that shows which areas of field to use, which to leave fallow, and what to plant where and when.

Capabilities profile: increasing profitability with reduced inputs

Using robots has the potential to reduce chemical usage in arable farming by as much as 95% and reduce cultivation energy and associated CO_2 emissions by 90% *(preservation)*. It can increase arable farming revenues by up to 40% and reduce production costs by up to 60%. *Performance? profitability, and productivity?*

Value Proposition

"We are building robots that will seed and care for each individual plant in your crop. They will only feed and spray the plants that need it, giving them the perfect levels of nutrients and support, with no waste" (*Small Robots website*).

Rationale

Robots are smaller, lighter, and more precise than the current farming systems using tractors. Using small robots instead of tractors will deliver greater yield from less inputs. Robots can also work in smaller and more unusually shaped fields; making the most of headlands (the corners of the fields, which are usually underused) while protecting hedgerows, biodiversity, and the British landscape longer term, it radically improves farming methods and food potential. *Productivity? Preservation?*

"Currently, 90% of cultivation energy is used ploughing which is only necessary because of heavy machinery crushing the soil," argues Professor Simon Blackmore, Head of Agricultural Robotics, Harper Adams University. And, if you treat the whole field the same, overuse of chemicals is inevitable. It is known that 95% of herbicides can be saved using a smart sprayer and small robots can help to make this technology work at scale. *Productivity? Preservation?*

Digitizing the field using artificial intelligence. Small Robot Company's AI-driven technology allows a level of autonomy, accuracy, and detail that now makes it possible for smaller farms to be profitable and for better decision-making. Eventually, each process, from knowing when to plant, all aspects of crop care, and knowing when to harvest, will be automated. Technology has changed almost every industry, but farming is lagging behind. By "digitizing" the field, the potential for efficiency is large. Data analysis from the field enables decisions which will consider agronomy, soil science, and market conditions. What to plant where and when to maximize yield; this means the ability to apply permaculture techniques at scale follows. *Performance?*

Gleadell Agriculture, an independent major trader of grain, seed, and fertilizer, is partnering with the Small Robot Company in its trials work, finding suitable trial sites and supplying seed in small bags for the trial growers through Dunns (Long Sutton) Ltd., one of the largest agricultural seed processors in the UK. Stuart Shand, Sales Director at Gleadell Agriculture, considers "Robot technology is potentially the most exciting agricultural development over the next few years, and it could take precision farming to a completely new level" and said "Given the pressures that UK farmers are under, they need to maximize efficiencies and reduce costs. Robot technology could help deliver on both counts and we look forward to working with the Small Robot Company to help achieve those goals," *Partnerships?*

Farming as a Serviceability Activity – "Pay-As-You-Grow Agribots." Small Robot Company has engineered a compact four-part robotic system to monitor plant and soil health and autonomously look after feeding, seeding, and weeding on farms. "If we can deliver a healthy crop at the end of the year – and for that we charge a certain amount per hectare – then there is no upfront purchase cost and the farmer doesn't have to worry about how reliable the robots are, they just pay per hectare," Ben Scott-Robinson, founder. *Profitability, Productivity.*

Small Robot Company offers its robots through a Farming as a Service (FaaS) model, which is both a hardware and a software service for farmers. Farmers pay a per hectare subscription fee for a robotic hardware service which digitizes the farm and delivers crop care at per-plant precision. The service comprises:

Tom: crop and soil monitoring robot

Dick: precision spraying and laser weeding robot

Harry: precision drilling and planting robot

Wilma: the operating system – The artificial intelligence-driven neural network

The robots take care of all the feeding, seeding, and weeding. When one of these robots is needed, it turns up and does its job. When it is finished, it is taken away. This makes it very low risk for farmers to trial and adopt. It also means that farmers no longer have to worry about buying, storing, or maintaining of most crop care machinery and equipment, which is often underutilized for most of the year.

Convergence …

The Small Robot Company vision is to replace tractors with lightweight agribots, in a bid to make farming more profitable and productive, more cost-efficient, and more environmentally friendly. Small Robot Company is at the "start-up" stage, but the support of its target market, arable farmers, many of whom have invested in the company, suggests it is a game changer in terms of reducing the need for expensive plant protection products, artificial fertilizers, and costly cultivations.

The Portsmouth-based Small Robot Company has secured £1 m through an online crowdfunding campaign, which attracted more than 1000 investors. It exceeded its original funding target of £500,000 within minutes of its launch in mid-December 2018, with the company saying that much of its success in raising money has been because of strong support from hundreds of farmers. It estimates that at least two-thirds of the initial £500,000 was raised from the farming community, with the largest single investment it has had to date; £60,000 coming from a farmer. It had previously received £300,000 in initial seed funding from farmers, including £90,000 raised through pre-sales to members of its farm advisory group. This group is made up of 20 farmers who are working with the business to develop the technology through hosting field trials for the robot prototypes. They have each paid a £5000 deposit, which will become £10,000 of credit for Small Robot Company services when it launches commercially. Their support comes despite the company being less than 2 years old and its robots – Tom, Dick, and Harry – still being at prototype stage.

This case study is based upon:

- Three agribots revolutionizing the farming industry, Maddy White, *The Manufacturer,* July 26, 2018
- Agri-tech Small Robot Company Scores Big Funding on Crowdcube with Days Remaining on Campaign, Erin Hobey, *Crowdfund Insider*, January 1, 2019

Shoes of Prey

Established in a Sydney in 2009, Shoes of Prey was created on the assumption that women possess a desire – to find the perfect pair of shoes. Acknowledging the individual and unique style of women globally, the brand developed their own 3D Designer software and became known for its creativity and innovation in a "one style fits all" market. Offering complete customization, the user decides on the shape, heel, color, and material to see their ultimate pair of shoes come to life. The Shoes of Prey value proposition offered not only individual "self-designed" but also a range of predesigned pieces. Shoes of Prey grew into a successful multinational company without leaving their value proposition of celebrating the personal style of women behind.

Partnering the brand's creative collaborations with fashion influencers, Shoes of Prey has continued to deliver well-crafted footwear at an affordable price point. The brand was seen to provide a wealth of options that are "sure to suit your individual preferences without breaking the bank." *Partnerships? Performance?*

In mid-October 2017, Shoes of Prey closed the physical design studios it had built in David Jones (a Sydney department store and those in outlets of Nordstrom the US department store group). The company, which at this time was backed by US venture capital funds Khosla Ventures and Greycroft Partners, ASX-listed alternative asset manager Blue Sky Funds, rich-list billionaire Mike Cannon-Brookes, and Nordstrom itself, announced its return to its roots as an online retailer. A co-founder said the company was leaving real-world stores and would focus on online sales, describing it as a natural evolution for the business arguing that while the physical stores have been hugely important, where they were at now as a brand, the most profitable and biggest opportunity for Shoes of Prey is online. Commentators suggested the change in direction represented a significant tactical retreat for the company and serving as a reminder that the path from fast-growing start-up to sustainable, mature business is rarely smooth and often uncertain. It was also argued that as a start-up company, problems occur, and the successful founders can navigate those ups and downs well because they can change direction and that the company had been pursuing a so-called "omnichannel" strategy. The strategy was designed to expose Shoes of Prey's offerings to lucrative shoppers and reduce the company's dependence on sales leads through advertising on Facebook and Google. It was also suggested that for a relatively young business that did not have a huge profile, the strategy of testing out a bricks and mortar channel made sense, and it provided credibility and an opportunity for customers that were maybe a bit nervous to touch and feel the shoes and get a sense for the quality of the product. Once credibility was established, the real opportunity for them was always online. Thus, the plan was to let people design and order custom shoes online, which were then manufactured at a dedicated factory in China and shipped to users within 2 weeks. In December 2016 the company raised from investors, including Nordstrom and Blue Sky to accelerate its US expansion. The company's physical retail operations represented 25% of its cost base, but just 15% of its sales. A source suggested the company was losing more money on the physical store presence than it had expected and that it did not match up with the realities of retail.

A director said the company was not yet profitable as it was seeking to scale up. It was suggested that cash flow was strong, the future appeared promising, and there were no plans to raise more funds. The company had a strong balance sheet, and by focusing on the most scalable and lucrative channel, namely, online, its long-term growth plans remained intact. Expectation was to reach profitability sometime next year (2017). *Performance? Profitability?*

In September 2017 the Company invested in a major expansion into sneakers. The business founder claimed the business still has plenty of funding. The move was the first unisex product for Shoes of Prey, which has raised $US24.6 million ($30.6 million) since its launch in Sydney in 2009 and its first major announcement since the closure of its physical shopfronts inside David Jones and Nordstrom stores in October 2016.

A director told *The Australian Financial Review* that the decision to return to the online business model had reduced cash burn at the company, which was reported by News Limited had been nearly $1 million a month by September 2016. A response from a director of the company suggested this was a "1-month snapshot" that did not give an accurate view into the company's 8-year growth history and that the spend in that particular month had been planned for. The accounts indicated Shoes of Prey had $7.7 million in cash remaining at the time, and it was argued that the improved cash position was evident as the business would not need to raise funds for the foreseeable future, despite an investment in product development that had seen it introduce 15 new styles in the past 12 months, as well as retooling to make sneakers. It was also said that their closure of the retail outlets had allowed the business to focus resources on the sneaker launch, described as "critical" given the growth of the "athleisure" fashion category, within which activewear and everyday wear are increasingly influenced by design. It was argued that it was a potential growth market but added that although sneakers were unisex, women continued to be the primary target customer for Shoes of Prey for now. *Performance? Profitability? Productivity?*

It was reported that the company owns its own factories in China, seen as important given the complexity of introducing sneakers to the customized manufacturing process. Shoe "lasts" were considered to be a significant investment but offered a larger range of size options, The company's dress shoes range was available in every size from a US "2" to a "15," including half sizes, as well as five different widths. The decision to offer sneakers as a female and male unisex product required them to build a whole new set of lasts. The director also told reporters that the concept of mass-customized shoes itself is still very early in its evolution and estimated the company needed at least a good 5 years where customer acquisition is as important a metric for us as lifetime value. Further expenditure was planned as strong traction in sneakers could put the company temporarily further into the red if it decided to invest to shore up its market position. Additional expenditure had been involved in investing heavily in logistics to reduce order cycle time which had been reduced to 2 weeks, from up to 8 weeks. The director declined to indicate when Shoes of Prey expected to become cash flow positive. The investment director of a major investor, Blue Sky Private Equity, expected the business would become profitable in 2017. *Profitability? Productivity?*

In May 2018 it was reported that Shoes of Prey was contemplating a sale of its business following a bridge financing round led by a Blue Sky Alternative Investments fund. Shoes of Prey, which reportedly lost $6 million last year, has been struggling to hit growth forecasts. The company was reported to be working on a new business strategy that included offering shoe sizes outside of standard ranges and using its manufacturing footprint to do private labels for other brands. After trading difficulties in 2018, it ceased trading in August. The company had been working on options aimed at avoiding a formal wind-down or liquidation. The company sold its manufacturing equipment to Estas Brands, which rehired its Chinese workers. This "ensured" that the manufacturing equipment, team, and know-how was preserved in the event a restart strategy was possible. In March 2019 Blue Sky told investors in its Blue Sky Private Equity Shoes of Prey fund it had begun the liquidation process in the USA and Australia and would de-register the business in China. It said there would be insufficient funds to repay its secured and unsecured lenders, "and therefore we do not believe investors will recover any capital." *Performance? Profitability? Productivity?*

A comment from one of the founding directors suggests some learning took place. It suggests difficulties in understanding that the mass-market customer didn't want to customize. If that had been clear, the company would not have gone down the path of raising venture capital and instead focused on building a strong but smaller business serving their selected niche. Other comments were made concerning the efficacy of The Shoes of Prey business model (delivering made-to-order shoes in just 2 weeks), which was a financially, legally, and ethically unsustainable business model. Customers who visited the retailer's website suffered "decision paralysis" when faced with the time-intensive task of producing their own shoes, which led to lower conversion rates. The business also had high fixed costs but could not fall back on scale to break even because each shoe was custom designed.

This case study is based upon:

- Shoes of Prey is shutting down its David Jones and Nordstrom design studios, John McDuling, *Australian Financial Review*, October 17, 2016.
- Shoes Of Prey to make sneakers in "athleisure" play, *Australian Financial Review,* Michael Bailey, September 19, 2017.
- Blue Sky backs Shoes of Prey, Sarah Thompson, Anthony MacDonald & Julie-anne Sprague, Australian Financial Review, May 22, 2018.
- Shoes of Prey founder reveals start-up is facing "struggles": ANOTHER day, another local retailer facing collapse — and this time, it's a popular shoe brand's future that hangs in the balance, Alexis Carey, News.com.au, August 28, 2018
- The final kick in Shoes of Prey collapse for Blue Sky backers, Jonathan Shapiro, *Australian Financial Review*, March 12, 2019.
- "Decision paralysis" the final straw for Shoes of Prey, Liz Main, *Australian Financial Review*, March 12, 2019.
- "Not a surprise": Investors lose millions in Shoes of Prey collapse, Cara Waters, *Sydney Morning Herald*, March 12, 2019.

Marley Spoon: Rapid Value Migration and Capability Responses

Marley Spoon is a home-meal-kits group that sends recipes and fresh ingredients directly to customers' homes making it easier to cook selected meals. However, the company has been underperforming since listing on the Australian Securities Exchange in July 2018, its value having more than halved during a generally poor performance of the "Market" and questions being asked concerning the efficacy of its business model. The business was founded in 2014 in Berlin and expanded to Australia the following April. But it stumbled in Britain, launching there in late 2014 and closing in 2016, being unable to make the economics work in the crowded British market.

The company raised AUD$70 million in fresh capital before listing on the ASX in July at AUD$1.42. The Marley Spoon instruments that trade on the ASX (Australian Stock Exchange) are called Chess Depositary Interests, which have the same characteristics as shares but are used by international companies wanting to trade on the Australian exchange. The Marley Spoon CDIs had sunk to as low as AUD39.5¢ on December 17. On January 3, 2019, they were listed at AUD40¢. *Profitability?*

Technology-driven online food-delivery businesses (such as Uber Eats and Deliveroo) where meals are picked up from restaurants and cafes and delivered to consumers willing to pay for the convenience for home delivery suggest that DIY meal preparation is not of interest to the "Millennials," despite emerging criticisms about the pay rates and conditions for the harried deliverers.

Marley Spoon co-founder Fabian Siegel warned Coles and Woolworths off the fast-growing $300 million meal-kit market, saying selling meal kits on supermarket shelves or online is "unviable" for the major chains, arguing major retailers in Australia and overseas have finally realized meal kits are not a short-term fad but an increasingly popular and efficient way for consumers to buy groceries for meals they cook at home. He argues that consumers who have discovered meal kits don't stop going to the supermarkets, but they are shifting a big chunk of sales away from the supermarkets to services like such as Marley Spoon. Fabian Siegel believes supermarkets don't understand the meal-kit business model, because it's not a retail business where you source things to stock and underestimate the high level of waste if meal kits, which typically contain a mix of protein, fresh produce, and portioned dry groceries such as herbs and spices, passed their use by dates. It's a much more sustainable way of cooking compared with a supermarket. The major grocery chains risked taking their eyes off their core business to pursue a meal-kit market worth between $250 million and $300 million a year in Australia, a fraction of the total $100 billion food and grocery market.

Mr. Siegel suggested there is a big difference between a retail business and a manufacturing business (which Marley Spoon sees itself), such as making just in time deliveries, sourced to order, based on weekly changing menus – it's a unique way of manufacturing and it's not as trivial as it looks. He argued that supermarkets have very complex and expensive infrastructure, and if you try to force a different product category into that infrastructure, it's going to be hard to make it economically viable. *Performance?*

A positive for Marley Spoon is the global trend for health and well-being and environmental sustainability. Marley Spoon delivers the exact quantity of ingredients needed arguing there are no excess ingredients remaining in a refrigerator for too long after being bought from a supermarket and not used. *Performance? Preservation?*

A negative for Marley Spoon is that the heavy sell-off in global share markets, which began at the start of October 2018, has impacted upon these high-growth but high-risk stocks that require several years for a return and had plenty of optimism built into their share prices because they were in the initial phases of building businesses. Fabian Siegel, one of the founders who has 20.6% of the company, argued that the high profit margins of supermarket giants Woolworths and Coles had enabled Marley Spoon to find a niche in Australia. *Profitability?*

However, there is the promise of strong competition; the chief executive of Coles, Steven Cain, said his company viewed pre-prepared meals as a big growth opportunity. Coles unveiled its first Coles Local convenience store format; the first small format store under the Coles Local banner opened in the well-heeled inner-city Melbourne suburb of Surrey Hills, and more stores are set to open in Victoria, NSW, and Queensland over the next 5 years. The Coles Local is half the size (1200 square meters) of a traditional full-service supermarket and carries about 8000 products – one-third of the usual range but twice the number in competitor Woolworths' Metro convenience stores, thereby enabling customers to do a full grocery shop. More than half the floor space is dedicated to fresh foods, including 100% Australian sourced fruit and vegetables and chilled ready-to-eat meals, including prepared meals and pre-cut produce under the Coles Local. Like IGA and Harris Farm stores regional independent competitive chains, the range of packaged groceries is heavily skewed to the local demographic. The Surrey Hills store carries a larger selection of Asian, premium, vegan, and vegetarian foods as well as local brands including meat. According to market research company IRI, the convenience store channel is growing faster than traditional supermarkets, fueled by time-poor customers doing top-up or impulse shops several times a week. Coles chief executive, Steven Cain, suggests "consumer needs are changing, and we are moving more towards convenience and we have to supply that demand." *Performance?*

Profitability?

More direct competition may come from an Uber Eats/Coles partnership. Uber Eats has established a partnership with supermarket chain Coles to deliver ready-to-eat and ready-to-heat meals. Coles and Uber Eats, the largest player in the online take-away food-delivery market, are evaluating a pilot program in a Sydney to order ready-made and semi-prepared meals from Coles through the Uber Eats app. A dedicated team in Coles will pick the orders from supermarket shelves and hand them to Uber Eats drivers, who will deliver the orders to homes or workplaces in less than 30 minutes for a $5 delivery fee. Coles director of fresh food, Alex Freudmann, reported shoppers were demanding more convenience, while Uber Eats general manager Jodie Auster noted that customers were seeking more choice, including healthier and cheaper meal and snack options. Rather than ordering a whole barbecue chicken from the neighborhood charcoal chicken shop ($17.90), for example, or chicken and avocado nigiri ($15.80) from the local sushi restaurant, customers would be able to order similar items from Coles for half the price.

However, in an earlier project (May 2018), Coles launched a trial using Uber drivers to deliver items missing from online grocery orders or items that needed to be replaced. However, the service was not rolled out, and Uber closed the express delivery service, known as Uber Rush, in May 2018. *Performance? Profitability?*

Productivity?

Case note: The Millennials, the label given to those born between about 1981 and 1996, are causing disruptive shifts around the world; it is a change; share market investors are trying to factor into their thinking. At a Sohn Investment Conference in New York in April 2018, two hedge funds predicted that online food-ordering companies would benefit as the Millennials' influence expands, with growth in restaurant sales outstripping supermarket sales (Evans, 2019)
    Based upon:

- Coles unveils first Coles Local convenience store format, Sue Mitchell, *Australian Financial Review*, November 13, 2018.
- Marley Spoon founder Fabian Siegel warns supermarkets off meal-kit market, Sue Mitchell, *Australian Financial Review*. November 14, 2018.
- Marley Spoon misery as Millennials head for the door, not the kitchen, Simon Evans, *Australian Financial Review*, January 4, 2019.
- Marley Spoon secures more than $30 m in funds to support meal-kit growth, Sue Mitchell, *Australian Financial Review*, January 29, 2019.
- Coles and Uber Eats test ready-to-eat meal deliveries, Sue Mitchell, *Australian Financial Review* January 29, 2019.

Repositioning in the Value Chain Network: Icon Aerospace Technology

Many supplier partners within value chain networks are locked into low value output positions, competing for orders alongside competitors in a typical situation where their only differentiation is price; over time they may be able to move into a favored supplier position with a customer, because of their consistent quality and reliability and because they have been able to meet a customer's requirements when a competitor supplier has failed to meet an order. Typically, they lack an RD&D capability to move away from low value positions. In an interview with Jonny Williamson of the *Manufacturer*, Icon's CEO recalled the company's journey within the aerospace value chain.

Prior to the global financial recession of 2008–2009, 70% of Icon Aerospace Technology's output was specific products for a specific customer. Although they could be described as unique solutions, they were, nevertheless, mostly at the lower end of the value chain. At that time the business used to mix all the compounds on-site, production which involved the use and storage of many different chemicals, some of them volatile. Management realized that nearly all the aerospace and defense primes (major companies) were looking for a supplier able to provide something of far greater engineering, high dollar value-added complexity, and made the decision that they were not going to be involved in the low value-added process anymore.

Icon Aerospace Technology has been on a 10-year journey to transition from low value output to products of a more complex nature; supported by a multimillion-pound investment in continuous improvement, the product range has mirrored the physical business' evolution, steadily producing more innovative and technologically advanced engineered products. An early decision was taken to outsource it the low value adding processes to what Icon considers to be one of the world's best compound suppliers. Icon has also invested in setting up on-site laboratory facilities in order to continually develop new materials and compounds. The valuable IP stays within Icon, whereas the lower-value work of mixing is handled elsewhere.

Icon Aerospace Technology is now a technology-led company, a world leader in highly engineered products utilizing polymers which are used in sealing, containment, propulsion, and protection systems worldwide. Specializing in solving complex technical challenges, the company creates future facing solutions which minimize weight and optimize performance and value for customers, who include the world's major names in aerospace and defense.

Icon practices *face2face engineering*, a unique approach to problem-solving. *face2face* reflects both what we do, creating strong bonds at the interface of different materials, and being face to face with customers. This helps build stronger bonds, gain deeper understanding, share ideas, and work better together. Strong links with academic and research institutions enable Icon to adopt the latest research and best practice into their work. Icon is continually pushing the development of new materials and processes – and how they "interfuse" with each other and within their customer's systems.

Active Industry Sectors

Aerospace; Approved by many of the world's leading aerospace companies, Icon designs, manufactures, tests, and verifies a wide range of standard and customized components and assemblies, including airframe seals, fire seals, ducts, bellows, and hoses. By deploying the latest materials and technologies to drive cost efficiencies and performance advantage, Icon has particular expertise in tackling technical complexity.

Defense; Icon has decades of experience supporting the defense sector. In aerospace, Icon has pioneered many products including Flexiflo™, a state-of-the-art in-flight refueling hose. By embracing innovative processes and technologies such as radar-absorbing materials, Icon's creative products deliver significant performance and weight optimization advantage in extreme operating environments, including NBC applications. The company has a proven track record in the manufacture, testing, and verification of high-performance components and assemblies for military land systems.

Industrial; By applying the knowledge and sophisticated skills acquired in aerospace and defense, Icon supports a number of specialist industrial sectors where similar levels of reliability and repeatable performance are required. Recent projects include eliminating vibration damage to parts in domestic wind turbines, specialist cable covering for use in underground transport application, and unique safety barriers used on railway applications.

Convergence …

Icon Aerospace is an example of an organization that understood the direction its customers' developments were taking and has matched its capabilities to meet their customers' expectations and have achieved their objective of repositioning in the value chain as a high-dollar value-adding supplier with a relevant capability portfolio including flexibility and speed of response *(performance)*. Icon invests in continuous improvement programs to increase capacity and remain competitive for *(productivity)* customers. This includes LEAN manufacturing *(profitability)* and developing company expertise and people through a range of initiatives – from national vocational qualifications NVQs and degree sponsorship to a "Knowledge Transfer Partnerships" with local universities *(partnerships)*. The company's comprehensive services range from design, development, and qualification through precision manufacturing and post-sales support *(producibility)*. Combined with scalable production facilities and a rapid prototyping service, Icon is able to resolve challenges and achieve a rapid turnaround that meets and exceeds expectations *(preservation)*.

Icon has the ambitious goal of doubling revenue by 2019. It is currently working through an order book worth upwards of £150 m (based on Teal Group data) and has created 100 new jobs over the past 18 months.

This case study is based on:

- Aerospace supplier advises how to move up the value chain, J. Williamson, *Manufacturer,*
- August 8, 2017.

Sustainability at Coca-Cola Amatil

The chief executive of Coca-Cola Amatil Alison Watkins, at a Committee for Economic Development of Australia forum, said that while some people would have called a "sustainability strategy" an oxymoron a few years back, it was now a "front and center" issue. Coca-Cola Amatil plan to use 50% recycled plastics in its bottles by the end of 2019, up from 26% currently.

The chief executive of beverages giant Coca-Cola Amatil has conceded that consumers might be "cynical" about the company's switch to put sustainability into its core strategy after lobbying hard against container deposit schemes several years ago. "It's about improving from where we are. We would not say we are perfect. We would not hold ourselves out to be on some sort of pedestal." At the forum on Friday, Coca-Cola Amatil announced that the shift came at a time when there was increasing consumer pressure for companies to take action on environmental issues like plastic waste. This pressure created a business risk that meant sustainability was no longer a "fringe" issue but at the "front and center" of business executives' minds. And that there is "a clear imperative to balance the delivery of results today with building a sustainable future for tomorrow." "It used to be the perception of businesses that the responsibility of dealing with plastics rested with government, but, from a consumer perspective, that's no longer the case." It's not a new pressure back in 2014, 80% of consumers supported the idea of a 10¢ deposit on cans and bottles to encourage recycling. *Preservation?*

By 2030, the company's aim is to have "no net new waste," which means that all its bottles will be made from recycled plastics. Some of that recycled plastic is bought from Australian-based companies, but a lot is shipped in from countries like China and Thailand.

The announcement comes at a time when Australian beverage volumes are in decline. She said proactive measures around both sugar content and single-use plastics were required lest Coca-Cola Amatil become a "sunset" business.

Nik Gowing, the founder and co-founder of Thinking the Unthinkable Project, contributed to the discussion. He suggested business leaders needed to grapple with the immense power of "pushbackism," suggesting that consumers and employees alike will increasingly "pushback" against companies that they believe are not taking environmental issues seriously. He quoted: "Employees are saying; we are stressed. We're coming under pressure working for this company," "We're in danger of angry consumers and angry citizens." Companies should use the resources they already have to better understand the speed at which sentiment shifts, he said, referring to younger employees who have a good grasp on public sentiment as well as the prevailing messages circulating on social media. *Preservation? Profitability? Productivity?*

This case study is based upon:

- Sustainability no longer a "fringe issue" for Coca-Cola Amatil, Natasha Gillezeau, *Australian Financial Review*, March 22, 2019

GE Looking for Better Times

BOSSES come in all shapes and sizes. One way to categorize them is to split them into two types: polishers and pickers (see Chap. 5). Polishers put their energy into products, improving and reimagining their design and production in a quest for perfection. Pickers, by contrast, are capital allocators, who stand back and decide unsentimentally how the firm should deploy resources. An example of this approach was Jeff Immelt.

Most chief executives would say they are more pickers than polishers. The task of creating a new product-service or honing a manufacturing process is best left to "geniuses." The BCG corporate "zoo" suggested firms split portfolios into four buckets: cash cows for milking, stars, dogs that should be put down, and question marks. Today BCG reckons that businesses shift between the buckets twice as fast as they did in the 1990s. *Performance, Profitability.*

Mr. Immelt remade GE partly because he had a difficult inheritance. GE's shares were overvalued, its earnings were inflated by gains from its pension scheme, and it had over expanded its financial arm, which later collapsed during the banking crisis. He globalized GE: 57% of sales now come from offshore (compared to 29% in 2001).

Mr. Immelt's main legacy will be as a capital allocator. He shrunk or disposed of mature or underperforming businesses or that were under margin pressure (plastics and kitchen appliances), or where GE has no value advantage, such as media. He eliminated most of its financial arm. And, he bought in areas with promising growth, where technology is becoming more important, aviation, power systems, and medical devices. Outside the financial arm, looking just at industrial operations, since 2001 GE disposed of businesses worth $126bn, or 167% of the capital employed in its industrial divisions. Counting capital expenditures, too, Mr. Immelt had redirected resources worth some 227% of GE's capital base. *Performance? Profitability?*

The results are not impressive. Annual free cash flow from GE's industrial business was around $10bn in 2001, and the figure has not risen, despite capital employed having increased from below $30bn to $75bn. Cash returns on capital have fallen to about 12%. GE's shares have lagged behind the S&P 500 index over most periods.

The cost of churning capital in predictable ways can be significant. It has been estimated that GE paid a multiple of 13 times gross operating profits for the businesses it has bought and received 9 times for those it sold. Some 90% of its industrial capital is now comprised of goodwill (the premium that a firm paid above book value for its acquisitions). A company's capital expenditure often is procyclical. For example, in 2010–2014 GE increased investment in its oil and gas business, when energy prices were high, and then cut back after they collapsed in 2015. *Performance? Profitability?*

For businesses in aggregate, and their investors, churning portfolios bring some benefits. Firms must respond to changes in customer tastes and technology (value migration). They may be able to improve their market shares for some products, permitting price increases. But it appears unlikely that hyperactive capital allocation greatly enhances wealth overall. It is impossible for every firm to own only outperforming businesses. And the fees lawyers and bankers charge are a tax on corporate activity that corrodes value.

For Mr. Immelt the jury is still conferring. GE's profits are rising even as its cash flows stall, as it logs the gains it expects to make on long-term infrastructure projects and servicing contracts. It has launched Leap (a gas turbine engine) and invested heavily in Predix (an open data platform) that it hopes will become an operating base for a host of industrial digital applications. It was buying new assets at the bottom of the cycle, with a planned merger of its energy business with Baker Hughes, an oil-services firm.

Part of Jeff Immelt inherited from his predecessor at GE was not in nearly as good a shape as Mr. Welch liked to pretend, share price was overvalued, pumped up by hype about Mr. Welch's talents. Its profits were inflated by gains from its pension scheme and its financial division, which had grown rapidly and contained big risks. Mr. Immelt tried to take GE back to its core as an industrial firm making sophisticated products such as power equipment and jet engines.

It is suggested these efforts were overshadowed by two mistakes. First, Mr. Immelt was slow to recognize just how dangerous GE's financial arm was. By 2007 it contributed 55% of profits and had racked up over $500bn of debt. When the Global Financial Crisis (GFC) struck, its funding dried up and profits collapsed. Mr. Immelt shut most of it down in 2015, but possibly too late. The second is less well understood but important: the performance of the nonfinancial business had been lackluster. As mentioned above, Mr. Immelt's reshuffling was huge, with disposals and acquisitions equivalent to 167% of its current capital employed. GE ditched its media arm, plastics division, and kitchen-appliances unit and bought into healthcare, energy, and power infrastructure. But they have been expensive; GE's capital employed has ballooned but not matched by returns. Weak operating performance, and the costs from the restructuring, means its cash flows are similar to what they were in 2001. GE has been running to stand still. *Performance? Profitability? Partnerships?*

Enter a New CEO

New CEO, Mr. Flannery, had been at GE since 1987, and the healthcare business he ran contributed 20% of profits. Two big tasks awaited him. Mr. Welch had been a pioneer of offshoring, and GE's supply chains are global, but the firm will have to guard against a protectionist backlash at home and abroad. That requires diplomatic and communication skills. The other task is to deal with GE's unspectacular financial performance. Trian, an activist hedge fund, owns a stake in GE and, behind the scenes, has probably been agitating for change, possibly exerting pressure for GE to break itself up. Observers at *The Economist* suggested that would be a bad idea: arguing what it now needs is less reengineering and more consistent execution.

Mr. Immelt was forced to concede during an interview (in 2016) with Jeff Cunningham, of *Chief Executive* (article December 8, 2017) that "In many ways, the environment today is nothing like what I thought it would be when I became CEO."

Performance? Profitability? Partnerships? Productivity?

Mr. Flannery unveiled a new strategy in November 2017. This rested on cost control, cultural change, and cuts. He slashed the promised dividend by half, to make savings on top of the $2bn in annual cuts and intended using incentives to restore management's focus on financial returns by altering the pay structure to incentivize executives to generate free cash flow. He is also planned to slim down the board of directors, even placing an activist investor on the board. But while his approach to costs and culture looked sensible, big investors wondered at his efforts to shrink the firm. Mr. Flannery stated he would dispose of some $20bn in assets in the next 2 years.

The firm had been a member of the Dow Jones Industrial Average, for over a century. Problems alas had turned GE into a disorganized, debt-laden mess. GE's shares plunged to below a quarter of their peak value in 2000. On June 26, 2018, GE was ejected from the Dow index and replaced by Walgreens Boots Alliance, a big healthcare firm. This action indicates the changes concerning the perceptions of investors at this time.

John Flannery announced details of a much-awaited restructuring plan. Over the next couple of years, GE planned to spin off its healthcare division and change its recent stake in Baker Hughes, a petroleum-services firm. He had previously confirmed the sale of its train locomotive division. Taken together, these three units generate roughly $40bn a year, about a third of the firm's annual revenues. *Performance? Profitability? Partnerships? Productivity?*

GE's share price rose on the news. The obvious reason was Mr. Flannery's renewed promise to reduce the size of the unwieldy conglomerate, including an intention to slash its net debt and pension obligations by $25bn. He also promised to cut an extra $500 m in costs, on top of previously announced cuts, by 2020. Beyond this willingness to wield the axe, Mr. Flannery's plan for fixing GE had three attractive elements: "spinners," "spin-offs," and "spinning down." First, the plan let management focus on the core businesses of power generation, aviation, and renewables. This produces about $70bn a year in revenues. GE's power division, which generates about half of those revenues, is in particularly deep trouble. A combination of poor management, poorly judged investments and weak global demand left it in crisis. The division planned to shed some 12,000 workers, nearly a fifth of its global workforce. *Productivity? Performance? Profitability?*

Second, the restructuring was to be undertaken in a way that would produce shareholder value. Rather than, say, sell the profitable healthcare unit to a strategic buyer in return for a rapid infusion of cash, Flannery planned to spin it off as a stand-alone firm. He proposed to give 80% of its shares to GE shareholders (who will capture any future financial gains) and sell the remaining 20% (Research by Emilie Feldman of the Wharton School shows that such spin-offs create value in two ways. Freed from overbearing parents, the new entities become more efficient at allocating capital. Intriguingly, her research shows that the divesting firms also improve their financial performance after a spin-off).

The third reason was Mr. Flannery's desire to reform GE's management culture. He launched a new "GE Operating System" which promised less centralized decision-making and a spinning down of resources from headquarters to business units. GE's bloated board of directors was replaced with a smaller, more relevant one, a co-founder of Trian Investments, an activist investor. Steven Winoker (of UBS, an investment bank) called this the most important of all the reforms, praising the new directors as "sharp, useful people."

Enter Another CEO

In September 2018 John Flannery was replace by Larry Culp, the first "outsider" to be appointed to the CEO role. Flannery's attempts to turn around the business – cutting costs by grounding the corporate jet fleet, selling off and shrinking business units, and slashing its dividend in half – were not enough to offset the troubles in its power division, which has taken a series of hits. Culp was brought on board in October to accelerate a turnaround centered on reducing debt, improving culture, and reshaping the portfolio. On March 13, 2019, it was reported GE rose 1.5% to $10.17 before regular trading in New York. The shares gained 32% in 2019 through Wednesday, recovering a portion of last year's 57% loss, the worst annual decline since at least the early 1970s.

Amid all the challenges, Culp also needs to position GE to grow again and avoid future crises. If he pushes ahead with the healthcare separation, GE will become more reliant than ever on its aviation division. One option is to sell more stock. That would both ease GE's debt burden and allow Culp to take his time with other asset sales. Culp has given mixed signals, concerning his willingness to do that. Otherwise, his most viable alternatives for raising cash are selling GE's jet-lessor unit, GECAS, or following through on a breakup plan Flannery unveiled earlier that would see the healthcare unit spun off to shareholders, with a stake sold via the public market. In the latter scenario, Culp would likely need to sell a bigger piece of the healthcare business than the 20% Flannery proposed to make it worthwhile. Still, paying down debt is really just a first step for GE; truly fixing the company will require a deep examination of the cultural instincts and market machinations that brought it to this point.

General Electric Co. will use up as much as $2 billion of cash in 2019, in an attempt to repair its balance sheet and return stability to the business. Free cash flow from the company's industrial businesses will decline in 2019 as restructuring costs hit at least $2.4 billion. "GE's challenges in 2019 are complex but clear," Culp said in a statement: "We have work to do in 2019, but we expect 2020 and 2021 performance to be significantly better."

Hindsight

What has happened? In a phrase, the twenty-first century. GE is a large organization and responsive to change. Its disruption is typical of what is happening. Not long ago, the S&P 500 dropped traditional organizations such as Bank of America,

Hewlett Packard, and Alcoa and replaced them with Goldman Sachs, Nike, and Visa. Brands, image, and networks beat assets. So, what do GE's lessons suggest for the rest of American industry? If change is the only constant, agility is a necessary skill for running a business, and strategy is more the immediate future, not a 5-year planning horizon. GE's long-term strategic thinking actually slows down the speed required to navigate abrupt shifts typical of global markets.

When interviewed in 2016 Jeff Immelt (shortly before the current crisis), his comments showed how fully aware he was of these challenges. Immelt said: "When I took over GE, the biggest surprise was the world twisted from one of relatively benign growth to one of just high volatility, high geo-political risk. In many ways, the environment today is nothing like what I thought it would be when I became CEO."

Immelt set about to reconstruct GE for this new world, and while he did everything a highly capable CEO could possibly do, he could not move fast enough given the chaotic environment. The problem wasn't Immelt; it was the legacy of how a major business operates, slowly and methodically.

Disruption today springs from many sources including the Chinese economic miracle, geopolitical disruption, and public discontent with major institutions which began not long ago with Occupy Wall Street and is now evident as Bitcoin savagely attacks our financial markets. Consumer tastes didn't change, they revolted. Tesla is now the most valuable automotive company in America with a fraction of the sales of GM or Ford. Yet, the company does no traditional advertising and does not have a dealer network. When that example was put to Immelt, he simply said, "business has no shelf life, and what was right 10 years ago may now be obsolete. You are always planning for the next disruption, and that means looking at new markets all the time."

So why didn't GE see those changes coming and plan for them? Or were they looking at the wrong "indicators"? The problem wasn't in the strategy, but in the time, it takes to implement it. Traditional strategic planning required months of approval and fails to stand up to Mark Zuckerberg's "hacker mentality," in which a team creates a solution in 6 days not 6 months. The Department of Labor isn't standing over their shoulders asking about equal opportunity or minimum wages. By the time the product is launched, Silicon Valley venture capitalists are organizing millions to fund them to the next stage. The changes put long-term planning itself in jeopardy. If it takes your team a year to develop a global competitive strategy, someone in Silicon Valley has developed an AI or Bitcoin replacement before your first brainstorming session. Immelt was aware of this when he said: "I think we used to think that if you had good leaders you could be in any business. We want to have good leaders, but we actually believe that this deep domain expertise is also critically important for the future. Business has no shelf life. You really can't be captivated by anything that's happened in the past. It's all about today and tomorrow in business."

The new era is teaching us that Immelt's comment that business has no shelf life, issued a painful edict to John Flannery and to his successor; that GE has no shelf life either. The abrupt changes he intended making were not revolutionary, they were just a realization of this fact of modern business life. *A role for Response Capabilities?*

Performance?   Profitability?   Partnerships?   Productivity?   Producibility?
Preservation?

This case study is based upon:

- Amid a whirlwind of deal making, GE's returns lag, *The Economist,* May 27, 2017.
- John Flannery will have more room to provide positive surprises. *The Economist.* June 17, 2017.
- The iconic American conglomerate has a new boss and a new strategy, *The Economist,* November 30, 2017.
- Did Jeff Immelt fumble the GE handoff?
- Jeff Cunningham, Chief Executive, December 8, 2017.
- Power failure, *The Economist,* June 28, 2018.
- General Electric CEO Larry Culp says the company has too much debt, *Alwyn Scott and Kate Duguid, Australian Financial Review,* November 13, 2018.
- GE sees cash burn up to $2 billion as turnaround costs weigh: CEO Larry Culp promises "significantly better" performance in 2020, 2021, *Richard Clough, Bloomberg,* March 14, 2019.

Ford Motor Company 2018/2019 and Beyond?

In March 2008 Ford Motor Company shed noncore brands in an effort to revive its ailing US business. Jaguar Land Rover (JLR), Aston Martin, and Volvo were more of a burden than an asset. Under Tata's ownership, JLR has gone on to rack up healthy profits and produce a string of popular sport-utility vehicles (SUVs) to widespread acclaim. Another former Ford marque, Aston Martin, last week (August 2018) announced plans to float on the stock market for up to £5bn. Under Ford, Volvo appeared not to fit. The Swedish carmaker was sold to Chinese firm Geely for $1.8bn. in March 2010, for a third of what Ford had paid for it. *Profitability and Productivity from a platform producibility approach.*

During the time since these events occurred, Europe has become a problem for Ford. The company has struggled to make consistent profits from its collection of factories and dealerships. *Productivity, Profitability.*

Analysts at Morgan Stanley suggest Ford has posted a pre-tax loss 12 times since 1999, adding up to a cumulative $3bn. of losses since then. Ford has persisted with a weary range of passenger vehicles, belatedly realizing and responding to the huge popularity of SUVs. This weak performance has had an impact on its share price, down by more than 40% in the past 5 years, versus the sales successes of General Motors and Fiat Chrysler Automobiles. *Performance, Profitability and Productivity.*

Ford is now promising action. In January 2019 James Hackett told employees the company would not accept last year's "mediocre" results and said the company was aiming to nearly double its annual operating profit. Ford had reported a 2018 operating profit of $7 bn with a profit margin of 4.4 percent, down from 6.1 percent in

2017. Adding that Ford is restructuring its global operations and has also announced an alliance in commercial vehicles with Germany's Volkswagen. The CFO said it would reallocate capital to focus on opportunities with higher returns and on "strategic partnerships," changes that will cost it $11bn. in restructuring charges, of which $7bn. will be in cash. *Profitability, Productivity, Partnerships, Producibility.*

Ford has yet to announce the changes, but they have added to the uncertainty of its UK operations; almost 12,000 are employed across plants and dealerships. These include two engine factories (one of which manufactures the controversial diesel engines) with a total of 3500 staff. The UK's imminent exit from the EU has added complications, with the risk of tariffs threatening to make them uncompetitive. *Productivity, Preservation.*

Other estimates suggest 14,000 European jobs could go in Ford's cuts (Evercore Analysts) and (Morgan Stanley analysts) predict even more cuts; about 24,000, assuming 60% of the £7bn. charges are redundancies, impacting 12% of its 202,000 global workforce. Having lost work it previously did for Volvo, and with work for Jaguar also ending, the Bridgend site risks being closed. Partnerships.

Ford's problem has been a long time in the making. Under its then chief executive, Alan Mulally, Ford entered the Global Financial Crisis (GFC) stronger than its US rivals, but with a $27bn debt pile at the end of 2007. It avoided the fate of GM and Chrysler (Chap. 11 bankruptcy and bailout by the US government). Mulally had started dismantling Ford's collection of premium brands by the time the crisis hit, selling luxury brand Aston Martin to a Kuwaiti consortium and Swedish marque Volvo to Geely of China. Selling the upmarket brands allowed it to focus on Mulally's One Ford plan. Crucially, it helped to ensure the Ford family's influential shareholding was not decimated by a cash call, for funding new Jaguar and Aston Martin models. *Performance (rationalizing brands).*

However, there was a cost. Ford has become increasingly narrow. Sales of its F-150 pickup/utility vehicle in America generate the bulk of its profits. Along with nearest competitors, Vauxhall, Renault, and Peugeot, Ford has suffered sustained competition from German premium carmakers in the UK and Europe. Low interest rate finance put premium marques within reach of an increasing number of households. *Performance.*

Ford also missed the shift in customer expectations (value migration), relying upon its reputation for reliability it largely ignored the boom in SUVs and has only recently started to compete. The dealership chain was both concerned and confused; "People want to buy SUVs, and they are more profitable." *Performance.*

Ford's Fiesta and Focus are the first and fourth best-selling new cars with British drivers, but its market share in Europe has been in steady decline. It slipped to 11.3% last year (2017) in the UK, down from 14.5% a decade ago and 19.2% in 2001. Meanwhile Mercedes, BMW, and Audi have all amassed shares of more than 6% each. Weak sales undermine the fragile economics of mass market car production, where margins are slender, and volumes are critical (scalability problems). *Performance, Productivity.*

Ford is also burdened with too many showrooms and is believed to be working on a strategy to reduce its UK dealerships. Industry watchers expect it to axe unpopular models. However, it is cost that is Ford's largest concern. Its car manufacturing facilities (notably Germany and Spain) are believed to be operating well below capacity. Ford has not made cars in Britain since 2002. "They have too much capacity in Germany," Evercore Analysts have observed and add, "Germany is the most expensive place in Europe to build cars and they're not a premium brand. There's a mismatch between the price of the product and their fixed cost base," (price/value equilibrium). *Performance, Productivity, Profitability.*

It has been suggested the best option would be for Ford to follow GM and exit Europe and focus on America. (GM sold the Vauxhall and Opel marques to the French automotive giant PSA for €2.2bn (£2bn) in 2017. Morgan Stanley suggest that Ford Europe represents the single-biggest risk to Ford's long-term financial health and success, an exit, while potentially expensive, may be the best option for Ford Motor Company and Ford Europe stakeholders. And, added a pessimistic comment concerning the segments they sell to being in decline; arguing the Ford brand has questionable to minimal value in most of the segments in Europe. Added to this they have too many dealers, too many nameplates, and too much capacity. And furthermore, the future diesel is an unknown, *Performance, Profitability, Productivity.* They have too many compliance costs. *Preservation.*

Exit would bring difficulties, and there are no obvious purchasers. Ford and Volkswagen have officially confirmed a partnership to build their next generation of vans and utility vehicles (utes) together. It will also likely spawn future shared products including Volkswagen's electric MEB platform underneath "blue oval-badged cars", autonomous vehicles, and other mobility services (Bloomberg January 24, 2019) or perhaps a strategic alliance with PSA or Fiat Chrysler. *Productivity, Partnerships.*

What happens next will be foretaste of what faces the mass market car industry. Continued excess capacity and talk of capacity rationalization through consolidation was a constant topic of the late Sergio Marchionne (Fiat/Chrysler CEO); the automotive labor unions have had concerns for some time. Government and Unions have been attempting to persuade Ford to upgrade the Bridgend factory, where 1725 staff are employed, to make parts for electric vehicles. Richard Parry-Jones, Ford's former chief technical officer, is believed to have been lobbying Ford's bosses in Dearborn on Bridgend's behalf. *Productivity.*

The pressure to build electric vehicles (another area where Ford has been behind its competitors) has been addressed. Ford is expanding production of all-electric vehicles after deciding that rapid growth in the market meant it needed to accelerate its plans. It is investing about $900mn. (£682mn.) in new production capacity at a plant in Michigan, creating about 900 jobs. Ford Head of global operations Joe Hinrichs said it had "taken a fresh look" at the growth in electric vehicles. Ford also announced it would start production of autonomous vehicles in about 2 years (BBC Business News, March 20, 2019). *Performance, Productivity, Profitability, Partnerships, Producibility.*

An (April 3, 2019) announcement from Volkswagen suggests the future competitive space for the automotive landscape. Volkswagen will be a device and software company, (Michael Jost, the VW brand's strategy chief), announced at a Berlin press conference introducing the car-sharing program. He emphasized that software and services, not just the car itself, will differentiate automotive brands in the future,

"To deal with this development, we need to reinvent the automobile." Volkswagen has announced it will invest 3.5 billion euros ($4 billion) by 2025 to build digital businesses and products including a cloud computing-based platform to connect vehicles and customers to offer services such as car sharing. VW plans to be a world leader in e-mobility services. VW said it expects to generate around 1 billion euros in sales by 2025 from offering new digital services including car-sharing, parking, and parcel delivery services. These digital services include smartphone applications like "We Park" into car infotainment systems and connecting vehicles with vendors like Amazon, so the e-commerce giant can unlock a VW vehicle and deliver a customer's order right to where their car is parked. VW will use "over-the-air" (OTA) software updates for cars if the operating system is designed in-house, rather than depending on third-party software supplied by the different vendors providing various sensors. OTA updates pushed directly to the vehicle save a customer from having to visit a dealer to have this performed.

A question arises: Is it time for the industry to consider consolidation or has VW built an unassailable competitive value advantage?

Performance, Profitability, and Productivity Opportunities Lost.

Jaguar Land Rover was sold for $2.3bn. to the Indian giant Tata in March 2008. After struggling during the financial crisis, the carmaker has generated hefty profits through an aggressive rollout of products, such as its Jaguar F-Pace sports utility vehicle. Profits hit a record £1.6bn. in 2016, and it sold 621,000 cars last year, up from about 290,000 in the final year of Ford ownership. Lately, 2018, however, it has started to struggle as uncertainty about its diesel engines, Brexit, and the tough Chinese market weigh on profits.

Aston Martin was sold to a Kuwaiti consortium headed by motor racing boss David Richards in March 2007 for £479mn. Italian private equity firm Invest Industrial later bought a stake in the company and hired Andy Palmer, the former Nissan No 2, to be chief executive. Palmer has revived Aston Martin by improving manufacturing and launching a range of new models, aiming to create stability after years of famine and feast. It now builds about 6200 cars a year, compared with 4200 in 2013. Aston last week said it planned to float on the London Stock Exchange, with an estimated value of about £5bn.

Volvo sold to Geely, and run by entrepreneur Li Shufu, now values the carmaker at $40bn. and is weighing up a stock market float. Volvo sold about 570,000 cars last year and has expanded into America with a new factory. Lost opportunities?

This case study is based upon:

- Rocketing Ford into the future, Laura Putre, *IndustryWeek*, May 10, 2017.
- Ford plans $11.5 billion in extra cuts, kills most US cars, Bloomberg, April 26, 2018.
- How the wheels came off Ford: The American car giant is looking to its home market for salvation – and away from Britain and Europe, J. Collingridge, *The Sunday Times,* September 2, 2018.
- Ford accelerates electric vehicle investment, BBC Business News, March 20, 2019.
- Ford Reboots European Strategy With SUVs, Mild Electrification, Ellen Proper & Elizabeth Behrmann, Bloomberg, April 3, 2019.

Printed by Printforce, the Netherlands